CORRIGENDA

The following corrections should be made in Appendix 1:

Table A1.2, line 3 $eE_0/\hbar$ should read $a_0E_0/\hbar$

Table A1.3, line 6 e/α_0 should read $e/\alpha_0{}^2$

        line 8 10^{-43} should read 10^{-41}

Table A1.4, line 4 $k^2{}_0$ should read $\kappa_0{}^2$

        $4.33226\ 10^{-31}$ should read $9.93764\ 10^{-14}$

        line 5 $10\ A^8$ should read $10^8\ A$

        line 6 α_0 should read $\alpha_0{}^2$

        line 7 $e\alpha_0{}^2$ should read $e^2\alpha_0{}^2$

THE INTERNATIONAL
ENCYCLOPEDIA
OF PHYSICAL CHEMISTRY
AND CHEMICAL PHYSICS

Topic 2. CLASSICAL AND QUANTUM MECHANICS

EDITOR: R. MCWEENY

Volume 2

QUANTUM MECHANICS:
METHODS AND BASIC APPLICATIONS

BY

R. McWEENY

THE INTERNATIONAL ENCYCLOPEDIA
OF PHYSICAL CHEMISTRY AND CHEMICAL PHYSICS

THE INTERNATIONAL ENCYCLOPEDIA
OF PHYSICAL CHEMISTRY AND CHEMICAL PHYSICS

Editors-in-Chief

D. D. ELEY

NOTTINGHAM

F. C. TOMPKINS

LONDON

List of Topics and Editors

1. Mathematical Techniques — H. JONES, *London*
2. Classical and Quantum Mechanics — R. McWEENY, *Sheffield*
3. Electronic Structure of Atoms — C. A. HUTCHISON, JR., *Chicago*
4. Electronic Structure of Molecules — J. W. LINNETT, *Cambridge*
5. Molecular Structure and Spectra — Editor to be appointed
6. Kinetic Theory of Gases — E. A. GUGGENHEIM (*deceased*)
7. Classical Thermodynamics — D. H. EVERETT, *Bristol*
8. Statistical Mechanics — J. E. MAYER, *La Jolla*
9. Transport Phenomena — J. C. McCOUBREY, *Birmingham*
10. The Fluid State — J. S. ROWLINSON, *London*
11. The Ideal Crystalline State — M. BLACKMAN, *London*
12. Imperfections in Solids — J. M. THOMAS, *Aberystwyth*
13. Mixtures, Solutions, Chemical and Phase Equilibria — M. L. McGLASHAN, *Exeter*
14. Properties of Interfaces — D. H. EVERETT, *Bristol*
15. Equilibrium Properties of Electrolyte Solutions — R. A. ROBINSON, *Washington, D.C.*
16. Transport Properties of Electrolytes — R. H. STOKES, *Armidale*
17. Macromolecules — C. E. H. BAWN, *Liverpool*
18. Dielectric and Magnetic Properties — J. W. STOUT, *Chicago*
19. Gas Kinetics — A. F. TROTMAN-DICKENSON, *Cardiff*
20. Solution Kinetics — R. M. NOYES, *Eugene*
21. Solid and Surface Kinetics — F. C. TOMPKINS, *London*
22. Radiation Chemistry — R. S. LIVINGSTON, *Minneapolis*

QUANTUM MECHANICS:
METHODS AND BASIC APPLICATIONS

BY

R. McWEENY

PERGAMON PRESS

OXFORD · NEW YORK · TORONTO

SYDNEY · BRAUNSCHWEIG

Pergamon Press Ltd., Headington Hill Hall, Oxford

Pergamon Press Inc., Maxwell House, Fairview Park, Elmsford,
New York 10523

Pergamon of Canada Ltd., 207 Queen's Quay West, Toronto 1

Pergamon Press (Aust.) Pty. Ltd., 19a Boundary Street, Rushcutters Bay,
N.S.W. 2011, Australia

Vieweg & Sohn GmbH, Burgplatz 1, Braunschweig

(A2JAQ)

First edition 1973

Library of Congress Catalog Card No. 73–175514

530.145
M 4

PRINTED IN GREAT BRITAIN BY BELL AND BAIN LTD., GLASGOW
ISBN 0 08 016794 2

INTRODUCTION

The International Encyclopedia of Physical Chemistry and Chemical Physics is a comprehensive and modern account of all aspects of the domain of science between chemistry and physics, and is written primarily for the graduate and research worker. The Editors-in-Chief have grouped the subject matter in some twenty groups (General Topics), each having its own editor. The complete work consists of about one hundred volumes, each volume being restricted to around two hundred pages and having a large measure of independence. Particular importance has been given to the exposition of the fundamental bases of each topic and to the development of the theoretical aspects; experimental details of an essentially practical nature are not emphasized although the theoretical background of techniques and procedures is fully developed.

The Encyclopedia is written throughout in English and the recommendations of the International Union of Pure and Applied Chemistry on notation and cognate matters in physical chemistry are adopted. Abbreviations for names of journals are in accordance with *The World List of Scientific Periodicals*.

CONTENTS

A*

PREFACE

THIS is the second of several short volumes covering the principles and methods of quantum mechanics, the selection of topics and techniques being based primarily on the needs of the chemical physicist. The only background assumed is a knowledge of the basic principles and formalism of the subject, equivalent to that provided by Volume 1*.

The present volume is concerned very largely with the development and application of general techniques for the solution, or approximate solution, of quantum mechanical problems. Topics covered have been chosen as much for their pedagogical importance, in illustrating the applicability and significance of quantum mechanical principles, as for their intrinsic value in quantum chemistry and chemical physics. Wherever possible, however, attention has been confined to the simplest imaginable system—one particle in a potential field. There are good reasons for this limitation. For a one-particle system, the analysis can usually be worked through in depth without excessive mathematical difficulty, and techniques or concepts can therefore be exposed in the simplest possible context. Secondly, surprisingly few wholly new areas are opened up by the consideration of second quantization (and the related techniques of diagrammatic perturbation theory) being, perhaps, the only obvious example. The general theory of angular momentum coupling, for instance, grows quite simply from the consideration of a one-electron system with spin–orbit coupling: and the application of group theory to a one-electron system with point group symmetry provides most of the basic ideas necessary for the study of, say, the permutation symmetry of many-electron systems. Finally, at the level of the independent-partical model (introduced in Chapter 3), the study of one-particle systems acquires sufficient relevance to provide a good general understanding of the applicability of quantum mechanics to complicated systems such as atoms and molecules. The treatment of many-electron systems, including the consideration of specifically many-body effects, is resumed in Volume 3.

Thanks are again due to Professor K. Ohno for his influence on some of the earlier chapters of this Volume, which were completed while the author was a guest of the Chemistry Department, University of Hokkaido; and to Mrs. S. P. Rogers for her patience and skill in preparing the manuscript for publication.

---

† Volume 1 of Topic 2 of the Encyclopedia

CHAPTER 1

METHODS OF APPROXIMATION IN QUANTUM MECHANICS

1.1. Preliminaries

The principles and formalism developed in Vol. 1[‡] are sufficient for all applications of quantum mechanics to low-energy many-particle systems (e.g. electrons in atoms and molecules) in which relativistic corrections, such as mass variation with velocity, can be ignored. Other effects, particularly those involving particle spins and the electromagnetic field, require a much deeper theory including a proper treatment of relativistic invariance requirements and a quantization of the field; a wholly satisfactory theory of this kind does not yet exist and the many observed effects that it would explain are still not fully understood. Nevertheless, it appears that the non-relativistic theory developed so far remains entirely adequate for the discussion of the structure and properties of atoms and molecules (though not nuclei), provided small phenomenological terms are added to the Hamiltonian to represent the observed effects of spins and other small interactions. Usually the form of these small terms may be inferred by semi-classical arguments, the spins being represented by point magnetic dipoles, and it is fair to say that when such terms are included the non-relativistic theory is extraordinarily successful in providing a coherent account of the structure and properties of matter. It is also true to say that so far, and in this field, the Schrödinger formulation of quantum mechanics (to which attention will be largely confined) is employed almost exclusively.

The actual implementation of the basic principles, on the other hand, is no easy matter. A many-electron molecule, for example, is a system of great complexity, for which there is little hope of obtaining an accurate solution to the Schrödinger equation—either time-dependent or time-independent. This is at once clear when we write down the

---

[‡] Volume 1 of Topic 2 of this series. Equations of Vol. 1 will be referred to by placing the letter I before the equation number. We continue to use the notation of Vol. 1: operators are indicated by sans-serif type (Gill Sans), while bold letters indicate *sets* of quantities (e.g. matrices, sets of vector components or variables) including many-variable volume elements (e.g. $\int d\mathbf{r}$ is a volume integral).

equation determining the stationary states of a system of N electrons moving in the field of a set of fixed nuclei, even with neglect of spins and with no external fields. This equation, which may be regarded as the starting-point of most applications of quantum mechanics in chemistry, is an *eigenvalue equation* (Vol. 1, Sections 3.1–3) and may be written

$$\mathsf{H}\Psi = E\Psi \tag{1.1}$$

where the Hamiltonian operator (cf. Vol. 1, p. 11) is[‡]

$$\mathsf{H} = \sum_i \left[-\frac{\hbar^2}{2m}\,\nabla^2(i) + V(i) \right] + \tfrac{1}{2}\sum_{i,j} e^2/\kappa_0 r_{ij}. \tag{1.2a}$$

Here $\nabla^2(i)$ indicates the Laplacian operator working on the variables of electron i, $V(i)$ is the potential energy of that electron in the given field, and $e^2/\kappa_0 r_{ij}$ is the mutual potential energy of two electrons at a distance r_{ij}. The wave function Ψ is a function of the positions (and spins) of all N particles, and (1.1) is therefore a partial differential equation in a large number of variables. In the case of many electrons, the interaction terms in (1.2) are a source of great difficulty; but even with only one electron, and the Hamiltonian

$$\mathsf{H} = -\frac{\hbar^2}{2m}\,\nabla^2 + V, \tag{1.2b}$$

the Schrödinger equation (1.1) is insoluble in closed form except for very special forms of the potential energy function V. The many-electron equation would be regarded by most mathematicians as totally intractable.

We must therefore be content with approximate solutions, and must place great emphasis on the development of approximation methods and on the application of any available mathematical techniques (e.g. those associated with possible symmetries of the system) that can help to simplify the problem. These methods are not new, and lean heavily on the mathematics used in Vol. 1. In the present chapter we survey the main approximation methods in current use, turning in later chapters to the study of systems of special importance, and to topics such as symmetry and the theory of angular momentum, which occupy a central role in quantum mechanics. Finally (Chaps. 6 and 7) we make a preliminary survey of some of the effects of spins and applied fields. All these topics are introduced, as far as possible, in the context of a *one*-particle system, in which many other mathematical complexities

---

[‡] $\kappa_0 = 4\pi\varepsilon_0$ (ε_0 = permittivity of free space) appears in the Coulomb repulsion terms when SI units are employed, but is absent in the mixed Gaussian system. Units are discussed in App. 1.

are absent. The detailed study of many-electron systems is deferred until later (Vol. 3) but most of the methods developed in the present volume are of general applicability.

1.2. Time-independent perturbation theory. Non-degenerate case

In this section we are concerned with solution of the stationary state Schrödinger equation (1.1) in the case where the Hamiltonian can be written in the form

$$H = H^{(0)} + \lambda H^{(1)}. \tag{1.3}$$

Here it is assumed that $H^{(0)}$ is the Hamiltonian of a *soluble* problem, with known eigenvalues and eigenfunctions, and that the parameter λ may be arbitrarily small. The term $\lambda H^{(1)}$ represents a *perturbation*, while $H^{(0)}$ is referred to as the "unperturbed" Hamiltonian. For example, when an external electric (or magnetic) field is applied to an unperturbed system, we may regard λ as the field strength and $-H^{(1)}$ as the electric (or magnetic) dipole moment operator. In this case λ has a direct physical meaning and it is appropriate to think of a process in which λ gradually increases its value from zero to a certain magnitude. There are cases, however, in which the Hamiltonian is split into two parts more or less artificially. What is important in such cases is that one part can be considered small in comparison with the other. Although we may finally discard the *perturbation parameter* (putting $\lambda = 1$) it is convenient to retain λ so as to keep track of the "order" of approximation, a term containing the factor λ^n being the "nth order" of smallness. The actual perturbation $H^{(1)}$ may then be "switched on", by increasing λ to its final value $\lambda = 1$.

We have assumed that we know all the eigenvalues and eigenfunctions $\{\Psi_n{}^{(0)}, E_n{}^{(0)}\}$ of the unperturbed Hamiltonian: these satisfy

$$H^{(0)}\Psi_n{}^{(0)} = E_n{}^{(0)}\Psi_n{}^{(0)}.$$

When a perturbation is added to the Hamiltonian, the eigenvalues and eigenfunctions will change, but their changes will be continuous in λ. In this section we also assume that the state on which attention is focused is non-degenerate; the degenerate case is considered in Section 1.3. Finally, we assume $\{\Psi_n{}^{(0)}\}$ to be a *complete* orthonormal set (Vol. 1, Section 4.7).

Let us expand both the eigenvalues and eigenfunctions in terms of λ:

$$E_n = E_n{}^{(0)} + \lambda E_n{}^{(1)} + \lambda^2 E_n{}^{(2)} + ..., \tag{1.4}$$

$$\Psi_n = \Psi_n{}^{(0)} + \lambda \Psi_n{}^{(1)} + \lambda^2 \Psi_n{}^{(2)} + ... \tag{1.5}$$

On substituting in the Schrödinger equation

$$\mathsf{H}\Psi_n = E_n\Psi_n$$

and observing that the coefficients of the same power of λ on both sides should be equal, we obtain equations for the zero-, first-, second- and nth-order corrections. The zero-, first-, second- and third-order equations are:

$$\mathsf{H}^{(0)}\Psi_n{}^{(0)} = E_n{}^{(0)}\Psi_n{}^{(0)}, \tag{1.6a}$$

$$(\mathsf{H}^{(0)} - E_n{}^{(0)})\Psi_n{}^{(1)} = (E_n{}^{(1)} - \mathsf{H}^{(1)})\Psi_n{}^{(0)}, \tag{1.6b}$$

$$(\mathsf{H}^{(0)} - E_n{}^{(0)})\Psi_n{}^{(2)} = E_n{}^{(2)}\Psi_n{}^{(0)} + (E_n{}^{(1)} - \mathsf{H}^{(1)})\Psi_n{}^{(1)}, \tag{1.6c}$$

$$(\mathsf{H}^{(0)} - E_n{}^{(0)})\Psi_n{}^{(3)} = E_n{}^{(3)}\Psi_n{}^{(0)} + E_n{}^{(2)}\Psi_n{}^{(1)} + (E_n{}^{(1)} - \mathsf{H}^{(1)})\Psi_n{}^{(2)}. \tag{1.6d}$$

The zero-order equation is the one we have already assumed satisfied by the eigenfunctions $\Psi_n{}^{(0)}$. The next equation determines the correction term $\Psi_n{}^{(1)}$. On multiplying both sides of (1.6b) by $\Psi_n{}^{(0)}$ and integrating over all space, the following relation is obtained:

$$\langle\Psi_n{}^{(0)}|\mathsf{H}^{(0)}|\Psi_n{}^{(1)}\rangle - E_n{}^{(0)}\langle\Psi_n{}^{(0)}|\Psi_n{}^{(1)}\rangle$$
$$= E_n{}^{(1)} - \langle\Psi_n{}^{(0)}|\mathsf{H}^{(1)}|\Psi_n{}^{(0)}\rangle. \tag{1.7}$$

Because $\mathsf{H}^{(0)}$ is a Hermitian operator and $\Psi_n{}^{(0)}$ satisfies (1.6a), the left-hand side vanishes. With the abbreviated notation

$$\langle\Psi_m{}^{(0)}|\mathsf{H}^{(1)}|\Psi_n{}^{(0)}\rangle = \langle m|\mathsf{H}^{(1)}|n\rangle$$

for matrix elements between the *unperturbed* eigenfunctions, (1.7) then yields

$$E_n{}^{(1)} = \langle n|\mathsf{H}^{(1)}|n\rangle. \tag{1.8}$$

This important result shows that the energy change due to application of a perturbation may be calculated, to first order, as the expectation value of the perturbation in the unperturbed state. A similar result is well known in classical mechanics.

The equations for the higher-order energy corrections involve the changes $\Psi_n{}^{(1)}$, $\Psi_n{}^{(2)}$, ... in the eigen*functions*, and to proceed further we must therefore first determine $\Psi_n{}^{(1)}$. We note that the first-order correction may be obtained in principle from the differential equation (1.6b), in which $\Psi_n{}^{(1)}$ is the only unknown function: this possibility is considered in Section 1.6. Here, however, we proceed in an alternative way by making the expansion

$$\Psi_n{}^{(1)} = \sum_i c_i\Psi_i{}^{(0)}. \tag{1.9}$$

Instead of solving the differential equation, we must then obtain the set of coefficients $\{c_i\}$. On substituting (1.9) in (1.6b) and making use of (1.6a), we find

$$\sum_i c_i(E_i{}^{(0)} - E_n{}^{(0)})\Psi_i{}^{(0)} = (E_n{}^{(1)} - \mathsf{H}^{(1)})\Psi_n{}^{(0)}.$$

Multiplication throughout by $\Psi_m{}^{(0)*}$ $(m \neq n)$, followed by integration over all space, then yields (remembering orthonormality, p. 3)

$$c_m(E_m{}^{(0)} - E_n{}^{(0)}) = -\langle m|\mathsf{H}^{(1)}|n\rangle$$

and consequently

$$c_m = -\frac{\langle m|\mathsf{H}^{(1)}|n\rangle}{E_m{}^{(0)} - E_n{}^{(0)}} \quad (m \neq n). \tag{1.10}$$

The coefficient c_n cannot be determined from (1.6b), but may be fixed from the normalization condition: thus

$$\langle\Psi_n|\Psi_n\rangle = \langle\Psi_n{}^{(0)}|\Psi_n{}^{(0)}\rangle + \lambda\langle\Psi_n{}^{(0)}|\Psi_n{}^{(1)}\rangle + \lambda\langle\Psi_n{}^{(1)}|\Psi_n{}^{(0)}\rangle + \cdots$$
$$= 1 + \lambda(c_n + c_n{}^*) + \cdots$$

and for this to be unity c_n must be purely imaginary. Hence the effect of admitting $c_n \neq 0$ is, to first order, merely to multiply the wave function by a phase factor $e^{i\theta}$. Since the phase is arbitrary there is no loss of generality in taking $c_n = 0$. From (1.10) it then follows that

$$\Psi_n{}^{(1)} = -\sum_{m(\neq n)} \frac{\langle m|\mathsf{H}^{(1)}|n\rangle}{E_m{}^{(0)} - E_n{}^{(0)}} \Psi_m{}^{(0)}. \tag{1.11}$$

The effect of the perturbation on the wave function $\Psi_n{}^{(0)}$ is thus to "mix in" small parts of all the functions $\Psi_m{}^{(0)}$ $(m \neq n)$ which are "connected" with it by non-zero matrix elements $\langle m|\mathsf{H}^{(1)}|n\rangle$.

After finding $\Psi_n{}^{(1)}$, we may use (1.6c) to find the *second*-order energy correction. On substituting (1.11) in (1.6c) and proceeding as in the derivation of $E_n{}^{(1)}$, we find easily

$$E_n{}^{(2)} = -\sum_{m(\neq n)} \frac{|\langle m|\mathsf{H}^{(1)}|n\rangle|^2}{E_m{}^{(0)} - E_n{}^{(0)}}. \tag{1.12}$$

The unperturbed states most important in determining the correction are again those for which $\langle m|\mathsf{H}^{(1)}|n\rangle$ are large, and which are closest in energy to the unperturbed state of interest.

The expansion coefficients in the second-order correction to the wave function,

$$\Psi_n{}^{(2)} = \sum_i d_i\Psi_i{}^{(0)} \tag{1.13}$$

may be determined essentially as in the first-order case. For $(m \neq n)$ we find

$$d_m = -\frac{\langle n|\mathsf{H}^{(1)}|n\rangle\langle m|\mathsf{H}^{(1)}|n\rangle}{(E_m{}^{(0)} - E_n{}^{(0)})^2}$$

$$+\frac{1}{E_m{}^{(0)} - E_n{}^{(0)}} \sum_{k(\neq n)} \frac{\langle m|\mathsf{H}^{(1)}|k\rangle\langle k|\mathsf{H}^{(1)}|n\rangle}{(E_k{}^{(0)} - E_n{}^{(0)})} \quad (m \neq n) \quad (1.14\text{a})$$

while d_n may again be determined from the normalization condition. The imaginary part of d_n may again be dropped, but there is now a non-zero real part

$$d_n = -\tfrac{1}{2} \sum_{k(\neq n)} \frac{|\langle k|\mathsf{H}^{(1)}|n\rangle|^2}{(E_k{}^{(0)} - E_n{}^{(0)})^2}. \qquad (1.14\text{b})$$

The second-order part of the wave function thus follows from (1.13) with the coefficients given by (1.14).

This procedure may be continued indefinitely. Here we note only that

$$E_n{}^{(3)} = \sum_{m(\neq n)} \sum_{k(\neq n)} \frac{\langle n|\mathsf{H}^{(1)}|m\rangle\langle m|\mathsf{H}^{(1)}|k\rangle\langle k|\mathsf{H}^{(1)}|n\rangle}{(E_m{}^{(0)} - E_n{}^{(0)})(E_k{}^{(0)} - E_n{}^{(0)})}$$

$$-\langle n|H^{(1)}|n\rangle \sum_{m(\neq n)} \frac{|\langle m|\mathsf{H}^{(1)}|n\rangle|^2}{(E_m{}^{(0)} - E_n{}^{(0)})^2}, \qquad (1.15)$$

a result which is sometimes useful, although evaluation of the sums usually leads to great difficulty. This difficulty becomes particularly acute when, as in many applications, the complete set of eigenfunctions of the unperturbed system includes a continuum: in this case (cf. Vol. 1, p. 86) summation must, over part of the range, be replaced by integration and careful convergence considerations become necessary.

EXAMPLE. *Stark effect in hydrogen. Ground state.* When an electric field is applied to a system the energy levels are usually shifted and degenerate levels split. This phenomenon is the Stark effect, which we now consider for the ground state of the hydrogen atom.

In actual calculations it is convenient to employ atomic units (fully discussed in App. 1) in which e, m and $\hbar$ provide the natural units of charge, mass and action, respectively. The units of length and energy are the "Bohr" and the "Hartree":

$$1\,\text{B} = a_0 = \frac{\hbar^2 \kappa_0}{m\,e^2}, \qquad 1\,\text{H} = E_0 = \frac{e^2}{\kappa_0 a_0}$$

where $\kappa_0 = 4\pi\varepsilon_0$ (ε_0 = permittivity of free space) and accordingly takes the numerical value unity when atomic units are used throughout. When all quantities are expressed in these units, the corresponding symbols (r, E, etc.) should be regarded as *dimensionless variables*; the fundamental constants e, m, $\hbar$, κ_0 take unit values and disappear from the equations; and any calculated quantity is understood to be the numerical measure of the corresponding physical quantity in the appropriate atomic units. For the present, however, we carry the units explicitly in order to indicate the correct physical dimensions of all quantities.

The first few unperturbed states of the hydrogen atom, anticipating the results of Chap. 2, are (using the notation Φ_n to avoid superscripts)

$$\Phi_1(\mathbf{r}) = N_1\, e^{-r/a_0} \qquad E_1 = -(\tfrac{1}{2})\,(e^2/a_0\kappa_0) \ \ (1s \text{ state}),$$

$$\Phi_2(\mathbf{r}) = N_2(r/a_0 - 2)\, e^{-r/2a_0} \qquad E_2 = -(\tfrac{1}{8})\,(e^2/a_0\kappa_0) \ \ (2s \text{ state}),$$

$$\Phi_3(\mathbf{r}) = N_2(r/a_0)\, e^{-r/2a_0} \cos\theta \qquad E_3 = -(\tfrac{1}{8})\,(e^2/a_0\kappa_0) \ \ (2p_0 \text{ state}),$$

$$\Phi_{4,5}(\mathbf{r}) = N_2(r/a_0)\, e^{-r/2a_0}\, 2^{-\frac{1}{2}} \sin\theta\, e^{\pm i\phi} \qquad E_{4,5} = -(\tfrac{1}{8})\,(e^2/a_0\kappa_0) \ \ (2p_{\pm 1} \text{ states}).$$

The last four states (Φ_4, Φ_5 corresponding to the upper and lower signs) are degenerate, but the non-degenerate perturbation theory applies to the state of interest, Φ_1. The normalizing factors are $N_1 = (\pi a_0{}^3)^{-1/2}$ and $N_2 = (32\pi a_0{}^3)^{-1/2}$.

When a uniform field F is applied (fixing, we suppose, a z-axis) the extra potential energy term in the Hamiltonian is[‡] Fze. If we use F in place of λ as the perturbation parameter, the perturbation is described by

$$\mathsf{H}^{(1)} = ez.$$

The first-order energy change is therefore, using (1.8),

$$E^{(1)} = \langle\Phi_1|\mathsf{H}^{(1)}|\Phi_1\rangle = 0$$

vanishing because every contribution to the integral from a volume element at point x, y, z is accompanied by a contribution of equal magnitude but opposite sign from one at point x, y, $-z$. The absence of a first-order change simply means the system has no dipole moment in its ground state.

There is, however, a first-order change in the wave function. If we include only the unperturbed states listed above, there is *one* non-zero off-diagonal element, H_{13}. The integral is easily evaluated to give

$$H_{13}(=H_{31}) = \langle\Phi_1|\mathsf{H}^{(1)}|\Phi_3\rangle = (2^8/3^5\,\sqrt{2})a_0\, e.$$

Since $E_3 - E_1 = \tfrac{3}{8}(e^2/a_0\kappa_0)$ the perturbed wave function, to first order, is

$$\Psi_1 = \Phi_1 - F(a_0{}^2\kappa_0/e)\,(2^{11}/3^6\,\sqrt{2})\Phi_3 + \dots$$

where the remaining terms arise from the higher unperturbed states. This result indicates that the perturbation causes a distortion of the wave function, symmetrical about the field axis, the centroid of the probability density moving in the negative z-direction (Fig. 1.1)—as would be expected intuitively since the electron carries a negative charge. The system is therefore *polarized* by the field. The natural unit of electric field intensity (see App. 1) is $e/a_0{}^2\kappa_0$, one unit corresponding to $5 \cdot 14220 \times 10^{11}$ V m^{-1}, and the actual distortion of the charge density by the fields normally encountered is therefore very small.

For the second-order energy change[§] we obtain, from (1.12),

$$E^{(2)} = -(2^{18}/3^{11})a_0{}^3\kappa_0 + \dots = -1 \cdot 480(a_0{}^3\kappa_0) + \dots$$

where again the remaining terms arise from admixture of the higher states. In

---

[‡] The arbitrary constant in the electrostatic potential is chosen so that the field potential energy vanishes at the origin.

[§] More correctly, the energy change is $F^2 E^{(2)}$; when F is expressed in units of $e/a_0{}^2\kappa_0$ this will appear in units of $e^2/\kappa_0 a_0$ (i.e. in Hartrees).

this simple example, the summation may be completed rigorously to give $E^{(2)} = -\tfrac{9}{4}(a_0{}^3\kappa_0)$. The result based on four excited functions (three of them ineffective) is therefore rather poor.

It is clear from the example just considered that a major disadvantage of perturbation theory, in the form presented so far, is the difficulty of evaluating the infinite sums over all unperturbed states. The highly excited functions are frequently unavailable, and usually they include a continuum of positive energy solutions (which may be mathematically awkward to handle).

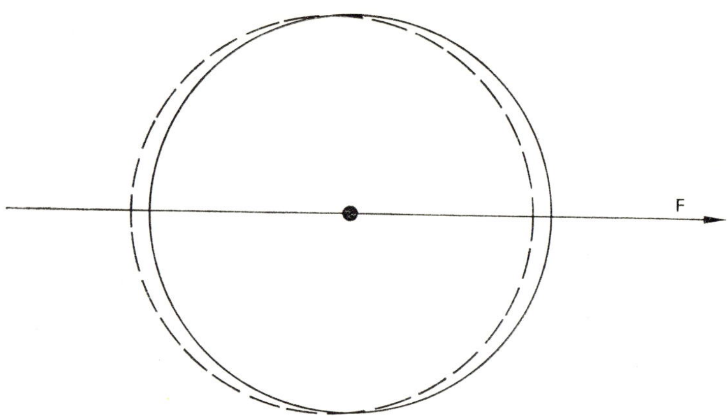

FIG. 1.1. Polarization of hydrogen $1s$ orbital. The contours of constant Φ are deformed by the field, probability density moving in the negative direction.

1.3. Time-independent perturbation theory. Degenerate case

We now consider the case in which the kth eigenvalue of the unperturbed Hamiltonian is n-fold degenerate. Let the linearly independent degenerate eigenfunctions be distinguished by a second subscript m, thus:

$$\mathsf{H}^{(0)}\Psi_{km}{}^{(0)} = E_k{}^{(0)}\Psi_{km}{}^{(0)} \quad (m = 1, 2, \ldots n). \qquad (1.16)$$

The eigenvalues and eigenfunctions of the unperturbed problem are expanded as before, as power series in λ;

$$E_{km} = E_k{}^{(0)} + \lambda E_{km}{}^{(1)} + \lambda^2 E_{km}{}^{(2)} + \ldots, \qquad (1.17)$$

$$\Psi_{km} = \Psi_{km}{}^{(0)} + \lambda\Psi_{km}{}^{(1)} + \lambda^2\Psi_{km}{}^{(2)} + \ldots \qquad (1.18)$$

where in general $E_{km}{}^{(1)}$, $\Psi_{km}{}^{(1)}$, etc., depend on m.

In (1.18) we see that when λ tends to zero, Ψ_{km} tends to $\Psi_{km}^{(0)}$, which is called a "correct zeroth-order eigenfunction". But as the kth eigenvalue is n-fold degenerate, *any* linear combination of the n eigenfunctions is also an eigenfunction associated with the same eigenvalue $E_k^{(0)}$. Consequently, if we are given a set of *arbitrarily chosen* eigenfunctions $\{\Phi_{km}^{(0)}\}$ we can hardly expect that these will coincide with the correct zeroth-order eigenfunctions. Our first task is therefore to find the correct zero-order eigenfunctions in terms of the arbitrary set.

Let us write

$$\Psi_{km}^{(0)} = \sum_{j=1}^{n} c_{mj}\Phi_{kj}^{(0)} \tag{1.19}$$

and then proceed to the separation of the orders as in the non-degenerate case. We substitute the expansions (1.17) and (1.18) into the full eigenvalue problem

$$(\mathsf{H}^{(0)} + \lambda\mathsf{H}^{(1)})\Psi_{km} = E_{km}\Psi_{km} \tag{1.20}$$

and obtain from the terms of first order in λ

$$(\mathsf{H}^{(0)} - E_k^{(0)})\Psi_{km}^{(1)} = (E_{km}^{(1)} - \mathsf{H}^{(1)})\Psi_{km}^{(0)}.$$

On multiplying by $\Phi_{kj}^{(0)*}$ from the left and integrating, the left-hand side of the above equation vanishes, i.e.

$$\langle\Phi_{kj}^{(0)}|\mathsf{H}^{(0)}|\Psi_{km}^{(1)}\rangle - E_k^{(0)}\langle\Phi_{kj}^{(0)}|\Psi_{km}^{(1)}\rangle = 0$$

because[‡] $\Phi_{kj}^{(0)}$ is an eigenfunction of $\mathsf{H}^{(0)}$ with eigenvalue $E_k^{(0)}$. When the same operations are applied to the right-hand side, and (1.19) is then substituted for $\Psi_{km}^{(0)}$, we obtain

$$E_{km}^{(1)}c_{mj} - \sum_{l=1}^{n} \langle\Phi_{kj}^{(0)}|\mathsf{H}^{(1)}|\Phi_{kl}^{(0)}\rangle c_{ml}. \tag{1.21}$$

On dropping the label km (of the required zero-order state) and putting

$$H_{jl}^{(1)} = \langle\Phi_{kj}^{(0)}|\mathsf{H}^{(1)}|\Phi_{kl}^{(0)}\rangle$$

it follows that the coefficients and first-order energy change should satisfy the following *secular equations*:

$$\sum_{l=1}^{n} (H_{jl}^{(1)} - E^{(1)}\,\delta_{jl})c_l = 0 \quad (j = 1, 2, \ldots n). \tag{1.22a}$$

These are simply linear homogeneous equations of the kind discussed

---

‡ Remember that $\mathsf{H}^{(0)}$ is a Hermitian operator, and may therefore be applied instead to the left-hand function in the matrix element, and that $E_k^{(0)}$ is real.

in Vol. 1, Section 3.7 and the n eigenvalues of $E^{(1)}$ are the roots of the secular determinant

$$\det \left| H_{jl}^{(1)} - E_{km}^{(1)}\, \delta_{jl} \right| = 0. \tag{1.22b}$$

If the roots are distinct, the mth being $E^{(1)} = E_{km}^{(1)}$, the ratios of coefficients $c_{m1}:c_{m2}: \ldots : c_{mn}$ are uniquely determined by (1.22a) and their absolute values by the normalization condition. All n solutions follow in this way.

When a root of (1.22b) remains s-fold degenerate, the corresponding eigenfunctions are not determined uniquely and they may be mixed by an arbitrarily unitary transformation. This arbitrariness may be removed if we perform a higher-order perturbation calculation which lifts the degeneracy: the calculation follows similar lines.

It should be noted that, since $H_{jl} = H_{jl}^{(0)} + \lambda H_{jl}^{(1)} = E^{(0)}\, \delta_{jl} + \lambda H_{jl}^{(1)}$, equation (1.22a) may be rewritten so as to give the total energy E_k to first order. In matrix form the resultant equation is

$$\mathbf{Hc} = E\mathbf{c}. \tag{1.23}$$

This is essentially the matrix form of the eigenvalue equation, as discussed in Vol. 1, Section 3.7. It differs from the corresponding equation for the coefficients in a complete set expansion in that the square matrix $\mathbf{H}$ and the column of coefficients are *finite*: the coefficients refer to a truncated expansion of the wave function in terms of the unperturbed functions referring *only to the degenerate level under consideration*. The relation between the exact solutions (complete set) and the first-order approximations (truncated set) is discussed further in a later section. The same method of solution applies in both cases.

Once we have the correct zeroth-order wave function $\Psi_{km}^{(0)}$ and the first-order corrections to the eigenvalue, $E_{km}^{(1)}$, we can apply the method of Section 1.1 to obtain $\Psi_{km}^{(1)}$, $E_{km}^{(2)}$, $\Psi_{km}^{(2)}$, $E_{km}^{(3)}$ and so on.

EXAMPLE. *Stark effect in hydrogen. Excited states.* We return to the perturbation problem discussed in Section 1.1 (p. 6), but now consider the first four excited states ($\Phi_2, \ldots \Phi_5$), which form a degenerate set. The perturbation operator $H^{(1)}$ and the forms of the unperturbed functions have been given already; but now we must determine the correct zeroth-order wave functions. To do this we need the matrix elements of $H^{(1)}$ with respect to the four functions, and from symmetry considerations it follows easily that the only non-zero element is

$$\langle \Phi_2 | H^{(1)} | \Phi_3 \rangle = 3a_0 e.$$

Thus the secular determinant (1.22b) becomes

$$\begin{vmatrix} -E^{(1)} & 3a_0 e & 0 & 0 \\ 3a_0 e & -E^{(1)} & 0 & 0 \\ 0 & 0 & -E^{(1)} & 0 \\ 0 & 0 & 0 & -E^{(1)} \end{vmatrix} = 0.$$

The four roots are $3a_0e$, $-3a_0e$, 0, and 0. The corresponding zero-order eigenfunctions and first-order energies $(E^{(0)} + FE^{(1)})$ are thus

$(\Phi_2+\Phi_3)/\sqrt{2}$	$(\Phi_2-\Phi_3)/\sqrt{2}$	Φ_4	Φ_5
$E^{(0)} + 3a_0eF$	$E^{(0)} - 3a_0eF$	$E^{(0)}$	$E^{(0)}$

In this case the functions Φ_4 and Φ_5 remain degenerate, even in higher order, as a consequence of axial symmetry. More complete discussions of this linear Stark effect may be found elsewhere (e.g. Kemble, 1937, pp. 403–8).

1.4.   The variation method

The perturbation method is sometimes difficult to apply: it requires, for example, the availability of an exactly soluble eigenvalue equation, to which the problem of interest is closely related; and it is not necessarily convergent. We now turn to an alternative method, in some ways more general, which does not suffer from such disadvantages. First we consider the simplest statement of the method, applicable to the calculation of a ground state wave function.

(a)   *Ground state*

The method to be developed depends on the following very simple theorem:

> For a physically eligible but otherwise arbitrary function Φ, the expectation value of the Hamiltonian is an upper bound to the lowest eigenvalue;
> $$\mathscr{E} = \frac{\langle \Phi | \mathsf{H} | \Phi \rangle}{\langle \Phi | \Phi \rangle} \geqq E_1, \tag{1.24}$$
> the equality applying when Φ is an exact solution.

A proof goes as follows. We number the eigenvalues and associated eigenfunctions of the Hamiltonian from below:

$$\Psi_1 \quad \Psi_2 \quad \Psi_3 \dots$$
$$E_1 \leqq E_2 \leqq E_3 \dots.$$

According to the Postulates (Vol. 1, Chap. 4) the set $\{\Psi_i\}$ is complete, and Φ can be expanded as

$$\Phi = \sum_i c_i \Psi_i.$$

From the eigenvalue property and the orthonormality of the functions $\{\Psi_i\}$ it then follows that

$$\langle\Phi|H|\Phi\rangle = \sum_i |c_i|^2 E_i \geqq E_0 \sum |c_i|^2.$$

But

$$\sum_i |c_i|^2 = \langle\Phi|\Phi\rangle$$

and on substitution the theorem follows immediately. The equality holds only when $c_i = 0$ for $E_i > E_1$; in this case $\Phi = \Psi_1$ (non-degenerate ground state) or is an arbitrary mixture of functions with $E_i = E_1$ (degenerate ground state).

The variation theorem (1.24) provides an immediate criterion for comparing two wave functions, the "better" approximation to the ground state wave function being taken as the one which yields the closer estimate of the eigenvalue. The "best" wave function (Φ) of any assumed form is therefore obtained by making the expectation value in (1.24) as small as possible (e.g. by varying any parameters which the function may contain). This method of constructing approximate wave functions by minimizing an energy expectation value is the *variation method*.

EXAMPLE. *The hydrogen atom.* As an example, let us use a Gaussian function of the form

$$\Phi(r) = \exp\left(-\mu^2 r^2\right),$$

where μ is a variational parameter, as an approximate wave function for the ground state of the hydrogen atom.

As Φ is independent of angles, the ∇^2 operator in the Hamiltonian reduces to give

$$H = -\frac{\hbar^2}{2m}\frac{1}{r^2}\frac{\partial}{\partial r}\left(r^2\frac{\partial}{\partial r}\right) - \frac{e^2}{r}.$$

By performing the integration, we obtain

$$\mathcal{E} = \frac{\langle\Phi|H|\Phi\rangle}{\langle\Phi|\Phi\rangle} = \frac{3\hbar^2}{2m}\left[\mu - \frac{2me^2}{3\hbar^2\kappa_0}\sqrt{\frac{2}{\pi}}\right]^2 - \frac{4}{3\pi}\frac{me^4}{\hbar^2\kappa_0^2},$$

which takes a minimum value when

$$\mu = \frac{2me^2}{3\hbar^2\kappa_0}\sqrt{\frac{2}{\pi}} = 0{\cdot}5319(1/a_0)$$

where $a_0 = \hbar^2\kappa_0/me^2$ is the Bohr radius. The minimum value, which is a closest upper bound to the exact ground state energy (-0.5 H), is

$$\mathscr{E}_{\min} = -\frac{4}{3\pi}\frac{me^4}{\hbar^2\kappa_0{}^2} = -0.4244 \text{ H}.$$

A single Gaussian function thus gives a poor variational energy, in error by some 15%.

To improve the result obtained in the above Example it would be necessary to use a more flexible functional form of Φ, possibly with several parameters. In fact, linear combinations of several Gaussian functions can give a close approximation to the exact function.

(b) *Excited states*

A procedure parallel to that just developed may be applied to excited states. It depends on the following theorem:

If Φ is orthogonal to the exact ground state[‡] eigenfunction Ψ_1 but is otherwise arbitrary, then

$$\mathscr{E} = \frac{\langle\Phi|\mathsf{H}|\Phi\rangle}{\langle\Phi|\Phi\rangle} \geqq E_2 \qquad (1.25)$$

where E_2 is the energy of the first excited state.

This theorem follows as for the ground state, on noting that orthogonality requires $c_1 = 0$ in the expansion

$$\Phi = \sum_i c_i\Psi_i.$$

For the second and higher excited states, similar theorems can be obtained. If we are interested in the kth excited state, we require that a trial function Φ be orthogonal to all the eigenfunctions $\Psi_1, \Psi_2, ..., \Psi_{k-1}$. Then the expectation value of H calculated with Φ is always equal to or larger than the true eigenvalue E_k.

These theorems for excited states have less utility than (1.24) since exact eigenfunctions are seldom available, even for the ground state. However, if Ψ_2 is *known* to be orthogonal to Ψ_1 by symmetry, then the variation theorem (1.24) becomes very useful in obtaining approximations for Ψ_2 and E_2.

---

‡ When the ground state is degenerate, Φ is required to be orthogonal to all the eigenfunctions with the eigenvalue E_1. The ground state is here assumed to be non-degenerate but extension to the degenerate case is straightforward.

(c) *Linear variation functions*

In quantum chemistry the wave function is very frequently approximated as a linear combination of a finite number of linearly independent basis functions $\{\Phi_i\}$, $i = 1, 2, ..., n$:

$$\Phi = \sum_{i=1}^{n} c_i \Phi_i. \tag{1.26}$$

The theory of such expansions is discussed in Vol. 1, Chap. 3, where it is shown that for complete set, with $n \to \infty$, the Schrödinger equation is equivalent to a *matrix* eigenvalue equation of the form (expansion functions not assumed orthonormal)

$$\mathbf{Hc} = E\mathbf{Sc} \tag{1.27}$$

in which $\mathbf{H}$ and $\mathbf{S}$ are infinite Hermitian matrices with elements

$$H_{ij} = \langle \Phi_i | \mathsf{H} | \Phi_j \rangle, \quad S_{ij} = \langle \Phi_i | \Phi_j \rangle \tag{1.28}$$

and the column $\mathbf{c}$ comprises an infinite set of expansion coefficients. The present problem, first discussed by Rayleigh and Ritz, is to find the optimum choice of coefficients in the *finite* expansion (1.26) which arises by "truncation" to the first n terms.

To obtain the optimum coefficients we insert (1.26) in the variational expression

$$\mathscr{E} = \frac{\langle \Phi | \mathsf{H} | \Phi \rangle}{\langle \Phi | \Phi \rangle}$$

and require that $\mathscr{E}$ be stationary against variation of the coefficients. Thus, on differentiating[‡] with respect to $c_k{}^*$, we require

$$\sum_i H_{ki}c_i - \frac{\partial \mathscr{E}}{\partial c_k{}^*} \sum_{j,i} c_j{}^* S_{ji} c_i - \mathscr{E} \sum_i S_{ki} c_i = 0. \tag{1.29}$$

This equation applies for all values of k, and the required stationary property of $\mathscr{E}$, namely $(\partial \mathscr{E}/\partial c_k{}^*) = 0$ $(k = 1, 2, ... n)$, therefore leads to the condition

$$\sum_{i=1}^{n} (H_{ki} - \mathscr{E}S_{ki})c_i = 0 \quad (k = 1, 2, ... n). \tag{1.30}$$

These equations, equivalent to (1.27) but with finite matrices, possess non-trivial solutions only for values of $\mathscr{E}$ which satisfy

$$\det |H_{ki} - \mathscr{E}S_{ki}| = 0. \tag{1.31}$$

This is essentially a polynomial equation with n roots. The lowest of

---

‡ Note that c_j and $c_j{}^*$ may be treated as independent variables (see Vol. 1, p. 49).

these ($\mathscr{E}_1$) is a best approximation to the ground state energy obtainable from a function of the form (1.26) with any given (i.e. fixed) basis functions. The corresponding wave function is found by inserting $\mathscr{E}_1$ into (1.30) and solving for the coefficients (cf. Vol. 1, Section 3.7). The resultant function is then a best approximation "according to the energy criterion" to the ground state wave function.

A very important advantage of using a linear variation function is that the kth root of (1.31) is an upper bound to the kth true eigenvalue, and therefore approaches, from above, the energy of the kth *excited state*. This property was found by MacDonald (1933) and was also

$$\mathscr{E}_k \geqq E_k, \tag{1.32}$$

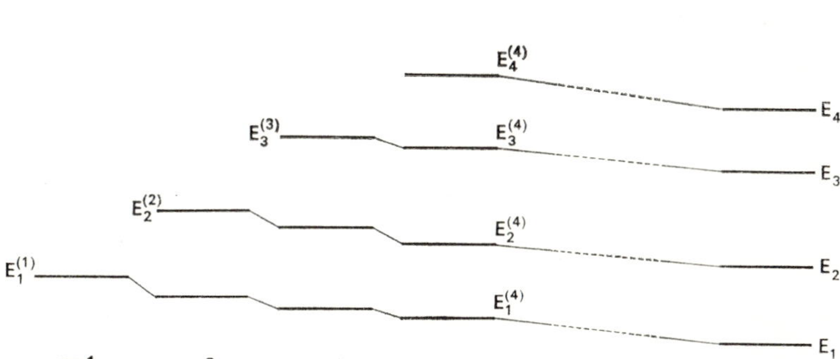

Fig. 1.2. Separation theorem. As the number of basis functions (n) is increased each $E_k^{(n+1)}$ falls between $E_{k-1}^{(n)}$ and $E_k^{(n)}$. Thus every E_k is approached monotonically from above.

implicit in work by Hylleraas and Undheim (1930). The theorem (1.32) originates from certain inequalities that can be established by regarding (1.31) as a bordered determinant and expanding accordingly. The effect of adding more and more linearly independent functions to the basis set, so that it approaches completeness, is indicated pictorially in Fig. 1.2 from which the theorem is self-evident.

These results are of great value not only from a practical point of view but also because they firmly establish the effects of truncation of the complete set expansions, postulated and discussed in Chapters 3 and 4 of Vol. 1. In short, the infinite matrices in (1.27), which is exactly equivalent to the Schrödinger equation, may be truncated to n rows and columns and can still provide useful approximations to the first n states. This is true even in the absence of *exact* solutions for any of the states, and the finite form of (1.27) thus provides a powerful approximation method.

1.5. Time-independent perturbation theory. The partitioning technique

We now consider an alternative version of the time-independent perturbation theory. This version provides a link between the perturbation theory treated in Sections 1.2 and 1.3 and the variation method discussed in the previous section, and is useful both in practical calculation and theoretical manipulations (e.g. Löwdin, 1963).

We start from (1.30), which (dropping the notational distinction between exact and approximate energies) may be written (cf. (1.27))

$$\mathbf{H}\mathbf{c} = E\mathbf{S}\mathbf{c} \tag{1.33}$$

where, for the n-term approximation (1.26). $\mathbf{H}$ and $\mathbf{S}$ are $n \times n$ matrices with elements given by (1.28). On choosing an orthonormal basis, this becomes

$$\mathbf{H}\mathbf{c} = E\mathbf{c}. \tag{1.34}$$

Now let us partition the basis set $\{\Phi_i\}$ into two sub-sets $\{\Phi_i{}^A\}$ and $\{\Phi_j{}^B\}$, where the A set might be regarded as a basis for a first approximation, while admission of the functions in the B set would represent a further refinement. We then rewrite (1.34) formally as

$$\begin{pmatrix} \mathbf{H}^{AA} & \mathbf{H}^{AB} \\ \mathbf{H}^{BA} & \mathbf{H}^{BB} \end{pmatrix} \begin{pmatrix} \mathbf{c}^A \\ \mathbf{c}^B \end{pmatrix} = E \begin{pmatrix} \mathbf{c}^A \\ \mathbf{c}^B \end{pmatrix},$$

which is equivalent to the two equations

$$\mathbf{H}^{AA}\mathbf{c}^A + \mathbf{H}^{AB}\mathbf{c}^B = E\mathbf{c}^A, \tag{1.34a}$$

$$\mathbf{H}^{BA}\mathbf{c}^A + \mathbf{H}^{BB}\mathbf{c}^B = E\mathbf{c}^B. \tag{1.34b}$$

If the expansion coefficients of the B-set functions were all zero, the second term in (1.34a) would vanish: we therefore regard it as a small correction term and estimate it from (1.34b) by solving formally for $\mathbf{c}^B$. Thus, using $\mathbf{E}$ for the *diagonal* matrix with $E_{ii} = E$ (all i),

$$\mathbf{c}^B = (\mathbf{E} - \mathbf{H}^{BB})^{-1}\mathbf{H}^{BA}\mathbf{c}^A \tag{1.35}$$

where $(\mathbf{E} - \mathbf{H}^{BB})^{-1}$ is defined as the inverse matrix of $(\mathbf{E} - \mathbf{H}^{BB})$. Substitution in (1.34a) then yields

$$[\mathbf{H}^{AA} + \mathbf{H}^{AB}(\mathbf{E} - \mathbf{H}^{BB})^{-1}\mathbf{H}^{BA}]\mathbf{c}^A = \mathbf{E}\mathbf{c}^A.$$

If we define a new matrix

$$\mathbf{H}^A{}_{\mathrm{eff}} = \mathbf{H}^{AA} + \mathbf{H}^{AB}(\mathbf{E} - \mathbf{H}^{BB})^{-1}\mathbf{H}^{BA} \tag{1.36}$$

the last equation can be written simply as

$$\mathbf{H}^A{}_{\mathrm{eff}}\mathbf{c}^A = E\mathbf{c}^A, \tag{1.37}$$

which has exactly the same form as the original equation (1.34) but with a truncated column and an *effective* Hamiltonian matrix.

By introducing the effective Hamiltonian $\mathbf{H}^A_{\mathrm{eff}}$ we have succeeded in reducing the whole problem to one which involves explicitly the states belonging to the sub-set A only. There is no approximation involved up to this stage and the reduction is only formal; however, it is clear that $\mathbf{H}^A_{\mathrm{eff}}$ involves not only $\mathbf{H}^{AB}$, $\mathbf{H}^{BA}$ and $\mathbf{H}^{BB}$ but also the required eigenvalue E, and that actual solution would therefore require some kind of iterative process, involving successive approximations to E.

From (1.36) and (1.37), various perturbation formulae can be derived. First we separate the diagonal and off-diagonal parts of $\mathbf{H}^{BB}$, denoting them by $\mathbf{D}$ and $\mathbf{B}$ respectively so that

$$\mathbf{H}^{BB} = \mathbf{D} + \mathbf{B}.$$

The inverse matrix $(\mathbf{E} - \mathbf{H}^{BB})^{-1}$ in (1.36) can then be expanded as follows:

$$
\begin{aligned}
(\mathbf{E} - \mathbf{H}^{BB})^{-1} &= \{(\mathbf{E} - \mathbf{D}) - \mathbf{B}\}^{-1} \\
&= \{(\mathbf{E} - \mathbf{D})[1 - (\mathbf{E} - \mathbf{D})\mathbf{B}^{-1}]\}^{-1} \\
&= [1 - (\mathbf{E} - \mathbf{D})^{-1}\mathbf{B}]^{-1}(\mathbf{E} - \mathbf{D})^{-1} \\
&= [1 + (\mathbf{E} - \mathbf{D})^{-1}\mathbf{B} + (\mathbf{E} - \mathbf{D})^{-1}\mathbf{B}(\mathbf{E} - \mathbf{D})^{-1}\mathbf{B} + \ldots](\mathbf{E} - \mathbf{D})^{-1}.
\end{aligned}
$$

$$(1.38)$$

It is usual to write

$$(\mathbf{E} - \mathbf{D})^{-1} = \frac{1}{(\mathbf{E} - \mathbf{D})}$$

though care must be taken to indicate the correct position of this factor in any product, and substitution of the expansion into (1.36) then determines the effective Hamiltonian:

$$\mathbf{H}^A_{\mathrm{eff}} = \mathbf{H}^{AA} + \mathbf{H}^{AB} \frac{1}{(\mathbf{E} - \mathbf{D})} \mathbf{H}^{BA} + \mathbf{H}^{AB} \frac{1}{\mathbf{E} - \mathbf{D}} \mathbf{B} \frac{1}{\mathbf{E} - \mathbf{D}} \mathbf{H}^{BA} + \ldots \quad (1.39)$$

Since $\mathbf{B}$ is simply the non-diagonal part of the $\mathbf{H}^{BB}$ block of $\mathbf{H}$ its elements are $H_{\alpha\beta}$ where we now use Greek letters to distinguish the functions of set B. Thus, a typical element of the effective Hamiltonian matrix, connecting Φ_i and Φ_j of sub-set A, is

$$H^A_{\mathrm{eff}\,ij} = H_{ij} + \sum_\alpha \frac{H_{i\alpha}H_{\alpha j}}{E - H_{\alpha\alpha}} + \sideset{}{'}\sum_{\alpha,\beta} \frac{H_{i\alpha}H_{\alpha\beta}H_{\beta j}}{(E - H_{\alpha\alpha})(E - H_{\beta\beta})} + \ldots \quad (1.40)$$

where $\alpha \neq \beta$ in the second sum (primed) because $\mathbf{B}$ is the *off*-diagonal part of $\mathbf{H}^{BB}$.

The relationship between the coefficients contained in $\mathbf{c}^A$ and $\mathbf{c}^B$ also follows from the expansion of the inverse matrix. We find

$$c_\alpha{}^B = \sum_i \frac{H_{\alpha i}}{E - H_{\alpha\alpha}} c_i{}^A + \sum_{\substack{\beta,i \\ (\beta \neq \alpha)}} \frac{H_{\alpha\beta} H_{\beta i}}{(E - H_{\alpha\alpha})(E - H_{\beta\beta})} c_i{}^A + \ldots \quad (1.41)$$

on substituting (1.38) in the matrix form (1.35).

Two typical applications indicate the usefulness of the present formulation:

Case 1. We take a single non-degenerate state Φ_k as the sub-set A and regard all other functions as comprising the sub-set B. Then from (1.40) $\mathbf{H}^A_{\text{eff}}$ becomes a single element

$$H^A_{\text{eff}\,ii} = H_{ii} + \sum_\alpha \frac{H_{i\alpha} H_{\alpha i}}{(E - H_{\alpha\alpha})} + \sum_{\alpha,\beta} \frac{H_{i\alpha} H_{\alpha\beta} H_{\beta i}}{(E - H_{\alpha\alpha})(E - H_{\beta\beta})} + \ldots \quad (1.42)$$

and (1.37) shows this is equivalent to

$$E = H^A_{\text{eff}\,ii} = f(E) \quad (1.42\text{a})$$

which may be used to determine an eigenvalue iteratively. We start from any first approximation $E_i{}^{(0)}$, and construct the sequence $E_i{}^{(0)}$, $E_i{}^{(1)}$, $E_i{}^{(2)}$, ... using

$$E_i{}^{(n+1)} = f(E_i{}^{(n)}).$$

This normally converges to E_i if the starting approximation is sufficiently close.

The best-known application of the procedure occurs in deriving the perturbation formula due to Brillouin (1933) and Wigner (1935). This follows on supposing that

$$\mathsf{H} = \mathsf{H}^{(0)} + \lambda\mathsf{H}^{(1)} \quad (1.43)$$

and assuming the basis functions $\{\Phi_i\}$ are the eigenfunctions of $\mathsf{H}^{(0)}$, with eigenvalues $E_i{}^{(0)}$. We readily find, as a formula giving the ith eigenvalue ($E = E_i$) of the perturbed Hamiltonian H,

$$E = E_i{}^{(0)} + \lambda H_{ii}{}^{(1)} + \lambda^2 \sum_{j(\neq i)} \frac{|H_{ij}{}^{(1)}|^2}{E - E_j{}^{(0)} - \lambda H_{jj}{}^{(1)}}$$

$$+ \lambda^3 \sum_{j,k(\neq i)}' \frac{H_{ij}{}^{(1)} H_{jk}{}^{(1)} H_{ki}{}^{(1)}}{(E - E_j{}^{(0)} - \lambda H_{jj}{}^{(1)})(E - E_k{}^{(0)} - \lambda H_{kk}{}^{(1)})} + \ldots \quad (1.44)$$

where the dummy indices have been indicated by j, k and the exclusion of i is indicated explicitly. The prime on the summation sign means, as usual, $j \neq k$ and the matrix elements refer only to the operator $\mathsf{H}^{(1)}$ in (1.43).

The Schrödinger perturbation formula, derived in Section 1.1, can be obtained from the Brillouin–Wigner expansion by iteration, putting $E = E_i{}^{(0)}$, as a first approximation to E in the right-hand side of (1.44) and expanding in powers of λ. The linear term in λ yields (1.8); the quadratic term gives (1.12); and so on.

The perturbed wave function is calculated by using (1.41). When $\mathbf{c}^A$ consists of only one element (let us take $c_i = 1$) the elements of $\mathbf{c}^B$ are clearly

$$c_j = \frac{H_{ji}}{E - H_{jj}} + \frac{1}{E - H_{jj}} \sum_{k(\neq i,j)} \frac{H_{jk}H_{ki}}{E - H_{kk}} + \dots$$

The reductions following from the assumption (1.43) then give

$$c_j = \frac{\lambda H_{ji}{}^{(1)}}{(E - E_j{}^{(0)} - \lambda H_{jj}{}^{(1)})}$$

$$+ \frac{\lambda^2}{(E - E_j{}^{(0)} - \lambda H_{jj}{}^{(1)})} \sum_{k(\neq i,j)} \frac{H_{ik}{}^{(1)}H_{kj}{}^{(1)}}{(E - E_k{}^{(0)} - \lambda H_{kk}{}^{(1)})} + \dots \quad (1.45)$$

Iteration, and expansion in powers of λ (not forgetting that normalization is necessary), finally lead to (1.10) and (1.14), which follow from the linear and quadratic terms, respectively.

Case 2. When the state in which we are interested is degenerate, the usual procedure is to diagonalize the Hamiltonian first by a unitary transformation of these degenerate functions (as in Section 1.3) and then to apply the usual perturbation method (Section 1.2) to the states so obtained. If we use the partitioning method, we may reverse the order and first construct the matrix $\mathbf{H}^A{}_{\text{eff}}$ which takes into account all the states outside the degenerate set (A). It is then only necessary to solve a truncated eigenvalue equation which, as in Section 1.3, refers formally to the degenerate set alone. This procedure is in some cases advantageous.

1.6. The Hylleraas variation principle

In Section 1.2 the first-order perturbation of the wave function was determined by using the unperturbed wave functions as the basis for an eigenfunction expansion. In principle, however, there is no reason why the differential equation for the first-order perturbation, namely (1.6b),

$$(\mathsf{H}^{(0)} - E_n{}^{(0)})\Psi_n{}^{(1)} = -(\mathsf{H}^{(1)} - E_n{}^{(1)})\Psi_n{}^{(0)} \quad (1.46)$$

should not be solved directly, i.e. without making the expansion (1.9). If this can be done, the second-order energy $E_n{}^{(2)}$ can be obtained from (1.6c) on multiplying from the left by $\Psi_n{}^{(0)}{}^*$ and integrating: this gives

$$E_n{}^{(2)} = \langle \Psi_n{}^{(0)} | (\mathsf{H}^{(1)} - E_n{}^{(1)}) | \Psi_n{}^{(1)} \rangle = \langle \Psi_n{}^{(0)} | \mathsf{H}^{(1)} | \Psi_n{}^{(1)} \rangle \quad (1.47)$$

—the final result following because, as will be seen presently, $\Psi_n^{(1)}$ may be assumed orthogonal to $\Psi_n^{(0)}$. We shall give an example to illustrate this method, which is of course feasible only in certain simple cases, before turning to a general procedure, due to Hylleraas (1930), for obtaining approximate solutions of the equation.

EXAMPLE. *Stark effect in hydrogen. Electric polarizability.* We again consider a hydrogen atom in a uniform electric field of strength F, directed along the z-axis. There is no first-order Stark effect (p. 6) but the second-order energy change may be associated with an induced electric moment. For small fields this is proportional to F:

$$\mu = \alpha F.$$

The factor α is the electric polarizability: the energy of the induced dipole, in the field, is $E_{\text{dip}} = -\frac{1}{2}\mu F = -\frac{1}{2}\alpha F^2$ and so may be calculated from the F^2 term in the energy. Solution of (1.46) thus yields an *exact* value of the polarizability.

The Hamiltonian in the presence of the field (taking the field potential as zero at the origin) is $\mathsf{H} = \mathsf{H}^{(0)} + F\mathsf{H}^{(1)}$ where $\mathsf{H}^{(0)}$ refers to the free hydrogen atom: thus

$$\mathsf{H}^{(0)} = -(\hbar^2/2m)\nabla^2 - e^2/\kappa_0 r, \qquad \mathsf{H}^{(1)} = ez.$$

We require the second-order energy, which is given by (1.47) in terms of $\Psi^{(1)}$.

The unperturbed eigenfunction and energy, for the ground state, are (p. 7)

$$\Psi^{(0)} = (1/\pi a_0^3)^{1/2} \exp(-r/a_0), \qquad E^{(0)} = -\frac{1}{2}(e^2/a_0\kappa_0).$$

The first-order energy change $E^{(1)} = \langle \Psi^{(0)}|ez|\Psi^{(0)}\rangle$ vanishes and equation (1.46) thus takes the form

$$\left(-\frac{\hbar^2}{2m}\nabla^2 - \frac{e^2}{\kappa_0 r} + \frac{e^2}{2a_0\kappa_0}\right)\Psi^{(1)} = -ez\Psi^{(0)}.$$

This may be solved by taking polar coordinates about the z-axis and separating the variables, r and θ (e.g. Vol. 1, App. 2). Solution of the θ-equation indicates

$$\Psi^{(1)} = f(r)\cos\theta$$

and the radial equation becomes

$$\frac{d^2f}{dr^2} + \frac{2}{r}\frac{df}{dr} - \left(\frac{2}{r^2} - \frac{2}{a_0 r} + \frac{2}{a_0^2}\right)f = \frac{2\kappa_0}{a_0\,e\,\sqrt{(\pi a_0^3)}}\,r\,e^{-r/a_0}.$$

It is easily verified, by comparing coefficients of each power of r, that

$$f(r) = (Ar + Br^2)\,e^{-r/a_0}$$

is a solution provided

$$A = 2a_0 B, \qquad B = -\kappa_0\{2e\,\sqrt{(\pi a_0^3)}\}^{-1}.$$

Thus

$$\Psi^{(1)} = -(a_0\kappa_0/e)z(1 + r/2a_0)\Psi^{(0)}.$$

The second-order term in the energy then follows from (1.47): in this particular case the integrations are trivial and the full result is, to second order,

$$E = E^{(0)} - (\tfrac{9}{4})a_0^3\kappa_0 F^2.$$

The polarizability is thus $\alpha = (\tfrac{9}{2})a_0^3\kappa_0$. This exact result is the one referred to in our earlier discussion (p. 8) of the same problem.

Direct solution of the differential equation, as in the above example, is not usually possible for more complicated systems. However, as was first noted by Hylleraas (1930), we can regard (1.46) as a stationary value condition for the function $\Psi^{(1)}$ which will minimize the energy expression up to second order; this suggests that instead of trying to solve (1.46) directly we may proceed in the spirit of variation theory (Section 1.3), obtaining an approximation $\tilde{\Psi}^{(1)}$, say, which will minimize the energy and give an upper bound to $E^{(2)}$. First we consider the ground state, dropping the subscript n in (1.46) and (1.47).

To establish the stationary value property we take as a variation function $\Psi = \Psi^{(0)} + \lambda X$, where X is a function orthogonal to $\Psi^{(0)}$, and obtain an expectation value

$$\mathscr{E} = E^{(0)} + \lambda E^{(1)} + \lambda^2 \tilde{E}^{(2)} + 0(\lambda^3) \tag{1.48}$$

where

$$\tilde{E}^{(2)} = \langle X | \mathsf{H}^{(0)} - E^{(0)}) | X \rangle + 2 \langle X | \mathsf{H}^{(1)} | \Psi^0 \rangle. \tag{1.49}$$

Now let us choose X to optimize the energy, neglecting the terms of order λ^3. For $X \to X + \delta X$ we obtain immediately, using Hermitian symmetry, the condition

$$\delta \tilde{E}^{(2)} = 2 \langle \delta X | [(\mathsf{H}^{(0)} - E^{(0)}) X + \mathsf{H}^{(1)} \Psi^{(0)}] \rangle = 0. \tag{1.50}$$

However, since X was assumed orthogonal to $\Psi^{(0)}$, there is a constraint

$$\langle \delta X | \Psi^{(0)} \rangle = 0. \tag{1.51}$$

The constraint is then taken into account by Lagrange's method of undetermined multipliers on multiplying the last equation by a parameter μ and adding it to (1.50). The constrained stationary value condition is then

$$\langle \delta X | [(\mathsf{H}^{(0)} - E^{(0)}) X + (\mathsf{H}^{(1)} + \mu) \Psi^{(0)}] \rangle = 0 \tag{1.52}$$

with δX unrestricted. This is satisfied, for *arbitrary* δX, only when

$$(\mathsf{H}^{(0)} - E^{(0)}) X = -(\mathsf{H}^{(1)} + \mu) \Psi^{(0)}.$$

The Lagrangian multiplier is eliminated on multiplying from the left by $\Psi^{(0)*}$ and integrating: thus $\mu = -\langle \Psi^{(0)} | \mathsf{H}^{(1)} | \Psi^{(0)} \rangle = -E^{(1)}$. The required stationary value is therefore achieved when X satisfies the equation

$$(\mathsf{H}^{(0)} - E^{(0)}) X = -(\mathsf{H}^{(1)} - E^{(1)}) \Psi^{(0)} \tag{1.53}$$

which coincides with (1.46), the equation defining the first-order correction to the wave function, i.e. the solution is $X = \Psi^{(1)}$. It is also evident that the constrained stationary value conditions (1.50) and (1.51) may be combined into the single condition

$$\tilde{E}^{(2)} = \langle X|(\mathsf{H}^{(0)} - E^{(0)})|X\rangle + 2\langle X|(\mathsf{H}^{(1)} - E^{(1)})|\Psi^{(0)}\rangle$$

$$= \text{stationary value} \qquad (1.54)$$

for, making the variation $X \to X + \delta X$, this yields exactly the condition (1.52) with $\mu = -E^{(1)}$. This form of the variation principle is convenient in what follows.

On substituting the solution $X = \Psi^{(1)}$ in (1.49) and using (1.53) we obtain

$$\tilde{E}^{(2)} = -\langle\Psi^{(1)}|(\mathsf{H}^{(1)} - E^{(1)})|\Psi^{(0)}\rangle + 2\langle\Psi^{(1)}|\mathsf{H}^{(1)}|\Psi^{(0)}\rangle \qquad (1.55a)$$

or, on account of the orthogonality of X and $\Psi^{(0)}$,

$$\tilde{E}^{(2)} = \langle\Psi^{(1)}|\mathsf{H}^{(1)}|\Psi^{(0)}\rangle \qquad (1.55b)$$

which (since $\mathsf{H}^{(1)}$ is Hermitian and $\tilde{E}^{(2)}$ is real) agrees exactly with $E^{(2)}$ defined in (1.47).

Suppose now X does *not* satisfy (1.53). In this case $\tilde{E}^{(2)}$ in (1.54) is not stationary for *general* variations (δX) and $\tilde{E}^{(2)} \neq E^{(2)}$. Nevertheless, X (a function of some restricted form) may serve as a *variational approximation* to $\Psi^{(1)}$. On denoting this function by $X = \tilde{\Psi}^{(1)}$, the corresponding variational energy (1.54) is

$$\tilde{E}^{(2)} = \langle(\tilde{\Psi}^{(1)}|(\mathsf{H}^{(0)} - E^{(0)})|\tilde{\Psi}^{(1)}\rangle + 2\langle\tilde{\Psi}^{(1)}|\mathsf{H}^{(1)} - E^{(1)}|\Psi^{(0)}\rangle. \qquad (1.56)$$

To show that this is an upper bound to $E^{(2)}$, so that a "best" approximation may be obtained as in Section 1.4, let us suppose that $\tilde{\Psi}^{(1)}$ differs from the exact $\Psi^{(1)}$ by some function Δ. On inserting $\tilde{\Psi}^{(1)} = \Psi^{(1)} + \Delta$ in (1.56) it then follows (remembering $\Psi^{(1)}$ is orthogonal to $\Psi^{(0)}$) that

$$\tilde{E}^{(2)} = \langle\Psi^{(1)}|(\mathsf{H}^{(0)} - E^{(0)})|\Psi^{(1)}\rangle + 2\langle\Delta|(\mathsf{H}^{(0)} - E^{(0)})|\Psi^{(1)}\rangle$$

$$+ \langle\Delta|(\mathsf{H}^{(0)} - E^{(0)})|\Delta\rangle + 2\langle\Psi^{(1)}|\mathsf{H}^{(1)}|\Psi^{(0)}\rangle + 2\langle\Delta|\mathsf{H}^{(1)} - E^{(1)}|\Psi^{(0)}\rangle$$

and on substituting the *exact* second-order energy, given by (1.54) with $X = \Psi^{(1)}$, we obtain

$$\tilde{E}^{(2)} - E^{(2)} = 2\langle\Delta|([H^{(0)} - E^{(0)})\Psi^{(1)} + (H^{(1)} - E^{(1)})\Psi^{(0)}]\rangle$$

$$+ \langle\Delta|(H^{(0)} - E^{(0)}|\Delta\rangle.$$

The first term vanishes, since $\Psi^{(1)}$ satisfies (1.46), while for the ground state the second is essentially positive by the usual variation theorem (p. 11). A best approximation to $\Psi^{(1)}$, in the usual sense, may therefore safely be obtained by minimizing the expression (1.54), the minimum being an upper bound to $E^{(2)}$.

EXAMPLE. *Polarizability of the hydrogen atom*. Let us consider again the hydrogen atom in an electric field, but now use a simple variational form of $\Psi^{(1)}$. The simplest way of recognizing a polarization is to insert a z dependence; we take

$$\Psi^{(1)} = Az\Psi^{(0)}$$

which is already orthogonal, by symmetry, to $\Psi^{(0)}$. From (1.56) we obtain

$$\tilde{E}^{(2)} = A^2 \langle z\Psi^{(0)}|\{-(\hbar^2/2m)\nabla^2 - e^2/\kappa_0 r + e^2/2a\kappa_0\}|z\Psi^{(0)}\rangle + 2A\langle z\Psi^0|ez|\Psi^{(0)}\rangle$$

which becomes, on evaluating the integrals,

$$\tilde{E}^{(2)} = \tfrac{1}{2}A^2(e^2a_0/\kappa_0) + 2A(ea_0^2).$$

This takes its minimum value, $\tilde{E}_{\min}^{(2)} = -2a_0^3\kappa_0$, when $A = -2a_0\kappa_0/e$. The approximate electric polarizability is thus $\alpha = 4a_0^3\kappa_0$, compared with $(\tfrac{9}{2})a_0^3\kappa_0$ from the exact solution of (1.46).

In the above example, the *exact* unperturbed wave function $\Psi^{(0)}$ was known. In case only an approximation $\Psi^{(0)}$ is available, the variational method based on (1.56) may still be used but will not in general give an upper bound to $E^{(2)}$.

The Hylleraas method may also be extended to apply to excited states and, provided the function $\Psi_n^{(1)}$ is suitably chosen, can sometimes be made to yield an upper bound to $E_n^{(2)}$. For a discussion of various extensions of the method the reader is referred to the literature (Brown *et al.*, 1964; see also Sinanoglu, 1961).

1.7. Double perturbation theory

In many instances it is necessary to recognize *two* perturbations; we may, for example, possess a set of eigenfunctions for a "model" Hamiltonian somewhat different from that of the system under consideration—the difference being formally equivalent to a perturbation—and may then inquire into the effect of an applied field, which is a second perturbation. This situation will in fact arise frequently in the discussion of electronic properties and their accuracy.

Here we consider only a non-degenerate ground state and shall proceed only to second order (for the general case see, for example, Brown *et al.*, 1964), writing

$$\mathsf{H} = \mathsf{H}^0 + \lambda\mathsf{H}^\lambda + \mu\mathsf{H}^\mu \tag{1.57}$$

where λ and μ are the two perturbation parameters which switch on the perturbations H^λ and H^μ as $\lambda, \mu \to 1$: formally H^λ is just the derivative $(\partial H/\partial\lambda)$. With a similar notation we expand Ψ and E as

$$\begin{aligned}
\Psi &= \Psi^0 + \lambda\Psi^\lambda + \mu\Psi^\mu + \tfrac{1}{2}\lambda^2\Psi^{\lambda\lambda} + \lambda\mu\Psi^{\lambda\mu} + \tfrac{1}{2}\mu^2\Psi^{\mu\mu} + \ldots, \\
E &= E^0 + \lambda E^\lambda + \mu E^\mu + \tfrac{1}{2}\lambda^2 E^{\lambda\lambda} + \lambda\mu E^{\lambda\mu} + \tfrac{1}{2}\mu^2 E^{\mu\mu} + \ldots.
\end{aligned} \tag{1.58}$$

Thus, $\Psi^{\lambda\lambda} = (\partial^2\Psi/\partial\lambda^2)_{\lambda,\mu=0}$, etc., and $\Psi^{\lambda\mu}$ is the mixed derivative $(\partial^2\Psi/\partial\lambda\,\partial\mu)_{\lambda,\mu=0}$. On substituting these expressions in the equation $\mathsf{H}\Psi = E\Psi$ and equating corresponding coefficients we obtain the usual first-order equations

$$(\mathsf{H}^0 - E^0)\Psi^\lambda + (\mathsf{H}^\lambda - E^\lambda)\Psi^0 = 0, \tag{1.59a}$$

$$(\mathsf{H}^0 - E^0)\Psi^\mu + (\mathsf{H}^\mu - E^\mu)\Psi^0 = 0, \tag{1.59b}$$

showing that Ψ^λ and Ψ^μ may be obtained separately, exactly as in ordinary perturbation theory. But the terms in $\lambda\mu$ give a new equation

$$(\mathsf{H}^0 - E^0)\Psi^{\lambda\mu} + (\mathsf{H}^\lambda - E^\lambda)\Psi^\mu + (\mathsf{H}^\mu - E^\mu)\Psi^\lambda - E^\lambda - E^{\lambda\mu}\Psi^0 = 0. \tag{1.60}$$

The energy terms except $E^{\lambda\mu}$ are determined as in ordinary perturbation theory; for example, (1.8) and (1.47) are replaced by

$$E^\lambda = \langle\Psi^0|\mathsf{H}^\lambda|\Psi^0\rangle,$$
$$E^{\lambda\lambda} = 2\langle\Psi^0|(\mathsf{H}^\lambda - E^\lambda)|\Psi^\lambda\rangle,$$

where the factor 2 in $E^{\lambda\lambda}$ arises only from the differences of notation ($\tfrac{1}{2}E^{\lambda\lambda}$ in (1.58) corresponds to $E^{(2)}$ in (1.4)).

To obtain the energy to second order, however, $E^{\lambda\mu}$ is also required. It may be obtained from (1.60) on multiplying by Ψ^{0*} and integrating: thus

$$E^{\lambda\mu} = \langle\Psi^0|(\mathsf{H}^\lambda - E^\lambda)|\Psi^\mu\rangle + \langle\Psi^0|(\mathsf{H}^\mu - E^\mu)|\Psi^\lambda\rangle$$

and it is evidently unnecessary to actually solve (1.60) to obtain the second-order energy. Since, as in Section 1.6, the perturbation functions Ψ^μ and Ψ^λ may be assumed orthogonal to Ψ^0, the above expressions may also be written

$$E^\lambda = \langle\Psi^0|\mathsf{H}^\lambda|\Psi^0\rangle, \qquad E^\mu = \langle\Psi^0|\mathsf{H}^\mu|\Psi^0\rangle, \tag{1.61a}$$

$$E^{\lambda\lambda} = 2\langle\Psi^0|\mathsf{H}^\lambda|\Psi^\lambda\rangle, \qquad E^{\mu\mu} = 2\langle\Psi^0|\mathsf{H}^\mu|\Psi^\mu\rangle, \tag{1.61b}$$

$$E^{\lambda\mu} = \langle\Psi^0|\mathsf{H}^\lambda|\Psi^\mu\rangle + \langle\Psi^0|\mathsf{H}^\mu|\Psi^\lambda\rangle \tag{1.61c}$$

where all terms are real and the bra and ket functions may be exchanged if desired.

An important application of double perturbation theory arises when one of the perturbations is much easier to deal with than the other. Let us therefore suppose Ψ^λ can be obtained easily (e.g. by the direct procedure of Section 1.6), but not Ψ^μ: suppose also that our main interest is in the property to which λ and H^λ refer. Then

$$E = E^0 + \lambda(E^\lambda + \mu E^{\lambda\mu}) + \dots$$

and the important quantity is the coefficient of λ. Thus, λ might represent an applied field strength and H^λ the corresponding electric moment of the electrons in the field, as in the example at the beginning of this section. H^μ might typically represent the difference between the *actual* Hamiltonian (without the applied field) and the "model" Hamiltonian H^0 whose exact eigenfunctions are available. Thus H^μ, which for a many-electron system contains inter-electronic variables, is much more difficult to handle than H^λ. The (permanent)‡ dipole moment of the system, taking account of the second perturbation of H^μ, is then determined by $(E^\lambda + \mu E^{\lambda\mu})$. It would appear that evaluation of the dipole moment will be difficult because $E^{\lambda\mu}$ in (1.61c) contains the perturbed state $|\Psi^\mu\rangle$. This is not the case, however, owing to the existence of an "interchange theorem".

The interchange theorem simply allows us to replace $H^\lambda|\Psi^\mu\rangle$ in the expression (1.61c) by $H^\mu|\Psi^\lambda\rangle$, thus leading to an alternative form:

$$E^{\lambda\mu} = 2\langle\Psi^0|H^\mu|\Psi^\lambda\rangle \tag{1.62}$$

which does not require knowledge of the perturbation Ψ^μ. To prove this result we consider the perturbation applied in two stages. First we use a Hamiltonian

$$\tilde{H} = H^0 + \lambda H^\lambda$$

and obtain perturbed wave function and energy

$$\tilde{\Psi} = \Psi^0 + \lambda\Psi^\lambda + \tfrac{1}{2}\lambda^2\Psi^{\lambda\lambda} + \ldots,$$

$$\tilde{E} = E^0 + \lambda E^\lambda + \tfrac{1}{2}\lambda^2 E^{\lambda\lambda} + \ldots$$

where the perturbation Ψ^λ satisfies (1.59a), and all quantities are similarly defined in either single or double perturbation theories. Then we go to the final Hamiltonian

$$H = \tilde{H} + \mu H^\mu$$

and obtain results good to the first order in μ as

$$E = \tilde{E} + \mu\langle(\Psi^0 + \lambda\Psi^\lambda)|H^\mu|(\Psi^0 + \lambda\Psi^\lambda)\rangle$$

$$= E^0 + \lambda E^\lambda + \tfrac{1}{2}\lambda^2 E^{\lambda\lambda} + \mu E^\mu + \lambda\mu[\langle\Psi^0|H^\mu|\Psi^\lambda\rangle + \langle\Psi^\lambda|H^\mu|\Psi^0\rangle] + \ldots.$$

Since $\langle\Psi^\lambda|H^\mu|\Psi^0\rangle = \langle\Psi^0|H^\mu|\Psi^\lambda\rangle$ (cf. (1.61) et seq.), the coefficient of $\lambda\mu$ gives $E^{\lambda\mu}$ at once in the form (1.52) and comparison with (1.61c) gives

$$\langle\Psi^0|H^\lambda|\Psi^\mu\rangle = \langle\Psi^0|H^\mu|\Psi^\lambda\rangle \tag{1.63}$$

which is the required interchange theorem.

---

‡ To evaluate the *induced* dipole moment, as in previous examples, it would be necessary to go to higher order.

Double perturbation theory may, of course, be carried to all orders and similar interchange theorems apply in higher order so long as Ψ^0 is a non-degenerate ground state. The interchange property has been widely exploited by Dalgarno and others (e.g. Dalgarno and Lewis, 1955; Dalgarno and Stewart, 1958) in discussing electric polarizabilities of atoms. For further discussions the reader is referred to the literature (e.g. Brown *et al.*, 1964).

1.8.   Time-dependent perturbation theory

In this section we turn to the case where the perturbation is a function of the time t. Suppose that at $t = 0$ the system is in one of the stationary states of the unperturbed Hamiltonian $\mathsf{H}^{(0)}$, which is assumed time-independent. The perturbation will in general have two effects; assuming that it ultimately becomes time-independent it will lead continuously to a new stationary state, which could of course be calculated alternatively using the stationary state theory already developed; but it may also cause discontinuous *transitions* from the initial state to any other state. Some of the more recent developments in perturbation theory have in fact used a time-dependent formalism to achieve results that could in principle have been obtained by stationary state methods, the required static perturbation being artificially "switched on" very gradually (e.g. starting at $t = -\infty$ and reaching its given form at $t = 0$). But the traditional applications of time-dependent theory have been to the discussion of *transitions* caused by actual perturbations, such as an applied field or the interaction of two systems during a collision process. We shall study the time-dependent theory mainly as a means of understanding transition phenomena. Here we discuss only the main features of the theory, specific applications to the absorption or emission of electromagnetic radiation being taken up in Chap. 7. The extensive applications in scattering and collision processes will be dealt with elsewhere (Vol. 4). Attention will also be confined to first-order theory. More general methods of treating time dependence start from the evolution operator (Vol. 1, Section 5.6) but lie outside the scope of this volume.

We start from the time-dependent Schrödinger equation (Vol. 1, Section 4.2)

$$i\hbar\,\frac{\partial\Psi}{\partial t} = \mathsf{H}\Psi \tag{1.64}$$

and assume that the Hamiltonian has the form

$$\mathsf{H} = \mathsf{H}^{(0)} + \mathsf{H}^{(1)}(t)$$

where we now suppress the parameter λ, previously used in separating the orders of small quantities. We also assume that the stationary state eigenfunctions of $H^{(0)}$, which satisfy the equation

$$i\hbar \frac{\partial \Psi_n^{(0)}}{\partial t} = H^{(0)} \Psi_n^{(0)}, \qquad (1.65)$$

are known. The unperturbed functions thus involve the time only through a phase factor,

$$\Psi_n^{(0)}(\mathbf{x}; t) = \Phi_n^{(0)}(\mathbf{x}) \exp\{-iE_n^{(0)}t/\hbar\} \qquad (1.66)$$

and the time-independent factor satisfies the usual stationary state equation

$$H^{(0)}\Phi_n^{(0)} = E_n^{(0)}\Phi_n^{(0)}. \qquad (1.67)$$

We assume, as usual, that $\{\Phi_n^{(0)}\}$ is a complete set,[†] and use $\mathbf{x}$ to denote all the variables in the wave function (in general both spatial and spin).

A solution of (1.64) may now be expanded in terms of the functions $\Phi_n^{(0)}$, or equivalently in the form

$$\Psi(\mathbf{x}; t) = \sum_i c_i(t)\Psi_i^{(0)}(\mathbf{x}; t) \qquad (1.68)$$

where the phase factors have been included with the expansion functions. Insertion of (1.68) into (1.64) and use of the relation (1.65) then gives

$$i\hbar \sum_i \frac{dc_i}{dt} \Psi_i^{(0)} = \sum_i c_i H^{(1)} \Psi_i^{(0)}.$$

On multiplying this equation by $\Psi_j^{(0)*}$ from the left and integrating over all space we obtain the fundamental equation

$$\frac{dc_j}{dt} = -\frac{i}{\hbar} \sum_i c_i H_{ji}^{(1)} e^{i\omega_{ji}t} \qquad (1.69)$$

in which $H_{ji}^{(1)} = H_{ji}^{(1)}(t)$ is a matrix element between stationary state eigenfunctions, without time factors,

$$H_{ji}^{(1)} = \langle \Phi_j^{(0)} | H^{(1)} | \Phi_i^{(0)} \rangle \qquad (1.70)$$

and the "frequency factor" in the exponential is

$$\omega_{ji} = (E_j^{(0)} - E_i^{(0)})/\hbar. \qquad (1.71)$$

If we can solve (1.69) and obtain the c_i, then the solution Ψ follows in the form (1.68).

---

† Normally the set is not wholly discrete; for the "continuum" states n must as usual be replaced by a continuous variable, and sums by corresponding integrals.

We shall consider the integration of the fundamental equation in one or two main types of application, starting with the system in a particular stationary state (m) and getting the probability that at time t it will be in some other state (j): according to basic principles (Vol. 1, p. 91) this will be determined by the value of $|c_j(t)|^2$.

(A) *Sudden change*

In this case we suppose that at $t = 0$ the system is in a stationary state $\Psi_m{}^{(0)}$ and that a perturbation $\mathsf{H}^{(1)}$ is instantaneously switched on. Thus, initially,

$$t = 0; \quad c_j{}^{(0)} = \delta_{jm}.$$

If there were no perturbation, the system would remain in the state $\Psi_m{}^{(0)}$ for all time; but as there is a perturbation the coefficients c_j $(j \neq m)$ will not in general vanish for $t > 0$.

To investigate what happens for t small, when all coefficients except c_m $(\cong 1)$ are small quantities, we separate the orders as usual[‡] (matrix elements of $\mathsf{H}^{(1)}$ defining the first order) and write

$$c_j = c_j{}^{(0)} + c_j{}^{(1)} + c_j{}^{(2)} + \dots. \tag{1.72}$$

To solve (1.69) to first order, we substitute (1.72) and discard terms of second and higher order. The coefficients $c_j{}^{(1)}$ must then satisfy

$$\frac{dc_j{}^{(1)}}{dt} = -\frac{i}{\hbar} \sum_i H_{ji}{}^{(1)}(t) c_i{}^{(0)} e^{i\omega_{ji}t}$$

or initially, since we assume $c_i{}^{(0)}(t = 0) = \delta_{jm}$,

$$\frac{dc_j{}^{(1)}}{dt} = -\frac{i}{\hbar} H_{jm}{}^{(1)}(t) e^{i\omega_{jm}t}.$$

The equations for the different coefficients are therefore "uncoupled" in first order and integration gives at once

$$c_j{}^{(1)} = -\frac{i}{\hbar} \int_0^t H_{jm}{}^{(1)}(t) e^{i\omega_{jm}t} dt, \tag{1.73}$$

which determines the solution to first order.

To complete the discussion it is necessary to make specific assumptions about the nature of the perturbation. Here we consider the case where $\mathsf{H}^{(1)}$ is switched on at time $t = 0$ and subsequently remains

---

‡ A parameter λ may again be included, if desired, and finally put equal to unity.

constant. The matrix element $H_{jm}{}^{(1)}$ can then be taken outside the integration sign to give

$$c_j{}^{(1)}(t) = -\frac{H_{jm}{}^{(1)}}{\hbar}\left[\frac{e^{i\omega_{jm}t} - 1}{\omega_{jm}}\right]. \tag{1.74}$$

According to the basic axioms (Vol. 1, p. 91) it follows that the probability of finding the system in state j is given by (after a simple reduction)

$$\left|c_j{}^{(1)}(t)\right|^2 = \frac{4\left|H_{jm}{}^{(1)}\right|^2}{\hbar^2\omega_{jm}{}^2}\sin^2\left(\tfrac{1}{2}\omega_{jm}t\right). \tag{1.75}$$

In particular, if the final state j has the same energy as the initial state m, we may consider the limit $\omega_{jm} \to 0$ and obtain

$$\left|c_j(t)\right|^2 \cong \left|H_{jm}{}^{(1)}\right|^2 t^2/\hbar^2. \tag{1.75a}$$

It must be stressed that, in deriving these results, we have assumed $H^{(1)}$ so small that it is legitimate to drop second- and higher-order terms in (1.69), and that $c_j{}^{(0)}$ is non-zero only for one state. Only in this case is it possible to separate the coupled differential equations and complete the integration. We have also assumed that $H^{(1)}(t)$ is of the special type that switches on at $t = 0$ and then remains constant, and that t is small enough, in (1.75), so as not to violate the assumption that $c_j{}^{(1)}(t)$ is a first-order quantity.

Physical interpretation. In Fig. 1.3 the factor $\sin^2\left(\tfrac{1}{2}\omega_{jm}t\right)/\left(\tfrac{1}{2}\omega_{jm}\right)^2$ is plotted against ω_{jm}. It is clear that by far the most likely transitions will be those to states with essentially the same energy as the initial state. In other words, there is a "conservation of energy"—at least within the limits associated with the uncertainty principle.[‡] At first sight, this makes the physical interpretation of the preceding results unclear. If the system considered is an atom or molecule with discrete energy levels, how can it ever make a quantum jump of several electron volts? Also, intuition and experience suggest that the probability of a transition taking place during time t will be proportional to t, whereas (1.75a) indicates proportionality with t^2.

The root of these difficulties is the fact that we have considered an isolated system, whereas a transition between discrete states of an atom or molecule essentially involves *exchange of energy* between two sub-systems; for example, energy emitted may be taken up by a crystal lattice, by the walls of a container, or by a radiation field. The system

---

[‡] As the system is developing in time the energy–time uncertainty relation (Vol. 1, p. 84) must be observed. From Fig. 1.3 a transition after time Δt is expected only if $\omega_{jm} < 2\pi/\Delta t$, or $E_j - E_m < h/t$ (i.e. within uncertainty principle limits).

receiving or supplying the energy characteristically has a large number of degrees of freedom and a virtually continuous energy spectrum; it forms an energy "reservoir". To clarify the general nature of the transition process, without detailed specification of the nature of either the system or the reservoir, it is useful to reformulate the preceding analysis, assuming the "system" comprises two weakly interacting subsystems—"molecule" and "reservoir".

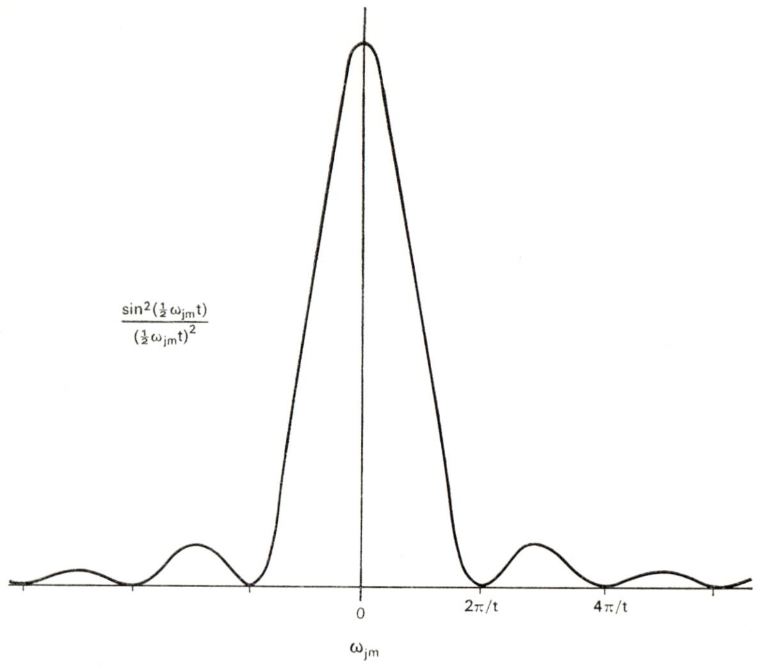

FIG. 1.3. Dependence of transition probability on energy difference. Probability of a transition during time t is proportional to the factor plotted, $\omega_{jm} = (E_j - E_m)/\hbar$, measuring the energy difference between initial and final states. The sharpness of the peak indicates conservation of energy, within limits imposed by the uncertainty principle.

Let Ψ_{mr} describe a system comprising the "molecule" in state m, energy E_m, and the "reservoir" in state r, energy ε_r. In the absence of interaction the Hamiltonian for the whole system would be $\mathsf{H}^{(0)} = \mathsf{H}^M + \mathsf{H}^R$, and the wave function Ψ_{mr} would be an eigenfunction of $\mathsf{H}^{(0)}$, with energy $E_{mr} = E_m + \varepsilon_r$. We now switch on an interaction term H' and expand the wave function in the form

$$\Psi = \sum_{m'r'} c_{m'r'} \Psi_{m'r'}. \tag{1.76}$$

The preceding analysis now applies and (1.73) becomes

$$c_{m'r'}{}^{(1)} = -\frac{i}{\hbar} \int_0^t H'_{m'r',mr}\, e^{i\omega(mr \to m'r')t}\, dt$$

except that now

$$H'_{m'r',mr} = \langle \Psi_{m'r'}|\mathsf{H}|\Psi_{mr}\rangle \tag{1.77}$$

and

$$\omega(mr \to m'r') = (E_{m'} + \varepsilon_{r'} - E_m - \varepsilon_r)/\hbar. \tag{1.78}$$

For a time-independent interaction we obtain, as before (cf. (1.75)),

$$\left|c_{m'r'}{}^{(1)}\right|^2 = \frac{4\left|H'_{m'r',mr}\right|^2}{\hbar^2} \frac{\sin^2 \frac{1}{2}\omega(mr \to m'r')t}{\omega(mr \to m'r')^2}. \tag{1.79}$$

But now there will be *many* reservoir transitions $r \to r'$ such that

$$\varepsilon_r - \varepsilon_{r'} \simeq E_{m'} - E_m$$

—giving a denominator close to zero and hence a non-negligible transition probability. As we are interested only in the "molecule", not the reservoir, the probability of the transition $m \to m'$ will be obtained by summing over all reservoir transitions that make $\omega(mr \to m'r') \simeq 0$.

Suppose there are $\rho(\varepsilon)\, d\varepsilon$ reservoir states in the range $(\varepsilon,\, \varepsilon + d\varepsilon)$, where $\rho(\varepsilon)$ is the "density of states": then if w is the total probability per unit time for realizing the transition $m \to m'$ we have

$$w = \frac{1}{t} \int \left|c_{m'r'}{}^{(1)}\right|^2 \rho(\varepsilon_{r'})\, d\varepsilon_{r'}.$$

We now take $\omega = \omega(mr \to m'r')$ as the integration variable, noting from (1.78) that $d\varepsilon_{r'} = \hbar\, d\omega$, and note that the first factor in the integrand varies extremely rapidly with ω, very much more so than the remainder. It is therefore permissible to give $H'_{m'r',mr}$ an appropriate mean value, and to put $\rho(\varepsilon_{r'}) = \rho_{\text{final}}$, withdrawing both factors from the integrand. The result is

$$w = \frac{4}{\hbar^2} \bar{H}\big|'_{m'm}\big|^2 \rho_{\text{final}} \int \frac{\sin^2 \frac{1}{2}\omega t}{\omega^2} \hbar\, d\omega$$

where $\left|\bar{H}'_{m'm}\right|^2$ denotes a mean value, over all final states of the reservoir in the vicinity of $\varepsilon_{r'} = (E_m - E_{m'} + \varepsilon_r)$, of the quantity $\left|\langle \Psi_{m'r'}|\mathsf{H}'|\Psi_{mr}\rangle\right|^2$. To evaluate this quantity it is necessary to specify the detailed nature of the particular system considered, but the general nature of the conclusions is always the same. Since contributions to the integral are negligible away from the peak (Fig. 1.3) the integration range may be taken as $(-\infty, +\infty)$ and on using the result $\int_{-\infty}^{+\infty} (\sin^2 \alpha x / x^2)\, dx = \alpha \pi$ we readily obtain the final formula

$$w = \frac{2\pi}{\hbar} \, |\bar{H}'_{m'm}|^2 \rho_{\text{final}}. \tag{1.80}$$

The transition probability per unit time is thus independent of t, and the general argument therefore shows that (in spite of equation (1.75a)), time proportional transitions to states of widely different energy can occur in the presence of a suitable reservoir, to supply or receive energy. The important factors determining the transition probability are essentially a matrix element connecting initial and final states of the system of interest, and a *density of states*—which may not refer to the system itself but rather to its environment. In specific applications the nature of these factors will become clear. The rather general formula (1.80) is often called "Fermi's golden rule".

(B)   *Slow change*

We now assume $H^{(1)}$ is initially zero but gradually approaches some finite limiting form. It will be of interest to follow the behaviour of the system for $t \to \infty$, but as it stands the fundamental equation (1.69) is unsatisfactory because the term with $i = j$ will give a divergent integral. This divergence may be removed by including a time-dependent phase factor in the expansion coefficients: for if we introduce

$$\bar{c}_i = c_i \exp\left(\frac{i}{\hbar} \int_0^t H_{ii}^{(1)}\, dt\right) \tag{1.81}$$

(1.69) becomes

$$\frac{d\bar{c}_j}{dt} = \sum_{i(\neq j)} \bar{c}_i H_{ji}^{(1)} \exp\left(\frac{i}{\hbar} \int_0^t \bar{\omega}_{ji}\, dt\right) \tag{1.82}$$

where

$$\bar{\omega}_{ji} = [(E_j^{(0)} + H_{jj}^{(1)}) - (E_i^{(0)} + H_{ii}^{(1)})]/\hbar. \tag{1.83}$$

The term with $i = j$ is thus eliminated, and with it the divergence difficulty. Clearly, $\bar{\omega}_{ji}$ is a measure of the difference between the *perturbed* energies, at time t, calculated according to the usual stationary state theory.

Again, for a system starting in state $\Psi_m{}^{(0)}$, the time-development equation separates into uncoupled first-order equations for the different coefficients and we obtain

$$\bar{c}_j{}^{(1)}(t) = -\frac{i}{\hbar}\int_0^t H_{jm}{}^{(1)} \exp\left(\frac{i}{\hbar}\int_0^t \bar{\omega}_{jm}\,dt\right) dt \quad (j \neq m). \qquad (1.84)$$

Suppose now that $H^{(1)}(t)$ has reached its final form at time t, but that all coefficients are still small enough for the first-order treatment to be valid. At such a time, writing (1.84) as

$$\bar{c}_j{}^{(1)}(t) = -\int_0^t \frac{H_{jm}{}^{(1)}}{\bar{\omega}_{jm}(t)}\frac{d}{dt}\exp\left(\frac{i}{\hbar}\int_0^t \bar{\omega}_{jm}\,dt\right) dt$$

and integrating by parts, we obtain[‡] on letting $t \to \infty$

$$\bar{c}_j{}^{(1)}(t) = \int_0^\infty \frac{\partial}{\partial t}\left(\frac{H_{jm}{}^{(1)}}{\omega_{jm}}\right)\exp\left(\frac{i}{\hbar}\int_0^t \bar{\omega}_{jm}\,dt\right) dt - \left[\frac{H_{jm}{}^{(1)}}{\omega_{jm}}\exp\left\{\frac{i}{\hbar}\int_0^t \bar{\omega}_{jm}\,dt\right\}\right]_0^\infty.$$

$$(1.85)$$

The final wave function, to first order, is then obtained by using (1.81) and substituting this result in (1.68). The final result may be written most conveniently in the form

$$\Psi = \bar{\Phi}_m \exp\left(-i\bar{E}_m t/\hbar\right) + \sum_{j(\neq m)} d_j \bar{\Phi}_j \exp\left(-i\bar{E}_j t/\hbar\right) \qquad (1.86)$$

where $\bar{\Phi}_j$ and $\bar{E}_j$ are the eigenfunctions and eigenvalues at $t = \infty$ (i.e. after application of the perturbation), as calculated by standard perturbation theory (Section 1.2), and

$$d_j = \int_0^\infty \frac{\partial}{\partial t}\left(\frac{H_{jm}{}^{(1)}}{\omega_{jm}}\right)\exp\left(\frac{i}{\hbar}\int_0^t \bar{\omega}_{jm}\,dt\right) dt, \qquad (1.87)$$

being the first term in (1.85).

These results are of great importance. They show that if a perturbation is applied infinitely slowly, so that the time derivative in the expression (1.87) for d_j is always infinitesimal, *no transitions will be induced* and the system will pass smoothly through the sequence of states corresponding to continuous change of perturbation parameter

---

[‡] It should be noted that $\bar{\omega}_{jm}$, defined according to (1.83), is a function of t.

in the stationary state theory, i.e. $\Psi_m \to \overline{\Psi}_m$ with no discontinuous jumps to other states. This result provides the quantum-mechanical basis for the interpretation of quasi-static processes in thermodynamics, and is referred to as the *adiabatic theorem*: it has been proved here only within the limits of first-order perturbation theory but its validity in fact may be established quite generally (Born and Fock, 1928; se also Messiah, 1962).

REFERENCES

BORN, M. and FOCK, V. (1928) *Z. Phys.* **51**, 165.

BRILLOUIN, L. (1933) *Actualités Sci. Ind.,* No. 71.

BROWN, W. B., EPSTEIN, S. T. and HIRSCHFELDER, J. (1964) in *Advances in Quantum Chemistry*, Vol. I, p. 255, Ed. LÖWDIN, P.-O., Academic Press, London and New York.

DALGARNO, A. and LEWIS, J. T. (1955) *Proc. Roy. Soc.* A **233**, 70.

DALGARNO, A. and STEWART, A. L. (1958) *Proc. Roy. Soc.* A **247**, 245.

HYLLERAAS, E. A. (1930) *Z. Phys.* **65**, 209.

HYLLERAAS, E. A. and UNDHEIM, B. (1930) *Z. Phys.* **65**, 759.

KEMBLE, E. C. (1937) *The Fundamental Principles of Quantum Mechanics with Elementary Applications*, McGraw Hill, New York (reprinted 1958 by Dover, New York).

LÖWDIN, P.-O. (1963) *J. Mol. Spectry*, **10**, 12.

MacDONALD, J. K. L. (1933) *Phys. Rev.* **43**, 830.

MESSIAH, A. (1962) *Quantum Mechanics*, North Holland, Amsterdam (Vol. 2, Ch. 17).

SINANOGLU, O. (1961) *Phys. Rev.* **122**, 491 (see also *J. Chem. Phys.* **34**, 1237).

WIGNER, E. (1935) *Math. Natur. Anz. (Budapest)* **53**, 477.

CHAPTER 2

THE CENTRAL FIELD PROBLEM. ATOMIC ORBITALS

2.1. The hydrogen atom as a central field system

In this chapter we consider the stationary states of a single particle moving in a central field, in which the potential energy depends only on the distance r from some fixed centre which we take as origin, $V = V(\mathbf{r}) = V(r)$. This provides further illustrations of the general principles discussed in Vol. 1 (Chap. 4) in a context which is of fundamental importance in atomic spectroscopy and quantum chemistry.

First, however, we show that any hydrogen-like system, comprising one nucleus and one electron, is indeed of central field type, i.e. we need consider formally only *one* particle moving in the field of a fixed nucleus, as assumed in Vol. 1. This means we must start from a two-particle problem and then try to separate the Schrödinger equation into parts describing respectively the internal motion and the translational motion of the mass centre through space.

From the postulates, it is a simple matter to write down the Schrödinger equation for two particles with masses m_1 and m_2. The classical Hamiltonian function is

$$H = \frac{1}{2m_1}\,\mathbf{p}_1{}^2 + \frac{1}{2m_2}\,\mathbf{p}_2{}^2 + V(\mathbf{r}_1, \mathbf{r}_2).$$

The Schrödinger Hamiltonian is obtained by replacing each momentum component by a differential operator, and the stationary states then arise as solutions of the partial differential equation[‡]

$$\mathsf{H}\Psi(\mathbf{r}_1, \mathbf{r}_2) = E\Psi(\mathbf{r}_1, \mathbf{r}_2) \tag{2.1}$$

where

$$\mathsf{H} = -\frac{\hbar^2}{2m_1}\,\nabla^2(1) - \frac{\hbar^2}{2m_2}\,\nabla^2(2) + V(\mathbf{r}_1, \mathbf{r}_2). \tag{2.2}$$

Here the Hamiltonian is assumed not to contain any terms depending on spin or external fields (the usual first approximation) and it is then

---

‡ Notation is referred to in the footnote on p. 1.

legitimate to consider a wave function depending only on the spatial variables $\mathbf{r}_1$, $\mathbf{r}_2$.

We now assume that the potential energy depends only on the distance $|\mathbf{r}_1 - \mathbf{r}_2|$ between the two particles and introduce new variables —the vector separation $(1 \to 2)$, and the position vector of the mass centre. With $M = m_1 + m_2$,

$$\mathbf{r} = \mathbf{r}_2 - \mathbf{r}_1, \quad \mathbf{R} = \frac{1}{M}(m_1\mathbf{r}_1 + m_2\mathbf{r}_2). \tag{2.3}$$

On introducing Cartesian coordinates and using the formulae of partial differentiation we find

$$\frac{\partial^2}{\partial x_1{}^2} = \frac{\partial^2}{\partial x^2} + \frac{2m_1}{M}\frac{\partial^2}{\partial x\,\partial X} + \frac{m_1{}^2}{M^2}\frac{\partial^2}{\partial X^2},$$

$$\frac{\partial^2}{\partial x_2{}^2} = \frac{\partial^2}{\partial x^2} - \frac{2m_2}{M}\frac{\partial^2}{\partial x\,\partial X} + \frac{m_2{}^2}{M^2}\frac{\partial^2}{\partial X^2}$$

with similar expressions for the other second derivatives. Equation (2.1) then takes the form

$$\left(-\frac{\hbar^2}{2\mu}\nabla_r{}^2 - \frac{\hbar}{2M}\nabla_R{}^2 + V(\mathbf{r})\right)\Psi(\mathbf{r},\mathbf{R}) = E\Psi(\mathbf{r},\mathbf{R}) \tag{2.4}$$

where μ is the *reduced mass*

$$\mu = \frac{m_1 m_2}{m_1 + m_2} = \frac{m_1 m_2}{M} \tag{2.5}$$

and

$$\nabla_r{}^2 = (\partial^2/\partial x^2 + \partial^2/\partial y^2 + \partial^2/\partial z^2),$$
$$\nabla_R{}^2 = (\partial^2/\partial X^2 + \partial^2/\partial Y^2 + \partial^2/\partial Z^2). \tag{2.6}$$

Equation (2.4) can be separated (Vol. 1, App. 2) by putting

$$\Psi(\mathbf{r},\mathbf{R}) = \phi(\mathbf{r})\Phi(\mathbf{R}) \tag{2.7}$$

and proceeding in the usual way. The separated equations are

$$\left(-\frac{\hbar^2}{2\mu}\nabla_r{}^2 + V(\mathbf{r})\right)\phi(\mathbf{r}) = E'(\mathbf{r})\phi(\mathbf{r}) \tag{2.8a}$$

for the internal motion and

$$-\frac{\hbar^2}{2M}\nabla_R{}^2\Phi(\mathbf{R}) = E''\Phi(\mathbf{R}) \tag{2.8b}$$

for the motion of the centre of mass. The energy of the whole system is $E = E' + E''$, the sum of energy eigenvalues for the two types of motion.

Equation (2.8b) shows that the motion of the centre of mass is the same as that of a free particle with mass M. Equation (2.8a), on the other hand, governs the relative motion of the two particles. If we consider the hydrogen atom, comprising one proton and one electron, and compare the resultant equation with I.(2.20), which referred to a fixed-nucleus system, the only difference is evidently that the reduced mass μ appears in place of the electron mass m. On setting $m_1 = M$ and $m_2 = m$ (proton and electron masses), the reduced mass $\mu = Mm/(M+m)$ is found to be very close to m: in fact, $\mu/m = 0 \cdot 999456$. When μ is used instead of m in the energy formula I.(2.23), the ground state energy of the hydrogen atom at rest is found to be

$$E_0 = -13 \cdot 598 \text{ eV.}$$

This value is much closer to the experimental ionization potential $(13 \cdot 595 \text{ eV})$ than the value $(13 \cdot 605 \text{ eV})$ obtained previously. The remaining discrepancy is due to small relativistic terms, which we do not consider at this point.

2.2. Angular momentum

When a classical particle moves in a central field the energy and the three components of angular momentum are all constants of the motion. The motion is an orbit whose orientation in space is fixed.

To discuss the corresponding situation in quantum mechanics we must examine the commutation properties of the associated operators (see, for example, Vol. 1, Chap. 4): if they all commute, then the quantities which they represent may all simultaneously be constants of the motion; if not, we must find as many mutually commuting operators as we can and obtain a maximal set of constants of the motion.

To obtain the commutation properties we may start from the classical angular momentum components (in Cartesian form) and set up angular momentum operators

$$\hbar \mathsf{L}_x = \mathsf{y}\mathsf{p}_z - \mathsf{z}\mathsf{p}_y, \quad \hbar \mathsf{L}_y = \mathsf{z}\mathsf{p}_x - \mathsf{x}\mathsf{p}_z, \quad \hbar \mathsf{L}_z = \mathsf{x}\mathsf{p}_y - \mathsf{y}\mathsf{p}_x \tag{2.9}$$

in accordance with the general principles.

Here we have written the operators in the form $\hbar \mathsf{L}_x$, $\hbar \mathsf{L}_y$, $\hbar \mathsf{L}_z$ so as to introduce the *dimensionless* angular momentum operators, as used in Vol. 1(p. 79). Thus, in Schrödinger language,

$$\mathsf{L}_x = \frac{1}{i} \left(y \frac{\partial}{\partial z} - z \frac{\partial}{\partial y} \right)$$

while $\hbar \mathsf{L}_x$ is the actual angular momentum operator, $\hbar$ itself having the dimensions of angular momentum. This convention avoids the repeated appearance of $\hbar$ in the basic equations, and is formally equivalent to working in atomic units (Appendix 1): it is easy to supply the $\hbar$ factors whenever results are required in absolute units, and $\hbar$ in fact turns out to be the natural unit in which to measure angular momentum.

We now show that the desired properties are completely determined by the basic commutation relations for position and momentum operators I.(4.30), namely

$$[\mathsf{x}, \mathsf{p}_x] = i\hbar\mathsf{I}, \qquad [\mathsf{y}, \mathsf{p}_y] = i\hbar\mathsf{I}, \qquad [\mathsf{z}, \mathsf{p}_z] = i\hbar\mathsf{I}. \tag{2.10}$$

We could, of course, use the Schrödinger representation in which x, p_x, etc., are multipliers and partial differential operators, but here we use the purely symbolic statements to emphasize that the conclusions are independent of the particular quantum-mechanical language in which they are expressed.

From (2.9), noting that the position and momentum operators commute unless they refer to the same axis in space, we see that

$$\hbar^2\mathsf{L}_x\mathsf{L}_y = [\mathsf{y}\mathsf{p}_x\mathsf{p}_z\mathsf{z} - \mathsf{y}\mathsf{x}\mathsf{p}_z{}^2 - \mathsf{z}^2\mathsf{p}_y\mathsf{p}_x + \mathsf{x}\mathsf{p}_y\mathsf{z}\mathsf{p}_z],$$

$$\hbar^2\mathsf{L}_y\mathsf{L}_x = [\mathsf{y}\mathsf{p}_x\mathsf{z}\mathsf{p}_z - \mathsf{x}\mathsf{y}\mathsf{p}_z{}^2 - \mathsf{z}^2\mathsf{p}_y\mathsf{p}_x + \mathsf{x}\mathsf{p}_y\mathsf{p}_z\mathsf{z}]$$

and operators from different angular momentum components therefore certainly do not commute. However, by subtraction

$$\hbar^2(\mathsf{L}_x\mathsf{L}_y - \mathsf{L}_y\mathsf{L}_x) = \mathsf{p}_z\mathsf{z}(\mathsf{y}\mathsf{p}_x - \mathsf{x}\mathsf{p}_y) - (\mathsf{y}\mathsf{p}_x - \mathsf{x}\mathsf{p}_y)\mathsf{z}\mathsf{p}_z$$

and using the last relation in (2.10) this becomes

$$\hbar^2(\mathsf{L}_x\mathsf{L}_y - \mathsf{L}_y\mathsf{L}_x) = i\hbar(\mathsf{x}\mathsf{p}_y - \mathsf{y}\mathsf{p}_x) = i\hbar^2\mathsf{L}_z.$$

Since the expressions in (2.9) differ only by a cyclic permutation of x, y, z we obtain three commutation relations:

$$[\mathsf{L}_x, \mathsf{L}_y] = (\mathsf{L}_x\mathsf{L}_y - \mathsf{L}_y\mathsf{L}_x) = i\mathsf{L}_z,$$

$$[\mathsf{L}_y, \mathsf{L}_z] = (\mathsf{L}_y\mathsf{L}_z - \mathsf{L}_z\mathsf{L}_y) = i\mathsf{L}_x, \tag{2.11}$$

$$[\mathsf{L}_z, \mathsf{L}_x] = (\mathsf{L}_z\mathsf{L}_x - \mathsf{L}_x\mathsf{L}_z) = i\mathsf{L}_y.$$

These are the basic commutation relations for angular momentum. We note that they are formally identical with those that refer to the spin components I.(4.40) giving some support to the interpretation of spin as a type of angular momentum. Equations such as (2.11) are often summarized, using a vector notation, in the form

$$[\mathbf{L} \times \mathbf{L}] = i\hbar\mathbf{L}. \tag{2.11a}$$

Here the cross indicates a "vector product" and the equation is simply a convenient shorthand for the three equations involving the component operators (the cyclic ordering of subscripts being given by the vector product rule, $[\mathbf{A} \times \mathbf{B}]_x = A_y B_z - A_z B_y$, etc.).

Although the three operators do not commute among themselves, and all three angular momentum components cannot therefore be simultaneously definite, we may inquire in a similar way about the *magnitude* of the angular momentum, whose square is

$$\mathbf{L}^2 = L_x^2 + L_y^2 + L_z^2$$

and which has a corresponding operator‡

$$\mathbf{L}^2 = L_x^2 + L_y^2 + L_z^2. \tag{2.12}$$

To simplify the derivation we note first that, for any three operators, A, B, C,

$$[A, BC] = ABC - BCA = ABC - BAC + BAC - BCA$$

where the middle terms are added to complete two new commutators $(AB - BA)C$ and $B(AC - CA)$, thus

$$[A, BC] = [A, B]C + B[A, C]. \tag{2.13}$$

On applying this result with $A = L_x$, taking $BC = L_y L_y$ and $L_z L_z$ in turn, we obtain

$$[L_x, L_y^2] = [L_x, L_y]L_y + L_y[L_x, L_y] = i\hbar(L_z L_y + L_y L_z),$$

$$[L_x, L_z^2] = [L_x, L_z]L_z + L_z[L_x, L_z] = -i\hbar(L_y L_z + L_z L_y).$$

Thus L_x commutes not with L_y^2 and L_z^2 individually but with their sum $L_y^2 + L_z^2$. Since it obviously commutes with L_x^2 we thus obtain $[L_x, \mathbf{L}^2] = 0$. The argument clearly implies

$$[L_x, \mathbf{L}^2] = [L_y, \mathbf{L}^2] = [L_z, \mathbf{L}^2] = 0 \tag{2.14}$$

since there is nothing to distinguish one component from the others.

The implication of the commutation properties of angular momentum is that in a state where the angular momentum has a definite magnitude, only one component (in any direction we choose) can be simultaneously definite. If we make a measurement and determine this component (let us adopt the usual convention and call it the z-component L_z), then one cannot obtain a simultaneously definite value of L_x or L_y. The situation is sometimes described pictorially, in classical language, by saying that the angular momentum vector with components L_x, L_y, L_z precesses

---

‡ Note that $\mathbf{L}^2$ is interpreted as one symbol, the operator associated with $\mathbf{L}^2 = \mathbf{L} \cdot \mathbf{L}$, not as an operator product, $\mathbf{LL}$. Also, since dimensionless operators are employed, the expectation value of $\mathbf{L}^2$ will be the square of the angular momentum, *in units of $\hbar^2$*.

about the z-axis (Fig. 2.1) so that L_z is definite but L_x and L_y may take values lying between two definite limits. Before trying to determine what these limits are, and to find what eigenvalues the angular momenta can take, we should verify that stationary states of definite angular momentum are indeed permitted in the case of motion in a central field. In order for this to be possible we know that the commuting angular momentum operators L^2 and L_z must also commute with the Hamiltonian.

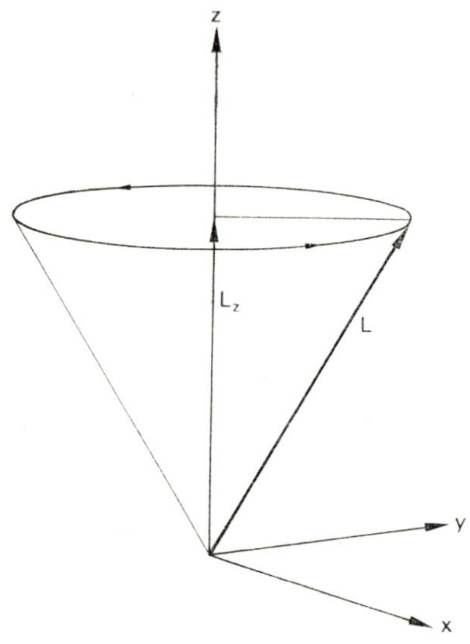

FIG. 2.1. Classical interpretation of angular momentum. Precession of the angular momentum vector about the z-axis (path indicated by the cone) corresponds to a definite value of L_z, but varying values of L_x, L_y.

The usual fixed-nucleus central field Hamiltonian has the form

$$\mathsf{H} = -\frac{1}{2m}\,\mathsf{p}^2 + V(\mathbf{r}) \tag{2.15}$$

and again the argument can be presented symbolically without reference to any particular representation. The characteristic feature of a central field is that $V(\mathbf{r})$ depends only on the magnitude of $\mathbf{r}$ and the associated operator thus involves the sum $\mathsf{x}^2 + \mathsf{y}^2 + \mathsf{z}^2$: if another operator commutes with this sum it will commute with any function of the sum and therefore with the potential energy operator.

It will be sufficient to consider the commutators

$$[\mathbf{L}_z, \mathbf{p}^2], \qquad [\mathbf{L}^2, \mathbf{p}^2], \qquad [\mathbf{L}_z, \mathbf{r}^2], \qquad [\mathbf{L}^2, \mathbf{r}^2].$$

First we note that, from the definition (2.9) together with (2.10),

$$[\mathbf{L}_z, \mathsf{p}_x] = \mathsf{p}_y/i, \qquad [\mathbf{L}_z, \mathsf{p}_y] = -\mathsf{p}_x/i, \qquad [\mathbf{L}_z, \mathsf{p}_z] = 0$$

and using (2.13) we obtain

$$[\mathbf{L}_z, \mathsf{p}_x{}^2] = (\mathsf{p}_x\mathsf{p}_y + \mathsf{p}_y\mathsf{p}_x)/i, \quad [\mathbf{L}_z, \mathsf{p}_y{}^2] = -(\mathsf{p}_x\mathsf{p}_y + \mathsf{p}_y\mathsf{p}_x)/i, \quad [\mathbf{L}_z, \mathsf{p}_z{}^2] = 0.$$

Hence $\mathbf{L}_z$ commutes with $(\mathsf{p}_x{}^2 + \mathsf{p}_y{}^2 + \mathsf{p}_z{}^2)$, and by symmetry

$$[\mathbf{L}_x, \mathbf{p}^2] = [\mathbf{L}_y, \mathbf{p}^2] = [\mathbf{L}_z, \mathbf{p}^2] = 0.$$

Also since the squares of the operators must likewise commute with $\mathbf{p}^2$ it follows at once that

$$[\mathbf{L}^2, \mathbf{p}^2] = 0.$$

The remaining commutators are determined by a similar argument; more simply, however, we need only note that the angular momentum components and all the commutators simply change sign if we interchange position and momentum coordinates. The vanishing of $[\mathbf{L}_z, \mathsf{p}_z]$ and $[\mathbf{L}^2, \mathbf{p}^2]$ therefore imply the vanishing of $[\mathbf{L}_z, \mathbf{r}^2]$ and $[\mathbf{L}^2, \mathbf{r}^2]$. Finally, therefore, we conclude that

$$[\mathsf{H}, \mathbf{L}_z] = 0, \qquad [\mathsf{H}, \mathbf{L}^2] = 0. \tag{2.16}$$

In other words, angular momentum eigenstates can occur as stationary states of a central field problem. In this case the energy, the squared magnitude of the angular momentum, and one—but not more than one —of its components may simultaneously be constants of the motion.

2.3. Eigenvalues of angular momentum

The possible eigenvalues of the angular momentum operators $\mathbf{L}^2$ and $\mathbf{L}_z$ are completely determined by the commutation rules. To find these eigenvalues it is expedient to introduce two new operators

$$\mathbf{L}^+ = \mathbf{L}_x + i\mathbf{L}_y, \qquad \mathbf{L}^- = \mathbf{L}_x - i\mathbf{L}_y \tag{2.17}$$

which, for reasons which will soon become clear, are called "step-up" and "step-down" operators respectively. It is then easy to derive, from (2.11), corresponding commutation relations

$$[\mathbf{L}_z, \mathbf{L}^+] = \mathbf{L}^+, \qquad [\mathbf{L}_z, \mathbf{L}^-] = -\mathbf{L}^-, \qquad [\mathbf{L}^+, \mathbf{L}^-] = 2\mathbf{L}_z \tag{2.18}$$

for the step-up and step-down operators.

Since there exist simultaneous eigenstates of $\mathbf{L}^2$ and L_z (Corollary 10, Chap. 4 of Vol. 1) we may postulate one such state, which we denote by[‡] $\phi_{x,m}$ satisfying

$$\mathbf{L}^2 \phi_{x,m} = x \phi_{x,m}, \tag{2.19a}$$

$$\mathsf{L}_z \phi_{x,m} = m \phi_{x,m}. \tag{2.19b}$$

Again the particular representation need not be chosen since we shall only use the commutation properties of the operators.

Let us operate on $\phi_{x,m}$ with the first commutator in (2.18). The result is, on rearranging and using (2.19b),

$$\mathsf{L}_z \mathsf{L}^+ \phi_{x,m} = \mathsf{L}^+ \mathsf{L}_z \phi_{x,m} + \mathsf{L}^+ \phi_{x,m},$$

$$= (m+1)\mathsf{L}^+ \phi_{x,m}.$$

In other words, $\mathsf{L}^+ \phi_{x,m}$ is either an eigenvector of L_z belonging to the eigenvalue $m+1$, or else the zero vector. We may therefore write

$$\mathsf{L}^+ \phi_{x,m} = c_m{}^+ \phi_{x,m+1} \tag{2.20a}$$

where $c_m{}^+$ is a constant to be determined presently.

The same argument applies if we use the second commutator in (2.18) and yields

$$\mathsf{L}_z \mathsf{L}^- \phi_{x,m} = (m-1)\mathsf{L}^- \phi_{x,m}.$$

We thus conclude that

$$\mathsf{L}^- \phi_{x,m} = c_m{}^- \phi_{x,m-1} \tag{2.20b}$$

where $c_m{}^-$ is another constant. Repeated use of L^+ and L^- thus allows us to generate a whole family of eigenvectors, with the m value changing in unit steps—hence the name step-up and step-down operators.

We now turn to (2.12), noting that $\mathbf{L}^2$ may be written in the alternative forms

$$\mathbf{L}^2 = \mathsf{L}^+ \mathsf{L}^- + \mathsf{L}_z{}^2 - \mathsf{L}_z,$$

$$\mathbf{L}^2 = \mathsf{L}^- \mathsf{L}^+ + \mathsf{L}_z{}^2 + \mathsf{L}_z. \tag{2.21}$$

Thus, for example,

$$\mathbf{L}^2 = \mathsf{L}_x{}^2 + \mathsf{L}_y{}^2 + \mathsf{L}_z{}^2$$

$$= (\mathsf{L}_x + i\mathsf{L}_y)(\mathsf{L}_x - i\mathsf{L}_y) + i(\mathsf{L}_x\mathsf{L}_y - \mathsf{L}_y\mathsf{L}_x) + \mathsf{L}_z{}^2$$

$$= \mathsf{L}^+ \mathsf{L}^- - \mathsf{L}_z + \mathsf{L}_z{}^2$$

where the final form follows on using (2.11).

---

‡ In this and subsequent chapters we adopt the convention of indicating specifically *one*-electron states by *small* Greek letters, reserving capitals for general many-electron states.

Equations (2.21) and (2.19) now imply that

$$\mathsf{L}^+\mathsf{L}^-\phi_{x,m} = (\mathbf{L}^2 - \mathsf{L}_z{}^2 + \mathsf{L}_z)\phi_{x,m}$$

$$= (x - m^2 + m)\phi_{x,m}.$$

Thus, forming scalar products with $\phi_{x,m}$ from the left we obtain

$$\langle\phi_{x,m}|\mathsf{L}^+\mathsf{L}^-|\phi_{x,m}\rangle = x - m^2 + m.$$

But since $\mathsf{L}^+ = \mathsf{L}_x + i\mathsf{L}_y$ is the adjoint of the operator L^- we may use the "turn over" rule (Vol. 1, p. 68) to obtain

$$\langle\phi_{x,m}|\mathsf{L}^+\mathsf{L}^-|\phi_{x,m}\rangle = \langle\mathsf{L}^-\phi_{x,m}|\mathsf{L}^-\phi_{x,m}\rangle \geqq 0$$

since in Hermitian vector space the length of any vector is non-negative. Thus

$$(x - m^2 + m) \geqq 0.$$

A similar procedure starting with the operator $\mathsf{L}^-\mathsf{L}^+$ instead of $\mathsf{L}^+\mathsf{L}^-$ leads to a similar inequality

$$(x - m^2 - m) \geqq 0$$

and addition of the two inequalities therefore gives

$$x \geqq m^2.$$

This shows that, for any given value of the squared magnitude of the angular momentum, the values that can be taken by any component are bounded both above and below.

Now let the upper and lower bounds of m be m_+ and m_-. The "top" state must then vanish if we try to step it up with L^+; for if the result were non-zero there would be a state with $m > m_+$. Hence

$$\mathsf{L}^+\phi_{x,m_+} = 0.$$

Also, from the second form of (2.21) we can say, putting $m = m_+$,

$$\mathbf{L}^2\phi_{x,m_+} = (m_+{}^2 + m_+)\phi_{x,m_+}$$

so that the eigenvalue of $\mathbf{L}^2$ in this state is

$$x = m_+(m_+ + 1).$$

The same argument applied to ϕ_{x,m_-} gives

$$x = m_-(m_- - 1);$$

from these results we obtain $m_+{}^2 + m_+ = m_-{}^2 - m_-$ or

$$(m_+ + m_-)(m_+ - m_- + 1) = 0.$$

The only acceptable solution is $m_- = m_+$, since $m_- = (m_+ + 1)$ would violate the assumption that m_+ is the upper bound.

We usually denote m^+, which determines the highest eigenvalue of L_z, by l. The eigenvectors may be labelled by the quantum numbers l, m, and the results so far established then become

$$\mathsf{L}^2\phi_{l,m} = l(l+1)\phi_{l,m}, \tag{2.22a}$$

$$\mathsf{L}_z\phi_{l,m} = m\phi_{l,m} \quad (m = l, l-1, \ldots -l). \tag{2.22b}$$

It is also clear that, since the eigenvalues change by unit steps, $l-(-l) = 2l$ is a positive integer. Hence l is either an integer or half an odd integer. It will appear shortly, on using the orbital angular momentum operators in Schrödinger form, that the differential equations corresponding to (2.22) have well-behaved solutions only for integral l. The preceding argument, based only on the commutation relations, naturally includes the case of spin angular momentum (Vol. 1, Section 4.9): we are already familiar with the "spin $\frac{1}{2}$" system where $l \to \frac{1}{2}$, $m \to +\frac{1}{2}$, $-\frac{1}{2}$ and later shall find that other half-odd integer values can be obtained by "coupling" spin and orbital angular momenta. With trivial changes of notation, however, the equations just derived apply to any type of angular momentum. We may also note in passing that since the different eigenvectors are orthogonal, so that $\langle\phi_{l,m}|\phi_{l,m\pm1}\rangle = 0$, then

$$\langle\phi_{l,m}|\mathsf{L}^\pm|\phi_{l,m}\rangle = 0$$

and it therefore follows that the expectation values of the L_x and L_y components in a state of given L_z are both zero (cf. Fig. 2.1).

Finally, it is useful to determine the coefficients $c_m{}^+$ in (2.20a) and $c_m{}^-$ in (2.20b) so as to preserve normalization when an eigenvector is stepped up or down. We assume that $\phi_{l,m\pm1}$ is normalized and therefore require

$$\langle\mathsf{L}^\pm\phi_{l,m}|\mathsf{L}^\pm\phi_{l,m}\rangle = |c_m{}^\pm|^2\langle\phi_{l,m\pm1}|\phi_{l,m\pm1}\rangle = |c_m{}^\pm|^2.$$

But the left-hand scalar product may be written, using (2.21),

$$\langle\phi_{l,m}|\mathsf{L}^\mp\mathsf{L}^\pm|\phi_{l,m}\rangle = \langle\phi_{l,m}|(\mathsf{L}^2 - \mathsf{L}_z{}^2 \mp \mathsf{L}_z)|\phi_{l,m}\rangle$$
$$= [l(l+1) - m^2 \mp m]\langle\phi_{l,m}|\phi_{l,m}\rangle$$

and the normalization condition is therefore

$$|c_m{}^\pm|^2 = l(l+1) - m^2 \mp m.$$

The coefficients $c_m{}^\pm$ are thus determined to within an arbitrary phase factor $e^{i\delta}$. We now adopt the phase convention $\delta = 0$, and the relative phases of all the eigenfunctions are then fixed. The resultant equations are, noting that $l(l+1)-m(m\pm1) = (l\mp m)(l\pm m+1)$,

$$\mathsf{L}^{+}\phi_{l,m} = \sqrt{\{(l-m)(l+m+1)\}}\,\phi_{l,m+1}, \qquad (2.23a)$$

$$\mathsf{L}^{-}\phi_{l,m} = \sqrt{\{(l+m)(l-m+1)\}}\,\phi_{l,m-1}. \qquad (2.23b)$$

In using the step-up and step-down operators it must be remembered that a choice of phases has been conventionally fixed; this is not a trivial matter and has sometimes caused much confusion.

2.4. Solution of the Schrödinger equation in polar coordinates

In dealing with central field problems it is natural to employ polar coordinates $(r,\ \theta,\ \varphi)$, in which case (Fig. 2.2) $V = V(r)$. The potential

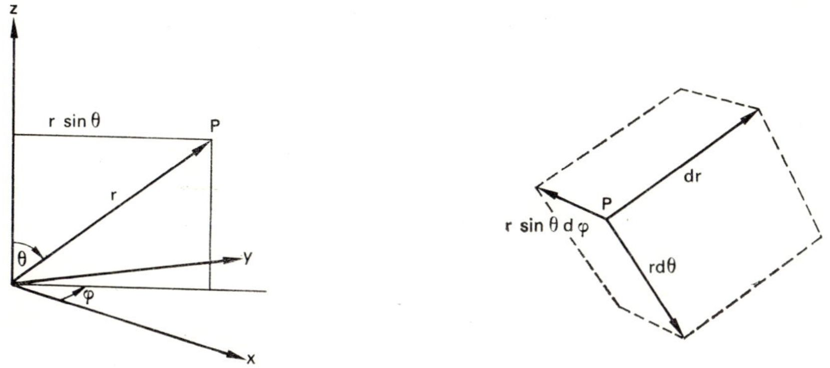

FIG. 2.2. Polar coordinates and volume element at point P.

energy function is then independent of θ and φ. The volume element takes the form $d\mathbf{r} = h_1h_2h_3\,dr\,d\theta\,d\varphi$ and with the scale factors $h_1 = 1$, $h_2 = r$, $h_3 = r\sin\theta$ the operator ∇^2 follows at once from I.(A1.15). The Schrödinger equation then becomes

$$\mathsf{H}\phi = -\frac{\hbar^2}{2m}\left\{\frac{1}{r^2}\frac{\partial}{\partial r}\left(r^2\frac{\partial}{\partial r}\right)+\frac{1}{r^2\sin\theta}\frac{\partial}{\partial\theta}\left(\sin^2\theta\frac{\partial}{\partial\theta}\right)\right.$$

$$\left.+\frac{1}{r^2\sin^2\theta}\frac{\partial^2}{\partial\varphi^2}\right\}\phi + V\phi = E\phi \qquad (2.24)$$

and this may be solved by the separation technique (Vol. 1, App. 2).

It is helpful to use the angular momentum properties from the start by noting that this Hamiltonian contains a term that can be identified as a multiple of the angular momentum operator $\mathbf{L}^2$. To show this we take the operators L_x, L_y, L_z in their Schrödinger form (p. 37)

$$\mathsf{L}_x = \frac{1}{i}\left(y\frac{\partial}{\partial z} - z\frac{\partial}{\partial y}\right), \quad \mathsf{L}_y = \frac{1}{i}\left(z\frac{\partial}{\partial x} - x\frac{\partial}{\partial z}\right), \quad \mathsf{L}_z = \frac{1}{i}\left(x\frac{\partial}{\partial y} - y\frac{\partial}{\partial x}\right)$$

$$(2.25)$$

and, by a straightforward application of the formulae for partial differentiation, re-express them in terms of r, θ, φ. The result is

$$\mathsf{L}_x = \frac{1}{i}\left(-\sin\phi\,\frac{\partial}{\partial\theta} - \cot\theta\cos\varphi\,\frac{\partial}{\partial\varphi}\right),$$

$$\mathsf{L}_y = \frac{1}{i}\left(\cos\phi\,\frac{\partial}{\partial\theta} - \cot\theta\sin\varphi\,\frac{\partial}{\partial\varphi}\right),$$

$$\mathsf{L}_z = \frac{1}{i}\frac{\partial}{\partial\varphi}\,. \tag{2.26}$$

From these results it is not difficult to obtain the operator $\mathbf{L}^2 = \mathsf{L}_x{}^2 + \mathsf{L}_y{}^2 + \mathsf{L}_z{}^2$. Thus,

$$\mathbf{L}^2 = -\frac{1}{\sin\theta}\frac{\partial}{\partial\theta}\left(\sin\theta\,\frac{\partial}{\partial\theta}\right) - \frac{1}{\sin^2\theta}\frac{\partial^2}{\partial\varphi^2}\,. \tag{2.27}$$

It is now clear that the Hamiltonian in (2.24) may be written

$$\mathsf{H} = \left\{-\frac{\hbar^2}{2m}\frac{1}{r^2}\frac{\partial}{\partial r}\left(r^2\frac{\partial}{\partial r}\right) + V(r)\right\} + \frac{\hbar^2}{2mr^2}\mathbf{L}^2 \tag{2.28}$$

and hence that, for eigenfunctions chosen to be simultaneous eigenfunctions of H and $\mathbf{L}^2$, the last term will be equivalent merely to a multiplicative factor $\hbar^2 l(l+1)/2mr^2$. Such eigenfunctions must therefore satisfy the much simpler equation

$$-\frac{\hbar^2}{2m}\left\{\frac{1}{r^2}\frac{\partial}{\partial r}\left(r^2\frac{\partial\phi}{\partial r}\right) - \frac{l(l+1)}{r^2}\,\phi\right\} + V\phi = E\phi. \tag{2.29}$$

The radial dependence of the wave function is thus determined; but since (2.29) contains only the *partial* differential operator $\partial/\partial r$ the solution may also exhibit a θ, φ-dependence and we therefore take

$$\phi(\mathbf{r}) = f(r)Y(\theta, \varphi). \tag{2.30}$$

Here $Y(\theta, \varphi)$ must be an eigenfunction of $\mathbf{L}^2$—which operates only on the variables θ, φ. We now introduce a new radial function

$$R(r) = rf(r) \tag{2.31}$$

and restate the basic equations for the radial and angular factors in the wave function. Substitution of (2.30) in (2.29) and use of (2.31) lead to

$$-\frac{\hbar^2}{2m}\frac{d^2R}{dr^2}+\left(V(r)+\frac{\hbar^2}{2m}\frac{l(l+1)}{r^2}\right)R = ER \qquad (2.32)$$

while the function $Y(\theta, \varphi)$ must satisfy

$$-\left\{\frac{1}{\sin\theta}\frac{\partial}{\partial\theta}\left(\sin\theta\frac{\partial Y}{\partial\theta}\right)+\frac{1}{\sin^2\theta}\frac{\partial^2 Y}{\partial\varphi^2}\right\} = l(l+1)Y \qquad (2.33)$$

which follows on substituting (2.27) in the equation $\mathsf{L}^2 Y = l(l+1)Y$. The possibility of separation into radial and angular factors is thus directly linked with the occurrence of commuting operators for radial and angular motion. We now discuss the radial and angular equations separately, taking the angular equation first.

The angular equation

It is clear from (2.26) and (2.27) that

$$\mathsf{L}^2 = -\frac{1}{\sin\theta}\frac{\partial}{\partial\theta}\left(\sin\theta\frac{\partial}{\partial\theta}\right)+\frac{1}{\sin^2\theta}\mathsf{L}_z{}^2$$

and when operating on simultaneous eigenfunctions of H, L^2 and L_z the last term will thus be equivalent to multiplication by $m^2/\sin^2\theta$ where m is the eigenvalue of L_z. Since Y must be an eigenfunction of L_z

$$\mathsf{L}_z Y = \frac{1}{i}\frac{\partial}{\partial\varphi}\,Y = mY \qquad (2.34a)$$

and integration yields

$$Y(\theta, \varphi) = F(\theta)\,e^{im\varphi}.$$

Finally, the condition that this function be an eigenfunction of L^2 determines the factor $F(\theta)$. Thus

$$\mathsf{L}^2 F = -\frac{1}{\sin\theta}\frac{d}{d\theta}\left(\sin\frac{dF}{d\theta}\right)+\frac{m^2}{\sin^2\theta}\,F = l(l+1)F. \qquad (2.34b)$$

On introducing $x = \cos\theta$ as a new variable, and putting $F(\theta) = P(\cos\theta)$ this equation becomes

$$(1-x^2)\frac{d^2P}{dx^2}-2x\frac{dP}{dx}+\left(l(l+1)-\frac{m^2}{1-x^2}\right)P = 0 \qquad (2.35)$$

which is the "associated Legendre equation" treated in Appendix 3, Vol. 1. For integral values of l the solutions are everywhere well behaved provided m is an integer with $|m| \leqslant l$. Since changing the sign of m makes no difference to the equation we need only consider $m \geqslant 0$ and may denote the standard solution by

$$P_l{}^m(\cos\theta) = P_l{}^{-m}(\cos\theta) = P_l{}^{|m|}(\cos\theta).$$

This function is the "associated Legendre polynomial", a polynomial of the lth degree in $\cos\theta$.

For given eigenvalues l and m, the angular part of the wave function is evidently a product of the form $P_l{}^{|m|}(\cos\theta)\, e^{im\varphi}$ and as m takes the full range of values ($m = l, l-1, \ldots -l$) there will be just $2l+1$ distinct

TABLE 2.1

Normalized spherical harmonics[‡] $Y_{lm}(\theta,\varphi)$

l	m	$Y_{lm}(\theta,\varphi)$
0	0	$(1/4\pi)^{1/2}$
1	0	$(3/4\pi)^{1/2}\cos\theta$
1	± 1	$\mp(3/8\pi)^{1/2}\sin\theta\, e^{\pm i\varphi}$
2	0	$(5/16\pi)^{1/2}(3\cos^2\theta-1)$
2	± 1	$\mp(15/8\pi)^{1/2}\sin\theta\cos\theta\, e^{\pm i\varphi}$
2	± 2	$(15/32\pi)^{1/2}\sin^2\theta\, e^{\pm 2i\varphi}$
3	0	$(7/16\pi)^{1/2}(5\cos^2\theta-1)\cos\theta$
3	± 1	$\mp(21/64\pi)^{1/2}(5\cos^2\theta-1)\sin\theta\, e^{\pm i\varphi}$
3	± 2	$(105/32\pi)^{1/2}\sin^2\theta\cos\theta\, e^{\pm 2i\varphi}$
3	± 3	$\mp(35/64\pi)^{1/2}\sin^3\theta\, e^{\pm 3i\varphi}$

[‡] The normalization also includes the Condon–Shortley phase factors (see text).

and well-behaved simultaneous eigenfunctions of $\mathbf{L}^2$ and $\mathbf{L}_z$—as already inferred from general principles. These functions are usually normalized so that on squaring and multiplying by an element of solid angle (i.e. the angle dependent part, $\sin\theta\, d\theta\, d\varphi$, of the volume element) integration over θ and φ gives unity. It is also convenient to include a phase factor so that when the step-up and step-down operators are expressed in differential form the eigenfunctions $Y_{l,m}$ have the phase relations already agreed upon (p. 44). The standard spherical harmonics are in this way defined as

$$Y_{l,m}(\theta,\varphi) = (-1)^{(m+|m|)/2}\left[\frac{2l+1}{4\pi}\frac{(l-|m|)!}{(l+|m|)!}\right]^{1/2}P_l{}^{|m|}(\cos\theta)\, e^{im\varphi}. \tag{2.36}$$

Functions defined in this way are said to satisfy the Condon–Shortley phase conventions. When phase relations between different eigenfunctions are not important, however, the factor $(-1)^{(m+|m|)/2}$ is frequently dropped. The first few functions are listed in Table 2.1.

Before turning to the radial equation we note that the restriction to integral l and m values, in the case of orbital angular momentum, appears in the Schrödinger formulation from the condition that the solutions of (2.34a) and (2.34b), respectively, are normalizable and single valued. Such considerations do not apply of course in the case of spin and the half-odd integer values are then allowed.

The radial equation

To discuss the radial equation fully we must assume a particular form of potential energy function. It is possible, however, to reach important conclusions with comparatively slight restrictions on the form of $V(r)$; we shall defer until later the special case of the Coulomb potential.

First, it is clear that (2.32) might be regarded as the Schrödinger equation for a one-dimensional motion with r in the interval $(0, \infty)$ and that

$$U(r) \;=\; V(r) + \frac{\hbar^2}{2m}\frac{l(l+1)}{r^2} \tag{2.37}$$

would play the part of an effective potential energy. The l-dependent term in $U(r)$ represents the potential energy due to a fictitious "centrifugal" force, increasing in proportion to the angular momentum.

Next we examine the form of solution $R(r)$ in the region near the origin, supposing only that $V(r)$ does not go to $-\infty$ faster than $-1/r$. Near $r = 0$, the $1/r^2$ term in (2.32) dominates, giving the approximation

$$\frac{d^2R}{dr^2} - \frac{l(l+1)}{r^2}\,R \;=\; 0. \tag{2.38}$$

On putting $R = r$, the indicial equation (Vol. 1, p. 145) is

$$\lambda(\lambda-1) - l(l+1) \;=\; 0$$

and solutions occur for

$$\lambda = -l, \quad \lambda = l+1.$$

Thus, near the origin

$$R(r) \sim cr^{-l} \quad \text{or} \quad cr^{l+1}.$$

In order that the wave function shall be quadratically integrable, the radial factor f must behave at $r \to 0$ so that $\int_0^\infty f^2 r^2\,dr$ remains finite, i.e. f must go to infinity slower than $1/r$. Consequently, the solution

$R = cr^{-l}$ is unacceptable for $l > 1$. When $l = 0$ the two possible values of λ are 0 and 1. Substitution shows that the expansion for $l = 0$, namely $R = c_0 + c_1 r + c_2 r^2 + \ldots$, cannot satisfy the equation (2.38) for r unless $c_0 = 0$; but in this case the solution corresponds to $\lambda = l + 1$ with $l = 0$. Thus, for all l, we may discard the solution $R = cr^{-l}$ and keep only $R = cr^{l+1}$ as the acceptable form for $r \to 0$.

Finally, we discuss how the function behaves for larger values of r. For this purpose we write (2.32) in the form

$$\frac{d^2R}{dr^2} = \frac{2m}{\hbar^2}\{U(r) - E\}R \tag{2.39}$$

with U defined in (2.37). This tells us how the curvature of the function R depends on its value at any point. When E is large and negative the curvature will be everywhere positive, and taking the form cr^{l+1} near $r = 0$ with a positive value of the constant c, the function R will behave as indicated in curve 1 of Fig. 2.3a. This is not an acceptable solution and this means that E is not an eigenvalue. If we now decrease E gradually, there will appear a region in which $U(r) - E < 0$ and the function R will take the form indicated in curve 2 of Fig. 2.3a. As E is decreased further, we shall find a particular value of E for which $R(r) \to 0$ for $r \to \infty$ (curve 3). The wave function then becomes square integrable, no longer diverging at infinity, and the value of E is the lowest energy eigenvalue, E_1.

On decreasing E still further $R(r)$ will cross the r axis at some value of r; for larger r the curvature will become negative, and R will tend to $-\infty$ for $r \to \infty$ (curve 4). But when E reaches the next lowest eigenvalue E_2, then R will tend to 0 from below for $r \to \infty$ (curve 5). The corresponding wave function will then have one radial node.

If the potential $V(r)$ has a deep and wide enough negative region, the procedure can be repeated a number of times and a certain number of negative eigenvalues will appear. The important fact is that in going from one eigenvalue to the next, in ascending energy order, the number of radial nodes increases by 1: the function R_n with eigenvalue E_n has $n - 1$ radial nodes. The number of these discrete negative energy eigenvalues, and hence the number of bound states, may be either finite or infinite, depending on the shape of $V(r)$. One way of finding this number is to integrate the differential equation

$$\frac{\hbar^2}{2m}\frac{d^2R}{dr^2} + U(r)R = 0, \tag{2.40}$$

which is obtained by setting $E = 0$ in (2.39), assuming $R = cr^{l+1}$ near the origin. The function may diverge when r tends to infinity but the number of nodes of R will nevertheless determine the number of bound states for this potential.

Next we turn to the case of positive energies. When $r \to \infty$, $U(r) \to 0$, and it is clear from (2.39) that d^2R/dr^2 and R have opposite signs for large r. Thus when R is positive, d^2R/dr^2 is negative and the

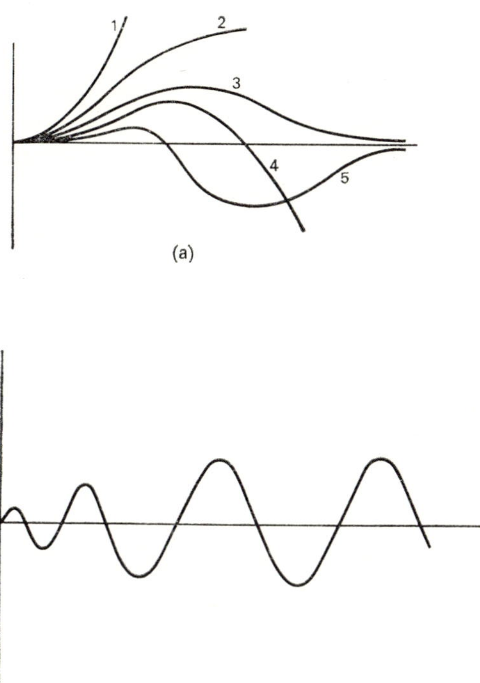

Fig. 2.3. Behaviour of radial function $R(r)$: (a) for negative values of the energy, (b) for positive values.

curve is convex towards the r-axis; but when R is negative, the curve is concave towards the r-axis (Fig. 2.3b). In both cases R is oscillatory and never tends to infinity. In fact, for large values of R, $U(r)$ can be neglected and equation (2.39) becomes

$$\frac{d^2R}{dr^2} = -k^2R, \quad (k^2 = 2mE/\hbar^2)$$

whose solution is

$$R = A \sin (kr + \alpha)$$

where A and α are constants. The radial wave function with the asymptotic form

$$f(r) = r^{-1}R(r) \sim (A/r) \sin (kr + \alpha) \qquad (2.41)$$

is not quadratically integrable as it stands, but it belongs to the class of functions discussed in Vol. 1, Section 2.4, from which normalizable solutions may be constructed by superposition of the functions corresponding to a full range of k values to give a wave packet.[‡] The wave function is thus acceptable in the extended sense appropriate in discussing free particles. Evidently for every positive value of E an acceptable wave function exists. There is a continuum of positive eigenvalues and we may use E itself as a quantum number specifying a state: the wave functions are then of the form

$$\phi_{E,l,m}(r, \theta, \varphi) = f_{E,l}(r)\,Y_{l,m}(\theta, \varphi). \qquad (2.42)$$

It should be noted that a wave packet constructed from functions in the vicinity of a given E value may produce a spherical pulse, possibly modulated by a spherical harmonic factor, proceeding outwards from the origin; such a pulse would describe a positive energy particle scattered by the potential field. Wave packets are fully discussed in Vol. 4.

Notation for central field states

Finally, we introduce the notation for labelling the bound states. This is borrowed from atomic spectroscopy in which the labels s, p, d, f, g are used to denote spectral series (sharp, principal, diffuse, etc.) arising from transitions to states with $l = 0, 1, 2, 3, 4, \ldots$. Thus with $l = 2$ we have a degenerate group of five "d-states" with m running from -2 to $+2$. The labels l, m in the spherical harmonic factor Y_{lm} are sometimes called the "azimuthal" and "magnetic" quantum numbers, respectively, terms again taken over from the early days of quantum theory.

For any given value of l there will, in general, be a number of states distinguished by the number of nodes in the radial factor. It is customary to label the "s states" (i.e. $l = 0$), $1s$, $2s$, $3s$, ..., ns, ... in which the principal quantum number n indicates the number of radial nodes plus one. For $l \neq 0$, however, there are both radial and angular nodes and the energy increases roughly with the total number: thus for $l = 1$, $m = 0$, there is one angular node, and the energy (for the more usual

---

[‡] Or alternatively by putting the system in a large spherical box and normalizing over the region of the box.

potentials) is closer to that of the $2s$ state (one radial node) than the $1s$ state (no nodes): it is then natural to call the corresponding state the $2p$ state (not $1p$). The long-accepted notation is that states of given l, m are labelled in ascending energy order by a principal quantum number n which takes values

$$n = l+1, l+2, \ldots .$$

Thus having solved (2.32) for a given value of l, the state with lowest eigenvalue will have $n = l+1$, the next state will have $n = l+2$, etc.

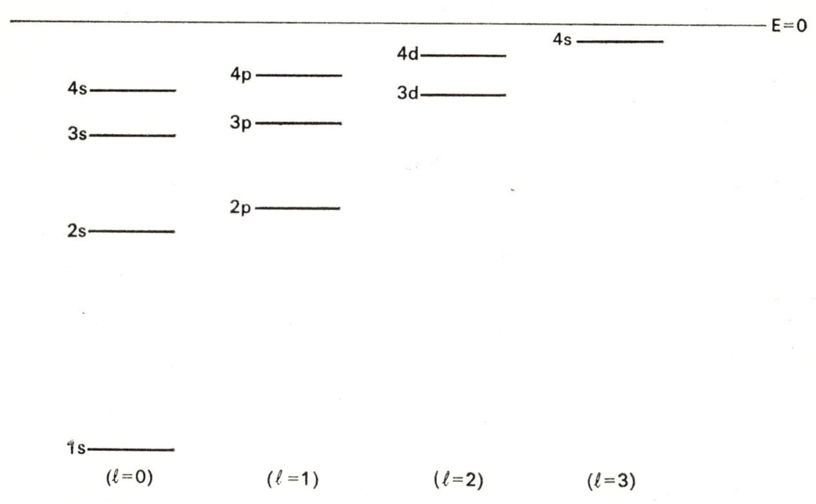

FIG. 2.4. Characteristic sequence of central field energy levels (not to scale).

In this way n indicates the total number of nodal surfaces in the wave function and

$$\text{Number of radial nodes} = (n-l-1).$$

Thus a $3d$ function has $3-2-1 = 0$ radial nodes, but the $3p$ function has $3-1-1 = 1$ radial node.

The sequence of energy levels for a central field of the type encountered in dealing with atoms (Chap. 3) is typically of the form indicated in Fig. 2.4. The corresponding wave functions are of the form

$$\phi_{nlm}(r, \theta, \varphi) = f_{nl}(r) Y_{lm}(\theta, \varphi). \tag{2.43}$$

The radial factors f_{nl}, and the energy E_{nl}, are independent of m, which distinguishes states which are essentially degenerate for any central

c

potential $V(r)$. For the special case, $V \propto 1/r$, the levels of given principal quantum numbers all coalesce and the principal quantum number then orders the levels unequivocally according to energy.

2.5. Atomic orbitals

Equation (2.32) can be solved in closed form for several choices of $V(r)$, one or two of which are suitable for representing electrons in atoms. The corresponding bound state wave functions, classified as in the last section, are then called *Atomic Orbitals* (AO's). In this section we deal with two important cases:

Hydrogen-like orbitals

First we consider a "hydrogen-like" system comprising a single electron in the field of a nucleus of atomic number Z. The potential energy function is then (using SI units with the abbreviation $\kappa_0 = 4\pi\varepsilon_0$)

$$V(r) \;=\; -Z\,e^2/\kappa_0 r \tag{2.44}$$

and the cases where $Z = 1, 2, 3, 4, \ldots$ correspond to H, He$^+$, Li^{++}, Be^{+++}, ... On inserting (2.44) into (2.32) and introducing atomic units[‡] the radial equation becomes

$$-\tfrac{1}{2}\frac{d^2 R}{dr^2} + \left\{ -\frac{Z}{r} + \frac{l(l+1)}{2r^2} \right\} R \;=\; ER. \tag{2.45}$$

Let us introduce the new variable ρ and a new energy parameter λ defined by

$$\rho = Zr, \quad \lambda = -2E/Z^2.$$

We then obtain the differential equation

$$\frac{d^2 R}{d\rho^2} + \left\{ \frac{2}{\rho} - \frac{l(l+1)}{\rho^2} - \lambda \right\} R \;=\; 0$$

which may be solved by the standard method (Vol. 1, App. 3). Its acceptable solutions are of the form $R = g(\rho)\exp(-\sqrt{\lambda}\cdot\rho)$, where the exponential factor is the asymptotic solution ($\rho \to \infty$). The equation for $g(\rho)$ is then satisfied by a finite polynomial (yielding a quadratically integrable radial function) provided $\lambda = 1/n^2$ where n is an integer.

[‡] Here we are neglecting as usual motion of the nucleus. The centre of mass correction (Section 2.1) can be allowed for by using a modified atomic unit with reduced mass $\mu \neq 1$ (instead of $m = 1$). Atomic units are fully discussed in Appendix 1.

In this case $g(\rho)$ is related to the associated Laguerre polynomial (Vol. 1, App. 3).

The eigenvalues and radial functions ($f_{nl} = R_{nl}/r$) may finally be expressed in the form

$$E_{nl} = E_n = -\frac{Z^2}{2n^2}, \tag{2.46a}$$

$$f_{nl}(r) = C_{nl}(2Z/n)^l \, e^{-Zr/n} L_{n+l}^{2l+1}(2Zr/n) \tag{2.46b}$$

TABLE 2.2

Hydrogen-like radial factors‡ $f_{nl}(r)$

n	l	$f_{nl}(r)$
1	0	$Z^{3/2} \, 2e^{-Zr}$
2	0	$Z^{3/2} \, (1/2\sqrt{2}) \, (2 - Zr) \, e^{-Zr/2}$
2	1	$Z^{3/2} \, (1/2\sqrt{6}) \, Zr \, e^{-Zr/2}$
3	0	$Z^{3/2} \, (2/81\sqrt{3}) \, (27 - 18Zr + 2Z^2r^2) \, e^{-Zr/3}$
3	1	$Z^{3/2} \, (4/81\sqrt{6}) \, (6Zr - Z^2r^2) \, e^{-Zr/3}$
3	2	$Z^{3/2} \, (4/81\sqrt{30}) \, Z^2r^2 \, e^{-Zr/3}$

‡ All expressed in atomic units. To convert to absolute units, replace r by r/a_0 and multiply each normalizing factor by $(a_0)^{-3/2}$.

where the factor C_{nl} ensures that the radial function is normalized "with weight factor r^2" (see I.(3.16) et seq.):

$$C_{nl} = \left[\frac{(n-l+1)!}{2n(n+l)_n!} \left(\frac{2Z}{n} \right)^3 \right]^{1/2}. \tag{2.46c}$$

On attaching the appropriate spherical harmonic factors (Table 2.1) we obtain normalized hydrogen-like atomic orbitals in the form (2.43). The first few radial functions are listed in Table 2.2 and are seen to contain nodes consistent with the general considerations of the last section.

Slater orbitals

In a many-electron atom, as we shall find presently, there is still some justification for a simplified "model" in which each electron moves in an "effective" central field, arising partly from the attraction of the nucleus and partly from repulsions exerted by the other electrons. This field no longer has simple Coulombic form and the atomic orbitals are no longer hydrogen-like: they are frequently approximated by linear combinations of "Slater orbitals" which are single-term functions of the form (2.43) but with a radial factor

$$f_{nl}(r) = C_n r^{n-1} e^{-\zeta r}, \quad C_n = (2\xi)^{n+1/2}/\sqrt{(2n)!}. \tag{2.47}$$

These radial functions are again normalized with weight factor r^2, and contain an adjustable parameter ζ referred to as the "orbital exponent" and in general dependent on the l value of the orbital. Normalized Slater orbitals are obtained by adding the spherical harmonic factors in Table 2.1.

It is easily shown, by substituting $R = rf_{nl}$ in (2.45) and equating to zero the coefficients of all powers of r, that the function (2.47) satisfies the general radial equation (2.39) with $U(r)$ given by (cf. (2.37))

$$U(r) = -\frac{\zeta n}{r} + \frac{n(n-1)}{2r^2}.$$

This means that a Slater orbital with radial part (2.47) satisfies the Schrödinger equation if the potential energy term is replaced by

$$V_{\text{eff}}(r) = -\frac{\zeta n}{r} + \frac{n(n-1) - l(l+1)}{2r^2} \tag{2.48}$$

corresponding to an effective field somewhat different from the actual Coulombic field of the nucleus. For large enough values of r, V_{eff} becomes essentially Coulombic and represents the potential energy of an electron in the field of a nucleus of effective atomic number $Z_{\text{eff}} = \zeta n$; and for $l = n-1$ (the maximum l-value for a given principal quantum number) the Slater orbital becomes identical with the corresponding hydrogen-like function. More generally, however, a single Slater orbital corresponds to a field which is non-Coulombic and strongly dependent on the value of l. When such an orbital is used to describe an electron in a *many*-electron atom, ζ is chosen to simulate a "screening" of the actual nuclear charge Z by the other electrons. Approximate rules for determining suitable Z_{eff} values have been given by Slater (1930) and improved values are listed by Clementi and Raimondi (1963).

It should be noted that the orbital exponents in general differ for orbitals with the same principal quantum number (i.e. the effectiveness of the "screening" may be different for, say, $2s$ and $2p$ electrons) and that since orbitals of different n and l are eigenfunctions corresponding to different potential energy functions, *they are not orthogonal*. In fact, the Hamiltonian operator with potential energy (2.48) has only *one* bound state eigenfunction: the second term in V_{eff} becomes large and positive near the nucleus and the negative region is then deep enough only to hold one bound state. The positive peak in V_{eff} is also responsible for the characteristic *nodeless* form of the radial factor (2.47). When Slater orbitals are used in atomic and molecular theory they are often orthogonalized by the Schmidt method (see Vol. 1, p. 47), starting with the orbital of lowest n value for each given l: the resultant functions then display essentially the same nodal behaviour as the corresponding hydrogen-like functions.

Relationship between real and complex orbitals

The simultaneous eigenfunctions of H, L^2 and L_z are essentially complex functions, owing to the presence of the factor $\exp(im\varphi)$ in the spherical harmonic. This factor arose from the requirement that the orbital should be an eigenfunction of L_z. However, a state of maximal knowledge in which all three operators have simultaneous eigenvalues is only entirely appropriate if L_z is measured by fixing some unique (z)-axis in space, e.g. by applying a magnetic field. This situation may arise in spectroscopic experiments but very commonly there is either no preferred axis or, on the other hand, the spherical symmetry is disturbed by environmental effects and there are two or three preferred axes (e.g. the x-, y-, and z-axes for an atom in a "cubic" environment). In such cases the degeneracy of functions of given n and l but differing in m may be used to set up more appropriate linear combinations.

The most commonly used real functions arise on replacing orbitals with spherical harmonic factors Y_{lm} and $Y_{l,-m}$ ($m \neq 0$) by their sum and difference; for then the factors $e^{im\varphi}$ and $e^{-im\varphi}$ will be replaced by $\cos m\varphi$ and $\sin m\varphi$ respectively. When Y_{lm} is defined according to (2.36) the *real* harmonics are defined by

$$
\begin{aligned}
S_{l0} &= Y_{l0} & (m = 0), \\
S_{lm} &= [(-1)^m Y_{lm} + Y_{l,-m}]/\sqrt{2} & (m > 0), \\
S_{lm} &= -i[(-1)^m Y_{lm} - Y_{l,-m}]/\sqrt{2} & (m < 0).
\end{aligned}
\tag{2.49}
$$

They are obtained from the complex harmonic Y_{lm} in (2.36) by the replacement

$$
\begin{aligned}
S_{lm}: \quad (-1)^{(m+|m|)/2} e^{im\varphi} &\to \sqrt{2}.\cos m\varphi & (m > 0), \\
S_{lm}: \quad (-1)^{(m+|m|)/2} e^{im\varphi} &\to \sqrt{2}.\sin m\varphi & (m < 0),
\end{aligned}
$$

the factor $\sqrt{2}$ restoring normalization. The real orbitals are then obtained by attaching the appropriate radial factors from (2.46b) or from Table 2.2.

The geometrical forms of the real harmonics are revealed most clearly in terms of Cartesian coordinates, each harmonic becoming a polynomial of degree l in x, y, z divided by r^l. The angular factors for the real s, p, and d functions are listed in this form in Table 2.3, which also shows the conventional notation for the resultant orbitals. The forms of the orbitals are indicated in Fig. 2.5.

The real orbitals have an interesting physical interpretation. Thus $2p_x$ is a mixture of $2p_{+1}$ and $2p_{-1}$ which describe two states, each with one unit of angular momentum around the z-axis but in opposite senses; these are combined with equal weights and the resultant state is thus

TABLE 2.3

Real harmonics‡ in Cartesian form

l	m	$H_{lm}(x, y, z)$	Orbital type
0	0	N_0	s
1	0	$N_1 \times \left\{ \begin{array}{l} z \\ x \\ y \end{array} \right.$	p_z
1	1	x	p_x
1	-1	y	p_y
2	0	$N_2 \times \left\{ \begin{array}{l} \frac{1}{2}(3z^2 - r^2) \\ \sqrt{3}\,xz \\ \sqrt{3}\,yz \\ \frac{1}{2}\sqrt{3}(x^2 - y^2) \\ \sqrt{3}\,xy \end{array} \right.$	d_{z^2}
2	1	$\sqrt{3}\,xz$	d_{xz}
2	-1	$\sqrt{3}\,yz$	d_{yz}
2	2	$\frac{1}{2}\sqrt{3}(x^2 - y^2)$	$d_{x^2-y^2}$
2	-2	$\sqrt{3}\,xy$	d_{xy}

‡ The real harmonics are conventionally defined so that positive m values correspond to $\cos m\varphi$ factors, negative m values to $\sin m\varphi$ factors. Multiplication by r^l yields an lth degree polynomial in x, y, z here denoted by $H_{lm}(x, y, z)$. The normalizing factors are given by $N_l = [(2l+1)/4\pi]^{1/2}$.

one in which L_z is uncertain, values ± 1 being equally probable. If the time factors $\exp(-iE_{2p}t)$ are added, it is clear that $2p_{+1}$ and $2p_{-1}$ become travelling wave solutions of the time-dependent Schrödinger equation, circulating in opposite senses around the z-axis. The real functions, $2p_x$ and $2p_y$, are simply standing wave solutions, set up by superposition of the travelling waves with suitable phases to give sine and cosine angular dependence.§ These conclusions apply of course for all values of m, irrespective of the form of the central potential.

---

§ The corresponding result for *linear* momentum is that combination of e^{ikx} and e^{-ikx} will produce a standing wave solution satisfying boundary conditions for a finite box. See, for instance, Vol. 1, Sections 2.1 and 2.4.

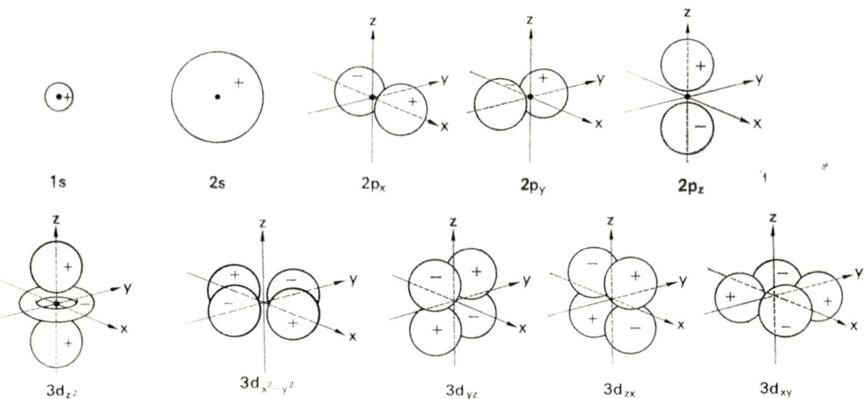

FIG. 2.5. Forms of real atomic orbitals (schematic).

2.6. Inclusion of electron spin

The effects of electron spin first came to light experimentally in the optical spectra of the alkali metals such as Na, K and Rb. These atoms all behave as if they possessed one loosely bound "outermost" electron, moving in the potential field of a "core", their spectra showing strong similarities with that of the hydrogen atom. The main difference is a lifting of the degeneracy between ns, np, nd, ... states, and this we already know is to be expected if there is any deviation from a strictly Coulombic central potential. The spectral series for the sodium atom, as inferred from experiment, are indicated in Fig. 2.6; the levels are labelled as representing the central field states (cf. Fig. 2.4) for a single "series electron" and the data indicate that in the ground state of the atom, the series electron is in a $3s$ state corresponding to a central field with $Z_{\text{eff}} \sim 2.5$. Similar data for K and Rb show that the series electrons are in $4s$ and $5s$ states respectively. All these assignments are compatible with the pre-1925 picture of the electronic structure of the alkali metals, based largely on the Periodic Table, the "series electron" corresponding to the "valence electron", and the "core" to an inert gas structure like that of the preceding element of the Periodic Table.

We therefore treat these atoms as one-electron systems, in a first approximation, and test the theory of spin, as developed in Vol. 1, Section 3.9, by trying to account for the observed "fine-structure" of the levels.[‡] The Hamiltonian for the series electron in the field of the core will be

$$H_0 = -\frac{\hbar^2}{2m} \nabla^2 + V(r) \qquad (2.50)$$

in the absence of spin terms. The spin terms are exceedingly small and, as they represent effects which are not present in a strictly classical

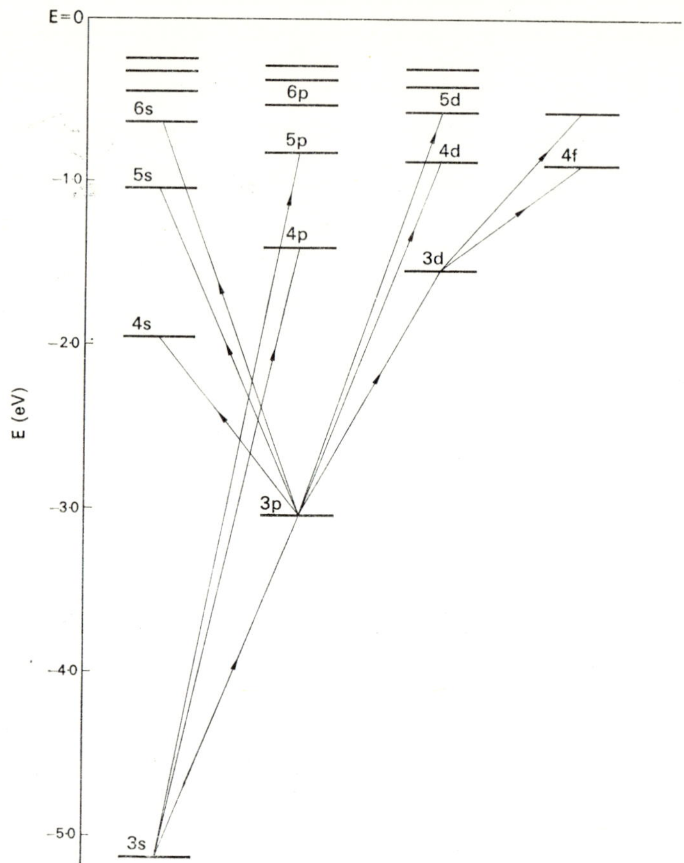

FIG. 2.6. Term diagram for the sodium atom. Transitions leading to various spectral series are indicated by the arrows (corresponding to absorption lines).

system, their forms cannot be directly inferred from the postulates in Chap. 4 of Vol. 1. We consider spin- and field-dependent effects more fully in Chap. 6, but for the moment proceed using classical arguments.

For an electron (charge $-e$) moving in an orbit with angular momentum $\hbar L$, the classically calculated interaction energy with a uniform magnetic field $\mathbf{B}$ turns out to be

$$(e/2m)\hbar\mathbf{L}\cdot\mathbf{B}.$$

‡ Historically, of course, it was the fine structure that compelled Goudsmit and Uhlenbeck to postulate the spin; the quantum-mechanical theory came later. The Stern–Gerlach experiment pointed only to non-zero magnetic moments of atoms and ions: it became clear only much later that this moment arose from an intrinsic *electronic* moment

This corresponds to the energy $-\mathbf{B}\cdot\boldsymbol{\mu}_L$ of a magnetic dipole of moment

$$\boldsymbol{\mu}_L = -(e\hbar/2m)\mathbf{L} = -\beta\mathbf{L} \qquad (2.51)$$

placed in the field. The proportionality constant β is called the "Bohr magneton": it has the dimensions of magnetic moment (since L is dimensionless) and the value

$$\beta = e\hbar/2m = 9\cdot2732 \times 10^{-24}\,\mathrm{JT^{-1}}.$$

Clearly 2β (the magnetic moment associated with 2 units of orbital angular momentum) is the natural (i.e. atomic) unit of magnetic moment. It might be expected that the spin angular momentum would give a similar interaction energy, corresponding to magnetic moment $-\beta\mathbf{S}$: this is not observed, the electron having an anomalous moment

$$\boldsymbol{\mu} = -2\beta\mathbf{S}. \qquad (2.52)$$

It would thus appear that the Hamiltonian operator (2.50) should be augmented by terms

$$\beta\mathbf{B}\cdot\mathbf{L} + 2\beta\mathbf{B}\cdot\mathbf{S}$$

representing the direct orbit–field and spin–field interactions, the angular momentum components now being regarded as operators (so that, for example, $\mathbf{B}\cdot\mathbf{S} = B_x S_x + B_y S_y + B_z S_z$) whose commutation properties have already been discussed.

There will be one more term, however, representing the spin–orbit interaction. The form of this term may be inferred easily, for an electron moving around a nucleus of charge Ze, by a classical argument. Seen from the electron, at point $\mathbf{r}$ and moving with velocity $\mathbf{v}$, the nucleus would be at $-\mathbf{r}$ and moving with velocity $-\mathbf{v}$. The electron would therefore experience a field $\mathbf{B}'$ due to an element of current $\mathbf{I} = -Ze\mathbf{v}$, located at $-\mathbf{r}$. By classical electromagnetic theory this field would be (App. 1)

$$\mathbf{B}' = \frac{-\mathbf{r}\times(-Ze\mathbf{v})}{\kappa_0 c^2 r^3} = \frac{Ze\hbar\mathbf{L}}{\kappa_0 c^2 m r^3}\,.$$

Since the electron itself has a magnetic moment $\boldsymbol{\mu} = -2\beta\mathbf{S}$ there should apparently be an interaction energy

$$-\mathbf{B}'\cdot\boldsymbol{\mu} = 2\beta\left(\frac{Ze\hbar}{\kappa_0 c^2 m}\right)\mathbf{L}\cdot\mathbf{S} = 4\beta^2(Z/\kappa_0 c^2 r^3)\mathbf{L}\cdot\mathbf{S}.$$

A more precise argument, however, including relativistic considerations, reduces this value by a factor $\frac{1}{2}$ (the "Thomas factor"). Here we shall simply assume that the essential features of the spin–orbit interaction

C*

are embodied in a corresponding operator $\lambda\mathbf{L}\cdot\mathbf{S} = (\mathsf{L}_x\mathsf{S}_x + \mathsf{L}_y\mathsf{S}_y + \mathsf{L}_z\mathsf{S}_z)$, where λ is a "spin–orbit coupling parameter".

The orbit–field, spin–field and spin–orbit interactions may thus be tentatively combined into a representative operator

$$\mathsf{H}' = \beta\mathbf{B}\cdot\mathbf{L} + 2\beta\mathbf{B}\cdot\mathbf{S} + \lambda\mathbf{L}\cdot\mathbf{S}. \tag{2.53}$$

We now regard this term as a small perturbation of H_0 in (2.50) and try to account for the fine structure of the observed energy levels, assuming first that $\mathbf{B} = 0$, i.e. that no external magnetic field is present.

2.7.　Classification and determination of the states, spin included

The term values for Na, K and Rb are given in Table 2.4 and it is seen that each of the p and d states is split into a doublet. To account for this fine structure we use the full Hamiltonian

$$\mathsf{H} = \mathsf{H}_0 + \mathsf{H}' = -\frac{\hbar^2}{2m}\nabla^2 + V(r) + \lambda\mathbf{L}\cdot\mathbf{S}. \tag{2.54}$$

The investigation of commutation properties (Section 2.2) must now be repeated in order to determine the constants of the motion when spin is admitted.

TABLE 2.4

Term values‡ for alkali metals Na, K, Rb

Na		K		Rb	
$4d$	$\begin{cases} 34548{\cdot}789 \\ 34548{\cdot}754 \end{cases}$	$6s$	$27450{\cdot}65$	$7s$	$26311{\cdot}46$
$5s$	$33200{\cdot}696$	$4d$	$\begin{cases} 27398{\cdot}11 \\ 27397{\cdot}01 \end{cases}$	$5d$	$\begin{cases} 25703{\cdot}52 \\ 25700{\cdot}56 \end{cases}$
$4p$	$\begin{cases} 30272{\cdot}51 \\ 30266{\cdot}88 \end{cases}$	$5p$	$\begin{cases} 24720{\cdot}20 \\ 24701{\cdot}44 \end{cases}$	$6p$	$\begin{cases} 23792{\cdot}69 \\ 23715{\cdot}19 \end{cases}$
$3d$	$\begin{cases} 29172{\cdot}90 \\ 29172{\cdot}85 \end{cases}$	$3d$	$\begin{cases} 21536{\cdot}75 \\ 21534{\cdot}42 \end{cases}$	$6s$	$20133{\cdot}6$
$4s$	$25729{\cdot}86$	$5s$	$21026{\cdot}8$	$4d$	$\begin{cases} 19355{\cdot}45 \\ 19355{\cdot}01 \end{cases}$
$3p$	$\begin{cases} 16973{\cdot}38 \\ 16956{\cdot}18 \end{cases}$	$4p$	$\begin{cases} 13042{\cdot}89 \\ 12985{\cdot}17 \end{cases}$	$5p$	$\begin{cases} 12816{\cdot}56 \\ 12578{\cdot}96 \end{cases}$
$3s$	0	$4s$	0	$5s$	0

‡ All values in cm^{-1} relative to the lowest s level of the series. Note that all levels other than those of s type are doublets.

It is at once clear that neither L_z nor S_z commutes with the term in $\mathbf{L} \cdot \mathbf{S}$. For example, since space and spin operators commute with each other,

$$[L_z, \mathbf{L} \cdot \mathbf{S}] = [L_z, (L_x S_x + L_y S_y + L_z S_z)]$$
$$= S_x[L_z, L_x] + S_y[L_z, L_y]$$
$$= i(S_x L_y - S_y L_x)$$

and similarly we find

$$[S_z, \mathbf{L} \cdot \mathbf{S}] = i(S_y L_x - S_x L_y).$$

But, by adding these results, it is clear that the *sum* $(L_z + S_z)$ will commute with $\mathbf{L} \cdot \mathbf{S}$, and hence (since both terms commute with H_0) with the Hamiltonian H. We refer to $L_z + S_z = J_z$ as the z component of *total* angular momentum, orbital plus spin, and thus have

$$[J_z, H] = 0.$$

The total angular momentum is symbolized by $\mathbf{J} = \mathbf{L} + \mathbf{S}$, which simply means that its components are the operators

$$J_x = L_x + S_x, \quad J_y = L_y + S_y, \quad J_z = L_z + S_z. \tag{2.55}$$

Thus, when spin is admitted in the Hamiltonian it is the z component of *total* angular momentum that is eligible as a constant of the motion.

We next examine $\mathbf{L}^2$ and $\mathbf{S}^2$ and find easily (e.g. by using (2.14) and their spin analogues) that

$$[\mathbf{L}^2, H] = [\mathbf{S}^2, H] = 0.$$

Also, using (2.55) and regrouping the terms in $J_x^2 + J_y^2 + J_z^2$ we observe that

$$\mathbf{J}^2 = \mathbf{L}^2 + \mathbf{S}^2 + 2\mathbf{L} \cdot \mathbf{S} \tag{2.56}$$

and obtain (since $\mathbf{L} \cdot \mathbf{S}$ commutes both with itself and with H_0, and hence with H)

$$[\mathbf{J}^2, H] = 0.$$

It thus appears that when spin–orbit interaction is admitted the maximal set of commuting operators is

$$H, \mathbf{J}^2, J_z, \mathbf{L}^2, \mathbf{S}^2.$$

The *magnitudes* of $\mathbf{S}$ and $\mathbf{L}$ may thus be constants of the motion, but the actual vectors are not: in the early days of quantum mechanics it was supposed that the vectors $\mathbf{L}$ and $\mathbf{S}$ precessed around their resultant $\mathbf{J}$, while $\mathbf{J}$ itself precessed around the z-axis (Fig. 2.7). Classical interpretations should not be taken literally, of course, but as a formal indication of how angular momenta may be coupled such "vector models" are often useful. It is easily verified that $\mathbf{J}$ has the same commutation properties as $\mathbf{L}$ and $\mathbf{S}$, namely, $\mathbf{J} \times \mathbf{J} = i\mathbf{J}$ and it follows that the eigenvalues of $\mathbf{J}^2$ must be of the form $J(J+1)$, while those of $\mathbf{J}_z$ will be $M = J, J-1, \ldots, -J$.

To determine the energies of the various states we employ perturbation theory (Section 1.3) using appropriate space–spin product functions as the eigenfunctions of the unperturbed Hamiltonian $\mathbf{H}_0$: to first order,

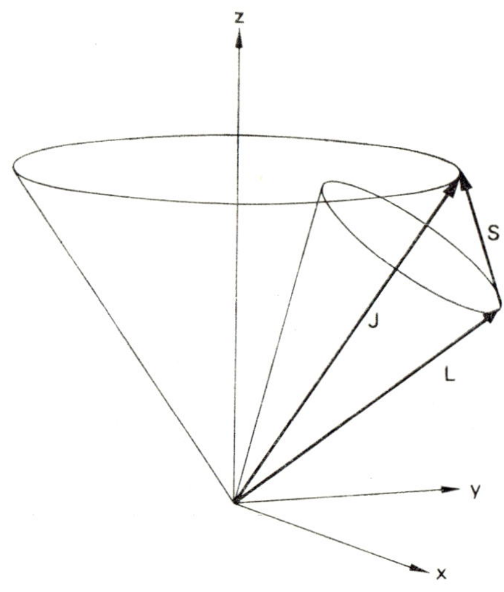

Fig. 2.7. Classical interpretation of spin–orbit coupling. The cones indicate precession of J about the z-axis, and of L about J. The magnitudes of J, L and S are constants of the motion; J_z is also constant, but not L_z and S_z.

we need only those which, in the absence of $\mathbf{H}'$, are degenerate. Thus, when the electron is in the $3p$ orbital we use a basis of *spin-orbitals*

$$3p_{+1}\alpha, \quad 3p_{+1}\beta, \quad 3p_0\alpha, \quad 3p_0\beta, \quad 3p_{-1}\alpha, \quad 3p_{-1}\beta.$$

These functions may be indicated by the values of the quantum numbers m_l and m_s, where we now add subscripts l and s to distinguish orbital and spin quantum numbers: it is therefore convenient to use a ket notation $|m_l m_s\rangle$ where it is understood that $m_l = +1, 0, -1$ ($l = 1$), and $m_s = \pm\frac{1}{2}$ ($s = \frac{1}{2}$). The secular determinant (1.22b) becomes

$$\det \left| \langle m_l m_s | \mathbf{H}' | m_l' m_s' \rangle - E^{(1)} \, \delta_{m_l m_{l'}} \, \delta_{m_s m_{s'}} \right| = 0 \tag{2.57}$$

and its roots determine the perturbation energies and perturbed states.

Instead of solving the secular equations, however, we shall obtain the required simultaneous eigenfunctions of $\mathbf{J}^2$ and $\mathbf{J}_z$ directly by "coupling" the angular momenta as in Fig. 2.8. Let us classify the basis functions according to the sum $m = m_l + m_s$ as below, using a ket notation for the products:

$$
\begin{array}{c|l}
m = +\tfrac{3}{2} & |+1, +\tfrac{1}{2}\rangle \\
m = +\tfrac{1}{2} & |\ \ 0, +\tfrac{1}{2}\rangle, \quad |+1, -\tfrac{1}{2}\rangle \\
m = -\tfrac{1}{2} & |\ \ 0, -\tfrac{1}{2}\rangle, \quad |-1, \ \ \tfrac{1}{2}\rangle \\
m = -\tfrac{3}{2} & |-1, -\tfrac{1}{2}\rangle
\end{array}
$$

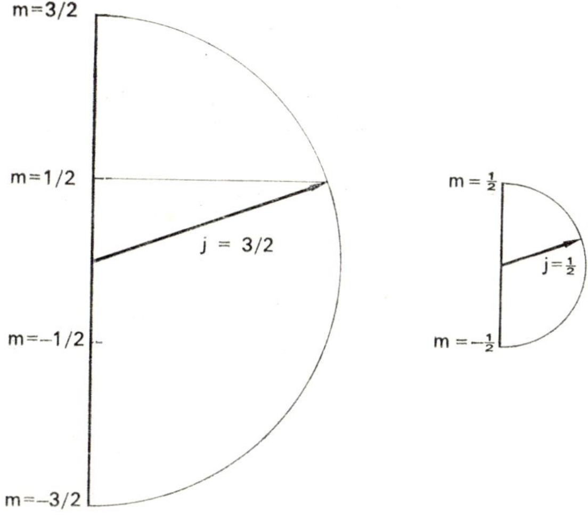

FIG. 2.8. Vector diagram for LS coupling ($l = 1$, $s = \tfrac{1}{2}$). The resultant J can take values with $j = \tfrac{3}{2}$, giving states with $m = \pm\tfrac{1}{2}$, $\pm\tfrac{3}{2}$ (left-hand diagram), and $j = \tfrac{1}{2}$, with $m = \pm\tfrac{1}{2}$ (right-hand diagram).

It is clear at once that any space–spin product is an eigenfunction of $\mathbf{J}_z = \mathbf{L}_z + \mathbf{S}_z$, with eigenvalue $m = m_l + m_s$. For example,

$$
\begin{aligned}
\mathbf{J}_z(3p_{+1}\alpha) &= \mathbf{L}_z(3p_{+1}\alpha) + \mathbf{S}_z(3p_{+1}\alpha) \\
&= 1(3p_{+1}\alpha) + \tfrac{1}{2}(3p_{+1}\alpha) \\
&= (1 + \tfrac{1}{2})(3p_{+1}\alpha).
\end{aligned}
$$

The *spin-orbitals* in each row of the table are thus eigenfunctions of $\mathbf{J}_z$ with eigenvalue $m = m_l + m_s$.

Now let us start at the "top" and examine the effects of J^2: rewriting the operator in a form analogous to (2.21) we obtain

$$\mathsf{J}^2|1, \tfrac{1}{2}\rangle = (\mathsf{J}^-\mathsf{J}^+ + \mathsf{J}_z{}^2 + \mathsf{J}_z)|1, \tfrac{1}{2}\rangle = j(j+1)|1, \tfrac{1}{2}\rangle$$

where $j = \tfrac{3}{2}$. This follows because J^+ destroys the state of maximum z component $m = \tfrac{3}{2}$ (which we now denote by j) and J_z, $\mathsf{J}_z{}^2$ merely multiply by m, m^2, respectively. Thus, $|1, \tfrac{1}{2}\rangle$ is an eigenstate of both J^2 and J_z, which we may denote by $|j = \tfrac{3}{2}, m = \tfrac{3}{2}\rangle$.

From general theory, however, we know that states of the same j but lower values of m can be reached by using the step-down operator. We therefore apply J^- to $|1, \tfrac{1}{2}\rangle$ and should obtain a state with $j = \tfrac{3}{2}$, $m = \tfrac{1}{2}$: the formula we need is essentially (2.23b), which gives in general

$$\mathsf{J}^-|m_l, m_s\rangle = \mathsf{L}^-|m_l, m_s\rangle + \mathsf{S}^-|m_l, m_s\rangle$$

$$= \sqrt{\{(l-m_l+1)(l+m_l)\}}|m_l-1, m_s\rangle$$
$$+ \sqrt{\{(s-m_s+1)(s+m_s)\}}|m_l, m_s-1\rangle.$$

Thus, with $l = 1$, $s = \tfrac{1}{2}$ and $m_l = 1$, $m_s = \tfrac{1}{2}$ we obtain

$$\mathsf{J}^-|j = \tfrac{3}{2}, m = \tfrac{3}{2}\rangle = \sqrt{2}|0, \tfrac{1}{2}\rangle + |1, -\tfrac{1}{2}\rangle.$$

This is a multiple of the normalized eigenvector $|j = \tfrac{3}{2}, m = \tfrac{1}{2}\rangle$, the proportionality factor being $\sqrt{\{(j-m+1)(j+m)\}}$ with $j = \tfrac{3}{2}$, $m = \tfrac{3}{2}$. Hence[‡]

$$|j = \tfrac{3}{2}, m = \tfrac{1}{2}\rangle = \sqrt{\tfrac{2}{3}}|0, \tfrac{1}{2}\rangle + \sqrt{\tfrac{1}{3}}|1, -\tfrac{1}{2}\rangle.$$

Repeated application of J^- gives all four states, with $m = \tfrac{3}{2}, \tfrac{1}{2}, -\tfrac{1}{2}, -\tfrac{3}{2}$. With the more explicit notation for the basis functions these are

$$|j = \tfrac{3}{2}, m = +\tfrac{3}{2}\rangle = 3p_{+1}\alpha,$$
$$|j = \tfrac{3}{2}, m = +\tfrac{1}{2}\rangle = \sqrt{\tfrac{2}{3}}(3p_0\alpha) + \sqrt{\tfrac{1}{3}}(3p_{+1}\beta),$$
$$|j = \tfrac{3}{2}, m = -\tfrac{1}{2}\rangle = \sqrt{\tfrac{2}{3}}(3p_0\beta) + \sqrt{\tfrac{1}{3}}(3p_{-1}\alpha),$$
$$|j = \tfrac{3}{2}, m = -\tfrac{3}{2}\rangle = 3p_{-1}\beta.$$

But there must be two more states, as we started with *six* linearly independent functions.

To find the remaining states, we note that $j = \tfrac{3}{2}$ is here a *maximum j* value since no higher value of m can be found, but that a lower half-

---

[‡] Alternatively, the function may be normalized from first principles—multiplying by a factor chosen so that the sum of the squares of the coefficients is equal to unity.

integer value, $j = \frac{1}{2}$, is permitted. The simultaneous eigenvalues of J_z could then only be $\pm \frac{1}{2}$, and the eigenvectors would have to be constructed from the table entries corresponding to $m = \frac{1}{2}$ and $m = -\frac{1}{2}$ respectively. Thus we expect, for example,

$$|j = \tfrac{1}{2}, \, m = \tfrac{1}{2}\rangle = a|0, \, \tfrac{1}{2}\rangle + b|1, \, -\tfrac{1}{2}\rangle.$$

To fix the coefficients it is sufficient to use the general theorem (Vol. 1, p. 88) that eigenfunctions of the same Hermitian operator (in this case J^2) but with different eigenvalues (in this case $\frac{3}{2}$ and $\frac{1}{2}$) must be orthogonal. Inspection of the coefficients in $|j = \frac{3}{2}, \, m = \frac{1}{2}\rangle$, and similarly for the case $m = -\frac{1}{2}$, then gives the two remaining eigenfunctions:

$$|j = \tfrac{1}{2}, m = \tfrac{1}{2}\rangle = \sqrt{\tfrac{1}{3}}(3p_0\alpha) - \sqrt{\tfrac{2}{3}}(3p_{+1}\beta),$$
$$|j = \tfrac{1}{2}, m = -\tfrac{1}{2}\rangle = \sqrt{\tfrac{1}{3}}(3p_0\beta) - \sqrt{\tfrac{2}{3}}(3p_{-1}\alpha). \tag{2.58}$$

This completes the determination of the "vector-coupled" states indicated schematically in Fig. 2.8. It should be noted in particular that, when spin–orbit coupling is admitted, the series electron is described in general by a "mixed" spin-orbital with both α and β components as in I.(4.41): it is only in special cases that a "pure" spin-orbital of α or β type is adequate.

2.8. Fine structure of the energy levels

Now that we have the eigenfunctions of J^2, J_z (and simultaneously L^2, S^2) we find it is unnecessary to solve a secular problem to obtain the energies of the states. For if we use as a basis the $|j, m\rangle$ (instead of the original product functions, of which they are linear combinations) a well-known theorem shows that all the off-diagonal elements must vanish. We need therefore only calculate the expectation values of H' (i.e. the diagonal elements $\langle j, m|H'|j, m\rangle$) to get the first-order energy changes.

The theorem needed (cf. Vol. 1, p. 89) states that if ψ and ψ' are eigenfunctions of an operator A, with different eigenvalues, and if B is an operator commuting with A, then $\langle \psi|B|\psi'\rangle$ will vanish: in the present case all six j, m-eigenfunctions differ in the eigenvalues of at least one operator commuting with H and hence all off-diagonal elements of H must vanish.

The expectation values are obtained very easily on noting that, *so long as we operate on simultaneous eigenfunctions of* L^2, S^2 and J^2 the perturbation operator is equivalent to a numerical multiplier; thus, making use of (2.56)

$$(\lambda\mathbf{L}\cdot\mathbf{S})|j, m\rangle = \tfrac{1}{2}\lambda(\mathbf{J}^2 - \mathbf{L}^2 - \mathbf{S}^2)|j, m\rangle$$

$$= \tfrac{1}{2}\lambda[j(j+1) - l(l+1) - s(s+1)]|j, m\rangle.$$

The first-order perturbation of the state ψ_{jm} is thus

$$E_{jm}^{(1)} = \tfrac{1}{2}\lambda[j(j+1) - l(l+1) - s(s+1)]. \tag{2.59}$$

For the sodium $3p$ level we therefore obtain the spin–orbit splitting shown in Fig. 2.9. The family of states with given eigenvalues of $\mathbf{L}^2$ and $\mathbf{S}^2$, but differing in eigenvalues of the resultant $\mathbf{J}^2$, is referred to in

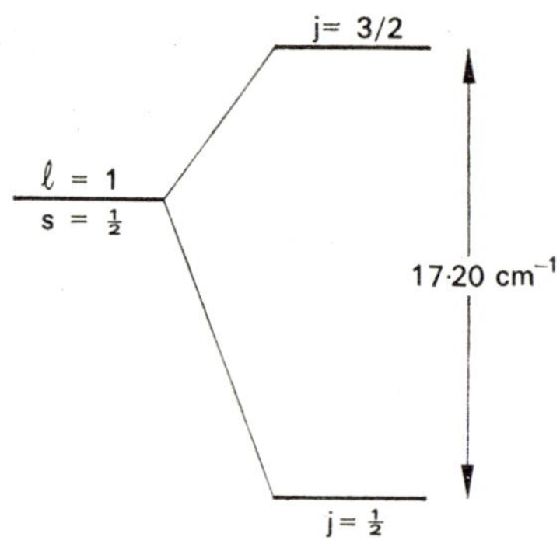

FIG. 2.9. Spin–orbit splitting of sodium $3p$ level.

general as a *multiplet*. The results for the sodium *doublet* are in virtually exact agreement with experiment if the parameter λ is given the value $\lambda = 11\cdot5$ cm⁻¹. Since the formula (2.59) does not depend on m, some degeneracy remains. The $j = \tfrac{3}{2}$ state is four-fold degenerate, while $j = \tfrac{1}{2}$ is two-fold degenerate. This is again in accord with experiment: if a magnetic field is applied the states of different m are resolved. This "Zeeman splitting" can be accurately accounted for by admitting the field-dependent terms in the perturbation operator $\mathbf{H}'$ given in (2.53). The study of effects due to external fields will be deferred, however, until Chap. 7.

It should be noted that the result (2.59) depends only on the *existence* of simultaneous eigenfunctions of $\mathbf{J}^2$, $\mathbf{L}^2$, $\mathbf{S}^2$ and that their explicit calculation was not actually required. This result is in fact more general than the present example might suggest: all the steps in the preceding discussion can be applied almost without change in the case of a *many*-electron atom, the quantum numbers then referring to *total* orbital and spin angular momenta (L, S) and their resultant J. The argument that leads to (2.59) still applies and this formula yields the "Landé interval rule" which states that in any atomic multiplet the separation between adjacent levels is proportional to the j value of the upper level.

REFERENCES

CLEMENTI, E. and RAIMONDI, D. L. (1963) *J. Chem. Phys.* **38**, 2686.
SLATER, J. C. (1930) *Phys. Rev.* **36**, 57 (see also *Quantum Theory of Atomic Structure*, McGraw-Hill, New York, 1960, Vol. I).

CHAPTER 3

ORBITAL PICTURE OF MANY-ELECTRON ATOMS, MOLECULES AND CRYSTALS

3.1. The independent particle model

For a one-electron system, an approximate solution of the Schrödinger equation may usually be obtained and interpreted without too much trouble. In Chap. 2 such solutions have been used extensively. The determination and interpretation of accurate electronic wave functions for a *many*-electron system is much more difficult and cannot be fully considered in this volume. In fact, however, the theory of the electronic structure of atoms, molecules and crystals is dominated by an "independent particle model" in which each electron is supposed to move in a common average field associated with the system as a whole. Each electron is then described by an orbital which is the solution of a one-electron eigenvalue equation containing an "effective potential", which includes not only the attraction between the electron and the nuclei but also certain terms that allow for its average interaction with all other electrons. A detailed discussion of one-electron systems in which the assumed potential field has a "shape" more like that of a molecule or crystal is therefore not an academic exercise but is a necessary step towards current applications of quantum mechanics in the theory of the electronic structure of matter. The independent particle model is a direct development from early work on the electronic structure of atoms, in which each electron is assigned to "its own" orbital: in its application to molecules and crystals, the main difference is that the atomic orbitals of Chap. 2 are replaced by "molecular orbitals" or "crystal orbitals". It is therefore of great interest to discuss some of the general features of orbitals describing the stationary states of an electron in a *poly*centric field. This provides further opportunities for applying the principles and methods of quantum mechanics, and will lead naturally to a further appreciation of the importance of symmetry—exhibited in rich variety by molecules and crystals—in readiness for a fuller discussion in the next chapter.

We begin with a brief descriptive account of the approach. For a many-electron system we start from a Hamiltonian operator (1.2a), namely[‡]

$$\mathsf{H} = \sum_i \mathsf{h}(i) + \tfrac{1}{2} \sum_{i,j}' g(i,j) \tag{3.1}$$

set up by the usual prescription (as, for example, in Section 2.1). Here the *one*-electron operator $\mathsf{h}(i)$ is the Hamiltonian for electron i *by itself* in the field of the nuclei, while $g(i,j)$ is the interaction energy between electrons i and j:

$$\mathsf{h}(i) = -(\hbar^2/2m)\nabla^2(i) + V(i), \quad g(i,j) = e^2/\kappa_0 r_{ij}. \tag{3.2}$$

$V(i) = V(\mathbf{r}_i)$ is the potential energy of electron i at point $\mathbf{r}_i$, in the field of the nuclei. Whenever we use atomic units (p. 6), the $\hbar$, m, e and κ_0 may each be replaced by unity.

Clearly, if the interaction terms could be discarded (i.e. if the particles were truly independent) H would reduce to

$$\mathsf{H}_0 = \sum_i \mathsf{h}(i). \tag{3.3}$$

For the moment, we ignore spin completely, and the eigenfunctions of H_0 can then be written without any approximation as orbital products (using lower-case letters for *one*-electron functions),

$$\Psi(\mathbf{r}_1, \mathbf{r}_2, \dots \mathbf{r}_N) = \phi_i(\mathbf{r}_1)\phi_j(\mathbf{r}_2) \dots \phi_p(\mathbf{r}_N). \tag{3.4}$$

This result follows easily by the separation technique (Vol. 1, App. 2), from which it also appears that the individual orbitals must then be eigenfunctions of the one-electron operator $\mathsf{h}(i)$ which refers to electron i by itself in the field of the nuclei. If ϕ_i corresponds to the *one*-electron energy ε_i, then

$$\mathsf{h}\phi_i = \varepsilon_i \phi_i \tag{3.5}$$

and it follows that the total electronic energy is simply

$$E = \varepsilon_i + \varepsilon_j + \dots + \varepsilon_p \tag{3.6}$$

exactly as we should expect for a set of independent particles.

The orbitals that occur in the wave function (3.4) are called the *occupied orbitals*; other solutions of (3.5), not occurring in Ψ defined in (3.4), are the *unoccupied orbitals*. The selection of orbitals used in describing a particular state of the many-electron system may be called an *orbital configuration*. Clearly the lowest energy state (i.e. the ground state) would result when all the electrons were assigned to the lowest

---

‡ For N electrons, the summation indices i, j run from 1 to N. The prime on the second sum indicates omission of the term with $i = j$. These summation ranges will be assumed in all that follows.

possible orbitals. It is known, however, that not more than two electrons can be described by the same orbital (for reasons that will be discussed presently) and it therefore follows that

$$E = n_1\varepsilon_1 + n_2\varepsilon_2 + \ldots \tag{3.7}$$

where the n's are "occupation numbers" and in the ground state take values $n_1 = 2$, $n_2 = 2$, ... when the orbitals are listed in ascending energy order ($\varepsilon_1 \leqq \varepsilon_2 \leqq \varepsilon_3 \ldots$).

For simplicity we are supposing at first that no degeneracies occur, so the ground state is described by a unique product. The *electron configuration* is then described by indicating the occupied orbitals and the numbers of electrons they contain, thus $\phi_1{}^{n_1}\phi_2{}^{n_2}\ldots$. For example, the electron configuration of the lithium atom ground state (3-electrons) is $1s^2 2s^1$, or simply $1s^2 2s$. The description of *excited states* is also simple: such states may be achieved by promoting an electron from an occupied orbital, ϕ_i say, to an unoccupied orbital ϕ_k and the excitation energy (i.e. the increase in total energy E) is then clearly

$$\Delta E(i \rightarrow k) = \varepsilon_k - \varepsilon_i. \tag{3.8}$$

The present approximation is thus extremely useful because it allows us to interpret many of the observed electronic transitions in atoms and molecules, which strictly involve changes in the whole electronic system, in terms of "jumps" of a *single* electron from one energy level to another (cf. Fig. 2.6). An important special case occurs when an electron is removed, from ϕ_i say, to infinity (with zero kinetic energy): the energy change is then the ionization energy, from ϕ_i, and we obtain

$$I_i = -\varepsilon_i. \tag{3.9}$$

The present approximation is thus extremely useful because it allows us to interpret ε_i physically as the negative of an ionization potential.

Finally, before asking how this simple picture is affected when electron repulsion terms are included, we note another important feature of the approximation. The interpretation of (3.4), according to the basic axioms, is that the probability of finding electron 1 in volume element $d\mathbf{r}_1$, electron 2 simultaneously in element $d\mathbf{r}_2$, and so on, is

$$\left|\Psi(\mathbf{r}_1, \mathbf{r}_2, \ldots \mathbf{r}_N)\right|^2 d\mathbf{r}_1 d\mathbf{r}_2 ,\ldots d\mathbf{r}_N$$
$$= \left|\phi_i(\mathbf{r}_1)\right|^2 \left|\phi_j(\mathbf{r}_2)\right|^2 \ldots \left|\phi_p(\mathbf{r}_N)\right|^2 d\mathbf{r}_1 d\mathbf{r}_2 \ldots d\mathbf{r}_N.$$

The probability of finding electron 1 in an element $d\mathbf{r}$ at the arbitrary point $\mathbf{r}$, *irrespective* of the positions of electrons 2, 3, ... N, is thus obtained by putting $\mathbf{r}_1 = \mathbf{r}$, $d\mathbf{r}_1 = d\mathbf{r}$ and "summing" over all possible positions of the other electrons 2, ... N. Since the orbitals are assumed normalized, the corresponding integrations each give unity and leave only $|\phi_i(\mathbf{r})|^2\,d\mathbf{r}$. In exactly the same way, the probability of finding electron 2 in element $d\mathbf{r}$ is seen to be $|\phi_2(\mathbf{r})|^2\,d\mathbf{r}$; and so on. Thus, if we define a one-electron probability density function $P(\mathbf{r})$ by

$$\begin{pmatrix}\text{Probability of an electron}\\ \text{(no matter which) in } d\mathbf{r}\end{pmatrix} = P(\mathbf{r})\,d\mathbf{r} \qquad (3.10)$$

it follows at once that a product function such as (3.4) in which we now suppose ϕ_1 occurs n_1 times, ϕ_2 occurs n_2 times, etc., will yield the particular form

$$P(\mathbf{r}) = n_1|\phi_1(\mathbf{r})|^2 + n_2|\phi_2(\mathbf{r})|^2 + \ldots \ . \qquad (3.11)$$

Now for many purposes, as a full discussion of many-electron systems shows, $P(\mathbf{r})$ may be regarded as the electronic "charge density" (measured in electrons per unit volume) in the system; it is, for example, the density of charge as inferred from X-ray scattering experiments. What we have shown is that, for an approximate wave function of product form, the charge density is a sum of orbital contributions. In particular, if ϕ_i is occupied there is a density term $|\phi_i|^2$ just as there would be with no other electrons present, while if ϕ_i is doubly occupied this density is doubled.

The above observations are in fact rather general and give the independent particle model its great conceptual appeal. It must be stressed, however, that at this point we are speaking of strictly non-interacting particles, representing a very crude model of the actual system. The real value of this approach is that, even when electron interactions are admitted and a function of the form (3.4) becomes only an *approximate* wave function, many of the features of a system of independent particles are preserved or only slightly modified. The term "independent particle model" thus refers to a particular kind of approximation in which the equations describing a real system bear a strong formal resemblance to those which would hold for an idealized system of non-interacting particles. As a full treatment of the independent particle model is outside the scope of this volume, we shall simply give an example, to indicate how the effects of electron interaction might be dealt with, and then pass to a brief discussion of the somewhat more general theory of Hartree and Fock.

EXAMPLE. *The helium atom.* Let us assign the two electrons of a helium atom to an orbital ϕ of the (normalized) form

$$\phi(\mathbf{r}) = \sqrt{\frac{\zeta^3}{\pi}}\,e^{-\zeta r}$$

and adopt the product function

$$\Psi(\mathbf{r}_1, \mathbf{r}_2) = \phi(\mathbf{r}_1)\,\phi(\mathbf{r}_2)$$

as an approximate wave function with which to perform a variation calculation. If the electrons were truly non-interacting this would be an exact wave function for $\zeta = Z\ (=2)$. As interaction is not neglected, the Hamiltonian (3.1) takes the form

$$\mathsf{H} = \mathsf{h}(1) + \mathsf{h}(2) + \mathsf{g}(1, 2)$$

and the introduction of ζ as a variational parameter will allow us to adjust the form of the wave function, to some extent, to allow approximately for the effect of the interaction term. The expectation value of the energy reduces at once to

$$E = 2\langle\,\phi|\mathsf{h}|\,\phi\,\rangle + \langle\,\phi\,\phi|\mathsf{g}|\,\phi\,\phi\,\rangle$$

where the integrals are readily evaluated (see, for example, Eyring *et al.*, 1944) to give, working in atomic units,

$$\langle\,\phi|\mathsf{h}|\,\phi\,\rangle = \int \phi^*(\mathbf{r}_1)\mathsf{h}(1)\,\phi(\mathbf{r}_1)\,d\mathbf{r}_1 = \tfrac{1}{2}\zeta^2 - z\zeta,$$

$$\langle\,\phi\,\phi|\mathsf{g}|\,\phi\,\phi\,\rangle = \int \phi^*(\mathbf{r}_1)\,\phi^*(\mathbf{r}_2)\,\frac{1}{r_{12}}\,\phi(\mathbf{r}_1)\,\phi(r_2)\,dr_1\,dr_2 = \tfrac{5}{8}\zeta;$$

the energy expression thus takes the form

$$E = \zeta^2 - 2Z\zeta + \tfrac{5}{8}\zeta$$

and proceeding in the spirit of the variation method (Section 1.3) we minimize E by requiring

$$(\partial E/\partial \zeta) = 2\zeta - 2Z + \tfrac{5}{8} = 0.$$

The best wave function of the assumed product form thus occurs when $\zeta = Z - \tfrac{5}{16}$,

$$E = -(Z - \tfrac{5}{16})^2\ \mathsf{H}$$

which give an estimate for helium of 77·45 eV compared with the known accurate value 78·98 eV.

Since we have used an orbital of hydrogen-like form, ϕ may be interpreted as an eigenfunction for an electron in the field of a "fictitious" nuclear charge corresponding to $Z' = Z - \tfrac{5}{16}$, with eigenvalue $\varepsilon_{1s} = -\tfrac{1}{2}(Z - \tfrac{5}{16})^2$, and the product function may be regarded as an exact eigenfunction of the independent-particle model with Hamiltonian

$$\mathsf{H}_{\mathrm{eff}} = \mathsf{h}_{\mathrm{eff}}(1) + \mathsf{h}_{\mathrm{eff}}(2)$$

where $\mathsf{h}_{\mathrm{eff}} = -\tfrac{1}{2}\nabla^2 - Z'/r$ is the Hamiltonian operator for one electron in the fictitious field with $V = -Z'/r$. The situation may be visualized by saying that each electron feels, besides the attraction due to nuclear charge Z, a repulsion from the second electron; and that on the average the effective field acting is closer to that of a "screened" nuclear charge of $(Z - \tfrac{5}{16})$.

3.2. The Hartree field. Atomic structure

The idea of an effective field, introduced in the above example, was developed essentially on intuitive grounds by Hartree (1928) and has been applied extensively in atomic structure calculations. Hartree assumed that the effective field felt by an electron in orbital ϕ_i could be derived by treating all the remaining electrons in the atom as "smeared-out" charge distributions. Thus for an electron configuration of doubly occupied orbitals ϕ_1, ϕ_2, ... ϕ_n, with wave function

$$\Psi(\mathbf{r}_1, \mathbf{r}_2, \dots \mathbf{r}_N) = \phi_1(\mathbf{r}_1)\phi_1(\mathbf{r}_2) \dots \phi_n(\mathbf{r}_N) \quad (n = N/2) \qquad (3.12)$$

the total charge density function (3.11) would become

$$P = 2|\phi_1|^2 + \dots + 2|\phi_i|^2 + \dots + 2|\phi_n|^2 \qquad (3.13)$$

while, for an electron in ϕ_i, that due to all the *remaining* electrons would be

$$P_i = 2|\phi_1|^2 + \dots + |\phi_i|^2 + \dots + 2|\phi_n|^2. \qquad (3.14)$$

Thus, if the electron in ϕ_i is at point $\mathbf{r}$ and we use $\mathbf{r}'$ to denote a variable point in the charge distribution P_i, the potential energy due to repulsion from the remaining electrons would be (atomic units)

$$\int \frac{P_i(\mathbf{r}')\, d\mathbf{r}'}{|\mathbf{r} - \mathbf{r}'|}$$

in which the integrand is the Coulomb repulsion between the electron at $\mathbf{r}$ and the smeared-out charge $P_i(\mathbf{r}')\, d\mathbf{r}'$ occupying a volume element $d\mathbf{r}'$. A plausible Hamiltonian operator for determining the wave function ϕ_i for this electron is thus

$$\mathsf{h}_{\text{eff}}^{(i)} = \mathsf{h} + \int \frac{P_i(\mathbf{r}')\, d\mathbf{r}'}{|\mathbf{r} - \mathbf{r}'|}. \qquad (3.15)$$

The effect of screening of the nucleus by the other electrons is now embodied in the last term, which is a function of position ($\mathbf{r}$) which may differ widely from Coulombic form except in certain ranges; thus when $\mathbf{r}$ is very small the electron will experience almost the full nuclear charge Z, but when $\mathbf{r}$ is very large the electron will be far from the atom and will experience an effective nuclear charge corresponding to a singly charged ion. The *orbital energy* of this electron is the eigenvalue associated with ϕ_i and follows on solving the equation

$$\mathsf{h}_{\text{eff}}^{(i)}\phi = \varepsilon\phi. \qquad (3.16)$$

It is a "fictitious" energy, referring to an electron in an artificially constructed field. The actual system is then replaced, in first approximation, by a "model" whose Hamiltonian is

$$\mathsf{H}_0 = \sum_i \mathsf{h}_{\text{eff}}^{(i)}(i)$$

just as in the example of Section 3.1 (p. 74).

Hartree's procedure for the calculation of atomic ground state wave functions may now be summarized essentially as follows:

1. From a plausible set of atomic orbitals (e.g. hydrogen-like orbitals with suitable values of the nuclear charge parameter) compute the charge density contributions and hence the Hamiltonian (3.15), containing the effective potential, for each orbital ϕ_i.

2. Solve each equation $h_{\text{eff}}^{(i)}\phi = \varepsilon\phi$ to get a revised approximation to the orbital ϕ_i. This will not generally agree with the ϕ_i used in constructing the potential energy function, which was merely a plausible starting point. Determine revised approximations to all the occupied AO's in this way.

3. Recompute the charge density and effective Hamiltonians and repeat the whole cycle, as in steps 1 and 2. Continue this process until the solutions of the eigenvalue equations agree, to any desired accuracy, with those arising from the previous cycle.

The orbitals and corresponding potential fields that finally emerge are then said to be "self-consistent" and the whole procedure is described as the *self-consistent field* (SCF) method. After obtaining the SCF AO's it is, of course, necessary to derive an expression for the total electronic energy, since this will not in general be a sum of orbital energies. By taking the expectation value of the Hamiltonian (3.1) in the state with wave function (3.12) we obtain easily (all summations confined to occupied orbitals)

$$E = 2 \sum_i \langle \phi_i | h | \phi_i \rangle + \sum_i \langle \phi_i \phi_i | g | \phi_i \phi_i \rangle + 4 \sum_{i<j} \langle \phi_i \phi_j | g | \phi_i \phi_j \rangle \tag{3.17a}$$

where the Dirac-type notation indicates, in general, the integrals (cf. the Example on p. 74)

$$\langle \phi_i | h | \phi_j \rangle = \int \phi_i^*(\mathbf{r}_1) h(1) \phi_j(\mathbf{r}_1) \, d\mathbf{r}_1,$$

$$\langle \phi_i \phi_j | g | \phi_k \phi_l \rangle = \int \phi_i^*(\mathbf{r}_1) \phi_j^*(\mathbf{r}_2) \frac{1}{r_{12}} \phi_k(\mathbf{r}_1) \phi_l(\mathbf{r}_2) \, d\mathbf{r}_1 \, d\mathbf{r}_2. \tag{3.18}$$

From (3.15), (3.16) and (3.14) it is not difficult to express each eigenvalue ε_i as an expectation value involving the one- and two-electron integrals and hence to rewrite (3.17a) in the form

$$E = 2 \sum_i \varepsilon_i - \sum_i \langle \phi_i \phi_i | h | \phi_i \phi_i \rangle - 4 \sum_{i<j} \langle \phi_i \phi_j | g | \phi_i \phi_j \rangle. \tag{3.17b}$$

The reason why the sum of the energies (ε_i) of the (pairs of) electrons in the doubly occupied orbitals must be reduced by the other terms in (3.17b) is simply that each ε_i is derived for an electron in a field provided by *all other* electrons: the sum of the orbital energies therefore counts all electron interactions *twice* and the correction terms simply remove half of this interaction energy. This correction may be regarded as the residual effect of electron interactions, calculated according to perturbation theory (Section 1.1) by averaging the full Hamiltonian over the unperturbed wave function of the model.

When Hartree's method is applied to the helium atom already considered in Section 3.1, the self-consistent $1s$ orbital is found to differ somewhat from the hydrogen-like form used previously (Fig. 3.1)

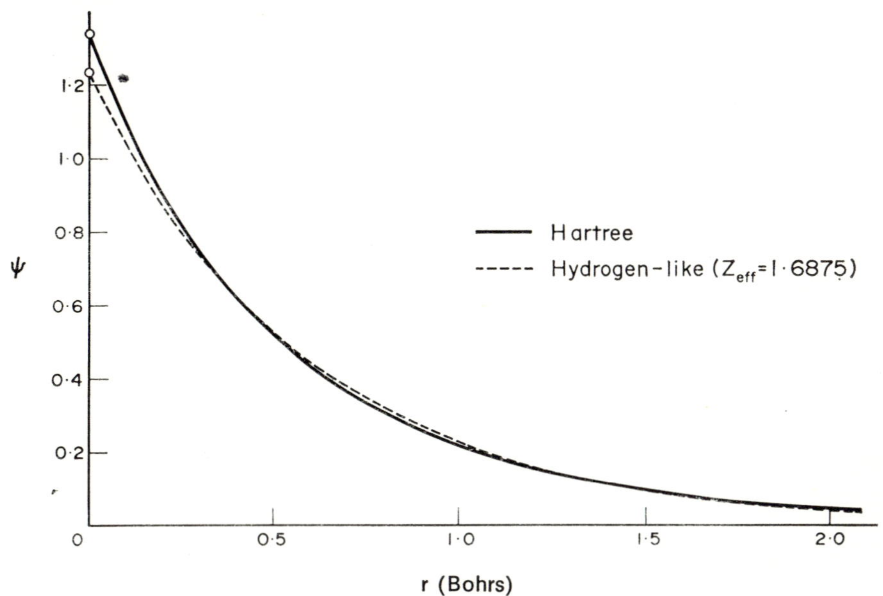

FIG. 3.1. Self-consistent orbital and potential for the helium atom. The $1s$ orbital is compared with a hydrogen-like orbital.

and the Hartree potential departs considerably from the Coulombic form (Fig. 3.2). The total energy obtained ($E = -2\cdot862$ H) is now within about 1% of the accurate value.

The technical details of the SCF method have deliberately been ignored; many of these refer to numerical methods of solution of the one-electron eigenvalue problem and, more particularly, of the radial equation obtained after separation of the variables (Section 2.4). One or two points, however, must be mentioned. In the first place, there is a different eigenvalue problem for determining each orbital; and each eigenvalue equation has a full infinite set of orthogonal solutions, the members of each set being in general non-orthogonal to the members of the other sets; care must then clearly be exercised in deciding which solutions to take as the occupied orbitals. Secondly, in the theory developed so far, it has been assumed that the ground state is non-degenerate, corresponding to a unique allocation of electrons to lowest energy orbitals. When the highest occupied orbital is degenerate,

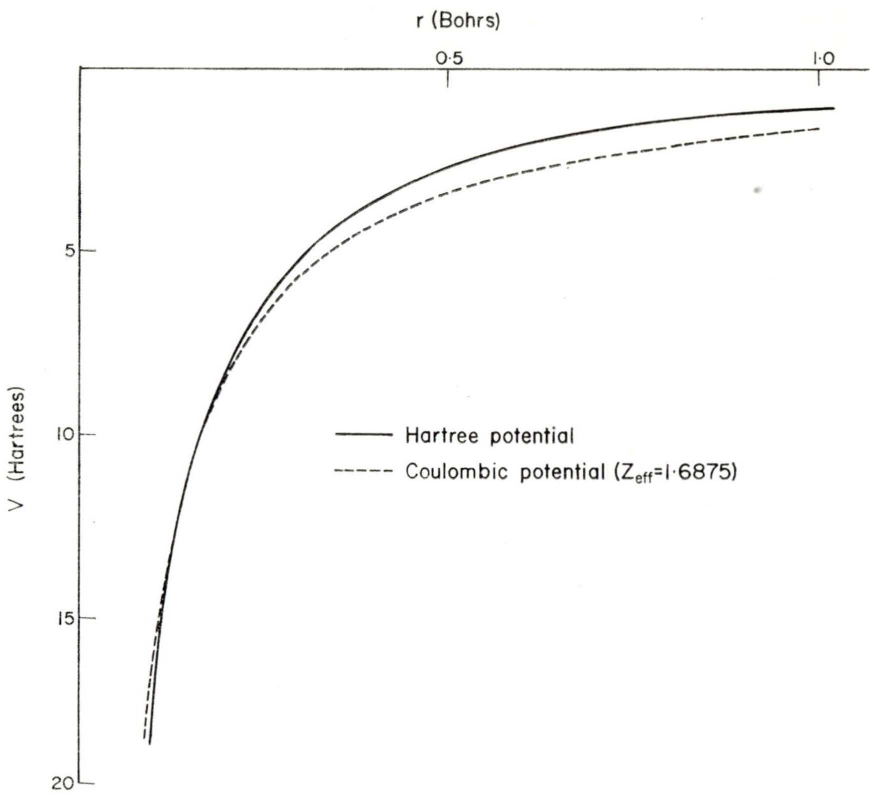

Fig. 3.2. Hartree field in the helium atom. Potential energy function compared with $V = -Z/r$ for an "effective" nuclear charge $Z = 1.6875$.

however, as in the $1s^2 2s^2 2p^2$ configuration[‡] of the carbon atom, where the sum of the orbital energies takes the same value for six distinct selections of the two $2p$ orbitals, modifications will clearly be necessary. In such cases the charge density is no longer spherically symmetric and

---

[‡] The term is here used in the sense of Condon and Shortley (1935), the specification of an occupation number referring not necessarily to a single orbital but to a degenerate set of orbitals; if we wish to refer to *particular* orbitals we continue to speak of an *orbital* configuration.

consequently the eigenvalue equations no longer refer to a central field problem. It is then usual to adopt a "spherical averaging" procedure by making just one allocation of electrons to orbitals (e.g. $1s^2 2s^2 2p_0{}^2$ for carbon) and then obtaining a radial density of charge, to use in Hartree's equations, by integrating over all angles. It is then easily shown that the charge density contribution from each occupied orbital should be replaced by an average over the full degenerate set,

$$\frac{1}{n_g} \sum_{\text{set}} |\phi_i|^2,$$

where n_g is the number of degenerate orbitals. Thus, for example, the $2p^2$ group in the carbon atom would give a spherically averaged charge density which could be represented by $2 \times \frac{1}{3}[|2p_{+1}|^2 + |2p_0|^2 + |2p_{-1}|^2]$ or, equivalently $\frac{2}{3}[|2p_x|^2 + |2p_y|^2 + |2p_z|^2]$: examination of the $2p$ angular factors listed in Table 2.3 confirms that this density depends on r only and is proportional to $r^2 e^{-2\zeta r}$.

As the description of the electronic structure of many-electron molecules and crystals is modelled on that of free atoms, it is worth considering one example in some detail to illustrate the main features of the approach.

EXAMPLE. *Self-consistent field for the sodium atom.* In Section 2.6 the sodium atom was discussed using a simplified model: it was supposed that all electrons but one provided a "core" and that outside the core was a single electron. The single electron, moving in a central field, appeared to be well described by an atomic orbital of $3s$ form in the ground state; and excitation to higher orbitals (e.g. $3p$, $3d$) provided a good interpretation of the spectral series—even to the point of accounting for fine structure due to electron spin. The success of this one-electron model may now be looked at from the point of view of the Hartree theory.

There are eleven electrons and in "filling up" the available orbitals according to the energy level diagram (Fig. 2.4) it appears that a plausible electron configuration for the ground state would be

$$\text{Na}[1s^2 2s^2 2p^6 3s].$$

From an assumed set of approximate atomic orbitals (e.g. the hydrogen-like or Slater-type orbitals of Section 2.5, with exponents chosen (p. 74) to simulate the "screening" of the nucleus by the inner electrons) the charge densities may be calculated and hence a first approximation to the Hartree field for each electron. Each differential equation is solved, yielding new orbitals and orbital energies, and the process continued until self-consistency is achieved.

The charge density contributions from the SCF orbitals are indicated in Fig. 3.3 along with the total density; there is a marked separation into "shells", the $1s^2$ "K-shell", the $2s^2 2p^6$ "L-shell", and finally one electron of the "M-shell". The $3s$ electron is the "valence electron" since it is responsible for the characteristic chemical properties of sodium, or the "series electron" since its excitations account for the observed series spectra.

The self-consistent field for the outermost electron is indicated in Fig. 3.4 and is not very different, over much of the range, from that of a suitably screened Coulomb potential. If the inert gas configuration $1s^2 2s^2 2p^6$ of the ion Na⁺ is regarded as a "core", and is kept fixed for various excitations of the valence electron, orbital energies for the latter may be computed readily. There is reasonable agreement between such orbital energies and the observed spectral term values, and this agreement may be improved very much by refinements described in the next section.

It is now clear that the simple *one*-electron model used in Section 2.6 may be rather realistic, provided the central field potential is carefully chosen so as to simulate the Hartree field of the many-electron system.

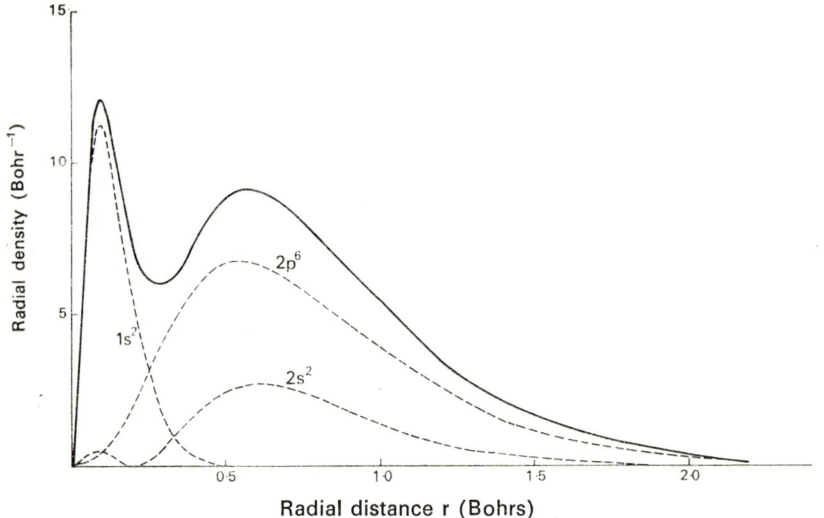

Fig. 3.3. Orbital contributions to the charge density in a sodium atom. The K- and L-shells, $1s^2$ and $2s^2 2p^6$, produce the two peaks. The $3s$ electron of the M-shell gives a more diffuse "valence electron" density (not shown); the peak occurs at about $3{\cdot}5$ B with a height of $\sim 0{\cdot}25$ B⁻¹.

In spite of the considerable successes of Hartree theory, as illustrated in the Example above, there are certain unsatisfactory features in the theory as developed so far. In the first place, the derivation has been intuitive and it is not clear whether the orbitals determined by the SCF procedure will be "best" orbitals in, for example, the sense of the variation method. Also the existence of electron spin has been completely ignored and the principle of allocating not more than two electrons to the same orbital has been treated as an *ad hoc* rule, translating Pauli's exclusion principle into wave mechanical form. The first uncertainty is easily cleared up (Slater, 1930a): if (3.12) is taken as a variational wave function, then the requirement that $E = \langle \Psi | H | \Psi \rangle$ shall be a stationary minimum against arbitrary variations of orbital form $\phi_1 \to \phi_1 + \delta\phi_1$, $\phi_2 \to \phi_2 + \delta\phi_2$, ... , consistent with preservation of normalization, is satisfied when the orbitals satisfy the equations

$$h_{\text{eff}}^{(i)}\phi_i = \varepsilon_i\phi_i \quad (i = 1, 2, \ldots n) \tag{3.19}$$

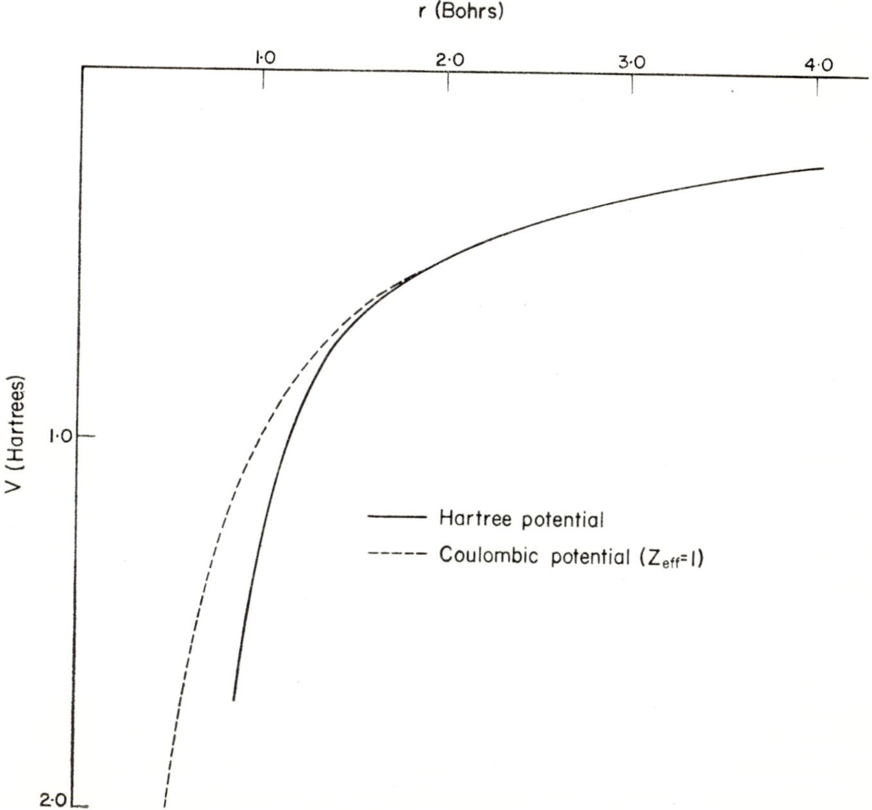

FIG. 3.4. Self-consistent potential for $3s$ electron in a sodium atom. The potential energy of the electron in the field of the "core" (K- and L-shells of Fig. 3.3) compared with the Coulombic form $V = -Z_{\text{eff}}/r$ with $Z_{\text{eff}} = 1\cdot0$, appropriate at large distances. The potential is accurately Coulombic in the region of the M-shell (see Fig. 3.2).

where the operator is the one defined in (3.15). The remaining uncertainties, associated with electron spin, are more fundamental and are discussed briefly in the next section. Here we note that only minor modifications of the model developed so far prove to be necessary, these amounting to little more than replacement of the $h_{eff}^{(i)}$ by a *single* effective Hamiltonian h^F (due to Fock, 1930) which determines *all* the orbitals. The reader may therefore omit the next section, on first reading, without loss of continuity.

3.3.   The self-consistent field with exchange

To remedy the omissions in Hartree's original theory of the self-consistent field we must first take account of electron spin. We know that knowledge of the spatial wave function of an electron is not sufficient to describe it completely; a *spin factor* must be added to distinguish the two possible "internal" states associated with the spin. When the Hamiltonian includes no spin terms, a state of definite energy and definite z-component of spin is then described by a *spin-orbital* indicated by $\phi\alpha$ or $\phi\beta$ according as the z-component is $\pm\frac{1}{2}$. When spin terms *are* included in the Hamiltonian, to allow for the magnetic moment associated with the electron spin, this simple "two-component" spin theory is capable of reproducing with high accuracy the results of observations on simple one-electron systems (Section 2.8).

To incorporate spin in many-electron theory we must first briefly return to the statistical interpretation. We know that with a conventional interpretation[‡] of the spin functions $\psi^*(\mathbf{x})\psi(\mathbf{x})\,d\mathbf{x}$ or $\phi^*(\mathbf{r})\phi(\mathbf{r})$, $\alpha^*(s)\alpha(s)\,d\mathbf{r}\,ds$ retains its normal significance: it is the probability of finding the particle in $d\mathbf{r}$ with spin in the interval $(s,\ s+ds)$. It is customary to introduce a probability density function $\rho(\mathbf{x})$ and to write this result in the form

$$(\text{Probability of electron in } d\mathbf{x}) = \rho(\mathbf{x})\,d\mathbf{x} = \psi(\mathbf{x})\psi^*(\mathbf{x})\,d\mathbf{x}. \qquad (3.20)$$

Since, in the particular example $\psi = \phi\alpha$, we are dealing with an "up-spin" electron, the density $\rho(\mathbf{x})$ (as it contains the factor $\alpha(s)$) will vanish unless $s \cong \frac{1}{2}$. The probability of finding the electron in a particular *spatial* volume element $d\mathbf{r}$, irrespective of the spin value, is of course obtained by "summing" (i.e. integrating) over all possible spin values: the δ-function normalization $\int \alpha^*(s)\alpha(s)\,ds = 1$ then yields, on introducing a spatial density function $P(\mathbf{r})$,

$$(\text{Probability of electron in } d\mathbf{r}) = P(\mathbf{r})\,d\mathbf{r} = \phi(\mathbf{r})\phi^*(\mathbf{r})\,d\mathbf{r} \qquad (3.21)$$

[‡] We continue to use $\mathbf{x}$ to denote space-spin variables collectively, reserving $\mathbf{r}$ for spatial variables and s for spin: the conventional interpretation of the spin function has been fully discussed elsewhere (Vol. 1, Section 4.9).

where the density functions with and without spin are connected simply by a spin integration

$$P(\mathbf{r}) = \int \rho(\mathbf{x}) \, ds \qquad (3.22)$$

which directly expresses their statistical relationship. Clearly $P(\mathbf{r})$ is, for one electron, the electron density function already used in Section 3.1, and addition of the spin factor to an orbital appears to make no difference whatever.

Now suppose there are two electrons. The statistical interpretation is

$$\begin{pmatrix}\text{Probability of electron 1 in } d\mathbf{x}_1 \\ \text{and electron 2 in } d\mathbf{x}_2\end{pmatrix} = \Psi(\mathbf{x}_1, \mathbf{x}_2)\Psi^*(\mathbf{x}_1, \mathbf{x}_2) \, d\mathbf{x}_1 \, d\mathbf{x}_2$$

$$(3.23)$$

and if we assumed a product form $\psi_1(\mathbf{x}_1)\psi_2(\mathbf{x}_2)$ and proceeded as in Section 3.1 we should easily infer that the probability of finding an electron (one *or* the other) in $d\mathbf{x}$ would be

$$\rho(\mathbf{x}) \, d\mathbf{x} = |\psi_1(\mathbf{x})|^2 + |\psi_2(\mathbf{x})|^2,$$

which is exactly analogous to (3.11) but with spin-orbitals instead of orbitals: integration over spin would then yield an expression just like (3.11)—again suggesting that the spin could be ignored completely. In Section 3.1, however, we glossed over an important point; the probability of finding electron 1 in $d\mathbf{x}$ and 2 anywhere (namely $|\psi_1(\mathbf{x})|^2 \, d\mathbf{x}$) is apparently different from that of finding electron 2 in $d\mathbf{x}$ and 1 anywhere (namely $|\psi_2(\mathbf{x})|^2 \, d\mathbf{x}$). But how do we know which electron is which ? If the electrons are indistinguishable we expect either of them to be found in any given volume element with *equal* frequency. The difficulty arises in its more general form if we consider the probability of a situation just like that referred to in (3.23) except that the electrons have changed places; the electron referred to *first* in Ψ then takes variables $\mathbf{x}_2$ (i.e. is at the point previously occupied by the second electron) and vice versa. The situation after exchanging electrons is indistinguishable from that referred to initially and we therefore expect an equal probability of occurrence

$$|\Psi(\mathbf{x}_2, \mathbf{x}_1)|^2 = |\Psi(\mathbf{x}_1, \mathbf{x}_2)|^2$$

even for an *exact* wave function. This is certainly not so for the product approximation $\psi_1(\mathbf{x}_1)\psi_2(\mathbf{x}_2)$. The implication appears to be that the wave function should possess a *symmetry property*: either

$$\Psi(\mathbf{x}_2, \mathbf{x}_1) = \Psi(\mathbf{x}_1, \mathbf{x}_2) \quad \text{or} \quad \Psi(\mathbf{x}_2, \mathbf{x}_1) = -\Psi(\mathbf{x}_1, \mathbf{x}_2).$$

This conjecture turns out to be correct.[‡] Moreover, although not yet proven on purely theoretical grounds, it is an empirical fact that for electrons the wave function is *antisymmetric*.

The antisymmetry principle for an N-electron system may be stated in the form

$$P_{ij}\Psi(\mathbf{x}_1, \mathbf{x}_2, ..., \mathbf{x}_N) = -\Psi(\mathbf{x}_1, \mathbf{x}_2, ..., \mathbf{x}_N)$$

where P_{ij} interchanges the variables $\mathbf{x}_i$, $\mathbf{x}_j$. Since any permutation can be expressed as a series of interchanges (whose number, even or odd, defines the *parity* of the permutation) this result also implies that

$$P\Psi(\mathbf{x}_1, \mathbf{x}_2, ..., \mathbf{x}_N) = \varepsilon_P\Psi(\mathbf{x}_1, \mathbf{x}_2, ..., \mathbf{x}_N) \qquad (3.24)$$

where $\varepsilon_P = +1$ (even parity), $= -1$ (odd parity). *All electronic wave functions are antisymmetric in the sense* (3.24).

It is now clear that a single product of spin-orbitals is in general unacceptable as an approximate wave function and that Hartree's method requires reappraisal. Fortunately, the essential ideas of the independent particle model are scarcely affected. For two electrons, assigned to spin-orbitals ψ_1 and ψ_2, it is clear that *two* products, $\psi_1(\mathbf{x}_1)\psi_2(\mathbf{x}_2)$ and $\psi_1(\mathbf{x}_2)\psi_2(\mathbf{x}_1)$, are equally eligible eigenfunctions of a model Hamiltonian $H_0 = h_{eff}(1) + h_{eff}(2)$, and it is also clear that (assuming the spin-orbitals orthonormal)

$$\Psi(\mathbf{x}_1, \mathbf{x}_2) = \frac{1}{\sqrt{2}}\left[\psi_1(\mathbf{x}_1)\psi_2(\mathbf{x}_2) - \psi_1(\mathbf{x}_2)\psi_2(\mathbf{x}_1)\right]$$

is acceptable as a normalized antisymmetric wave function. The connection with Pauli's exclusion principle is also transparent: if the two electrons occupy the same spin-orbital (i.e. "have the same space and spin quantum numbers") $\psi_1 = \psi_2$ and the wave function vanishes identically—no such state can be found.

For N electrons the situation is entirely similar. For a given selection of spin-orbitals there are $N!$ equally eligible product functions: these may be put together to yield an antisymmetric wave function in only one way, giving

$$\Psi(\mathbf{x}_1, \mathbf{x}_2, ..., \mathbf{x}_N) = \frac{1}{\sqrt{N!}} \sum_P \varepsilon_P P\psi_1(\mathbf{x}_1)\psi_2(\mathbf{x}_2) ... \psi_N(\mathbf{x}_N) \qquad (3.25)$$

where the summation is over all products obtained by permuting the variables and adding the factor ε_P $(= \pm 1)$ according to parity. If the

---

‡ The intuitive argument is not of course a proof: the result follows in fact from the form of the Hamiltonian, which involves all electrons in a symmetrical way, coupled with certain results from group theory (cf. Chap. 4).

individual orbitals are orthonormal (as we usually assume) the factor $1/\sqrt{N}\,!$ ensures normalization of the wave function Ψ. The sum in (3.25) is in fact the expansion of a determinant and is often written

$$\Psi(\mathbf{x}_1, \mathbf{x}_2, \ldots, \mathbf{x}_N) = \frac{1}{\sqrt{N!}}\begin{vmatrix} \psi_1(\mathbf{x}_1) & \psi_2(\mathbf{x}_1) & \ldots & \psi_N(\mathbf{x}_1) \\ \psi_1(\mathbf{x}_2) & \psi_2(\mathbf{x}_2) & \ldots & \psi_N(\mathbf{x}_2) \\ \cdots\cdots\cdots\cdots\cdots\cdots\cdots\cdots\cdots\cdots \\ \psi_1(\mathbf{x}_N) & \psi_2(\mathbf{x}_N) & \ldots & \psi_N(\mathbf{x}_N) \end{vmatrix}$$

$$= (N!)^{-1/2}\,\det\left|\psi_1(\mathbf{x}_1)\psi_2(\mathbf{x}_2)\ldots\psi_N(\mathbf{x}_N)\right| \qquad (3.26)$$

and referred to as a "Slater determinant" after Slater (1929) who first introduced antisymmetry in this way; we shall usually employ the abbreviated notation in which only the diagonal elements are shown. The exclusion principle is then seen as a consequence of the well-known result that a determinant vanishes identically if any two columns are the same ($\psi_i = \psi_j$). The antisymmetry requirement conveniently clears up the apparent ambiguity of how to assign N electrons to a given *selection* of spin-orbitals; there *is* no ambiguity; all $N!$ permutations are used but they must be put together to give the unique result (3.26).

It remains only to reformulate the considerations of Section 3.2. For simplicity, this will be done only for the case of a "closed-shell" ground state, for which the selection of lowest energy orbitals is unique and all are doubly occupied. For "open-shell" systems, where some orbitals are degenerate and are singly occupied, further small changes are necessary, but these are of little consequence for present purposes. Here we simply summarize the argument, noting first that the density function $\rho(\mathbf{x})$ appears as a sum of spin-orbital contributions

$$\rho(\mathbf{x}) = \sum_{i\,(\mathrm{occ})} \psi_i(\mathbf{x})\psi_i{}^*(\mathbf{x}) \qquad (3.27)$$

where i runs over all the occupied spin-orbitals. When each *orbital* appears twice (once with α factor, once with β), integration gives a spatial density function exactly the same as (3.11)

$$P(\mathbf{r}) = 2 \sum_{i\,(\mathrm{occ})} \phi_i(\mathbf{r})\phi_i{}^*(\mathbf{r}) \qquad (3.28)$$

where the sum is now only over distinct doubly occupied orbitals.

With the wave function Ψ of (3.26) the expectation value of the energy appears in the form (Slater, 1929)

$$E = 2\langle\phi_i|\mathsf{h}|\phi_i\rangle + \sum_i \langle\phi_i\phi_i|g|\phi_i\phi_i\rangle + 4 \sum_{i<j} [\langle\phi_i\phi_j|g|\phi_i\phi_j\rangle$$

$$- \tfrac{1}{2}\langle\phi_i\phi_j|g|\phi_j\phi_i\rangle], \qquad (3.29)$$

D

which differs from Hartree's form (3.17a) in the inclusion of the last term: the two-electron integrals are (cf. (3.18))

$$\langle\phi_i\phi_j|g|\phi_i\phi_j\rangle = \int \phi_i^*(\mathbf{r}_1)\phi_j^*(\mathbf{r}_2)\frac{1}{r_{12}}\phi_i(\mathbf{r}_1)\phi_j(\mathbf{r}_2)\,d\mathbf{r}_1\,d\mathbf{r}_2,$$

$$\langle\phi_i\phi_j|g|\phi_j\phi_i\rangle = \int \phi_i^*(\mathbf{r}_1)\phi_j^*(\mathbf{r}_2)\frac{1}{r_{12}}\phi_j(\mathbf{r}_1)\phi_i(\mathbf{r}_2)\,d\mathbf{r}_1\,d\mathbf{r}_2,$$

$$(3.30)$$

and are referred to as "Coulomb" and "exchange" integrals respectively.

The modified form of Hartree's eigenvalue problem defined by (3.15) and (3.16), with inclusion of the exchange terms, may now readily be obtained by the variation method. If each orbital is allowed to vary, $\phi_1 \rightarrow \phi_1 + \delta\phi_1$, $\phi_2 \rightarrow \phi_2 + \delta\phi_2$, ..., subject to preservation of normalization and mutual orthogonality (required in obtaining (3.29)), the condition for a stationary energy is that the orbitals are solutions of

$$\mathsf{h}^F\phi(\mathbf{r}) = \varepsilon\phi(\mathbf{r}). \tag{3.31}$$

Here the Fock operator h^F (Fock, 1930) is defined by

$$\mathsf{h}^F = \mathsf{h} + \int \frac{P(\mathbf{r}')}{|\mathbf{r}-\mathbf{r}'|}\,d\mathbf{r}' - \tfrac{1}{2}\int d\mathbf{r}'\,\frac{P(\mathbf{r};\mathbf{r}')}{|\mathbf{r}-\mathbf{r}'|}\,(\mathbf{r}\rightarrow\mathbf{r}')\,\ldots \tag{3.32}$$

where $P(\mathbf{r};\mathbf{r}')$ is a generalization of the function in (3.28), defined by

$$P(\mathbf{r};\mathbf{r}') = 2\sum_{i(\mathrm{occ})}\phi_i(\mathbf{r})\phi_i^*(\mathbf{r}') \tag{3.33}$$

while $(\mathbf{r}\rightarrow\mathbf{r}')$ means "change $\mathbf{r}$ to $\mathbf{r}'$ in any function standing to the right of the operator". The second term in (3.32) is similar to that in Hartree's operator (3.15) but represents the potential energy of an electron at $\mathbf{r}$ in the field of the *whole* electron distribution (i.e. without subtraction of a contribution from the electron in the orbital under consideration). The final term is described as an "exchange operator" and has a somewhat curious effect; operating on any function $f(\mathbf{r})$ it should be interpreted as "change $\mathbf{r}$ to $\mathbf{r}'$, multiply by $P(\mathbf{r};\mathbf{r}')/|\mathbf{r}-\mathbf{r}'|$ and integrate over $\mathbf{r}'$". The exchange operator thus produces a new function of $\mathbf{r}$; and it is readily verified that the operator is linear and Hermitian. The exchange operator is an *integral operator* (cf. Vol. 1, p. 110).

The SCF procedure, using the Fock operator h^F, is essentially the same as for Hartree's equations. However, there is only *one* eigenvalue problem to determine *all* the occupied orbitals and this removes the non-orthogonality difficulty encountered in the Hartree theory—for solutions of a given eigenvalue equation are automatically orthogonal.

Such improvements must be paid for in the extra labour incurred in handling the exchange term. Compromises have therefore been sought, in which the exchange operator is approximated by an ordinary potential term of suitable form (e.g. Slater, 1951; Lindgren, 1966): for atoms, these approximations lead to a *near*-optimum set of orbitals and have had considerable success. For present purposes, however, we are interested more in the *existence* of a one-electron effective Hamiltonian, whose eigenfunctions give the optimum orbitals to use in an independent-particle model, than in the precise form of the operator.

The physical structure of the Hartree model is also only slightly affected by inclusion of exchange. The orbital energies retain their significance as negative ionization potentials, to a "first order" in which the orbitals of the ionized system are assumed to differ little from those appropriate before ionization (Koopmans' approximation):

$$I_i = -\varepsilon_i = -\langle\phi_i|\mathsf{h}^{\mathrm{F}}|\phi_i\rangle = -[\langle\phi_i|\mathsf{h}|\phi_i\rangle + 2\sum_j (\langle\phi_i\phi_j|g|\phi_i\phi_j\rangle$$
$$-\tfrac{1}{2}\langle\phi_i\phi_j|g|\phi_j\phi_i\rangle)]. \tag{3.34}$$

And the transition energies associated with an electron jump from ϕ_i to an empty orbital ϕ_k turn out to be, instead of (3.8),

$$\Delta E(i \to k) = \varepsilon_j - \varepsilon_i - [\langle\phi_i\phi_k|g|\phi_i\phi_k\rangle - \langle\phi_i\phi_k|g|\phi_k\phi_i\rangle]$$
$$\pm\langle\phi_i\phi_k|g|\phi_k\phi_i\rangle \tag{3.35}$$

where the upper and lower signs refer to *two* possible excited states (triplet and singlet) now separated by an exchange term (absent in Hartree's theory). Finally the total electronic energy may be written in terms of orbital energies as (cf. (3.7))

$$E = 2\sum_i \varepsilon_i - \sum_i \langle\phi_i\phi_i|g|\phi_i\phi_i\rangle - 4\sum_{i<j} [\langle\phi_i\phi_j|g|\phi_i\phi_j\rangle$$
$$-\tfrac{1}{2}\langle\phi_i\phi_j|g|\phi_j\phi_i\rangle] \tag{3.36}$$

where, again, the two-electron terms correct for double counting of the interactions (as in (3.17b)), and the summations are confined to occupied orbitals.

This must conclude our study of the independent particle model. At this level of approximation, which is fundamental in many-electron quantum mechanics, the one-electron eigenvalue equation (3.31), with a well-defined effective Hamiltonian, continues to occupy a central position. Further aspects of many-electron theory will be dealt with elsewhere (Vol. 3); here we continue our study of one-electron systems, turning to potential functions of the polycentric form appropriate to molecules and crystals.

3.4. Molecular orbitals. General considerations

We have seen how the atomic orbitals describing the stationary states of a single electron in a central field provide a rather satisfactory basis for discussing, via the independent particle model, the electronic structure of complicated many-electron atoms. Electrons are allocated to the available orbitals, in ascending energy order and with two electrons in each (opposite spins), to yield a ground-state electron configuration. By suitable adjustment of the orbital forms, the corresponding one-determinant wave function can give a good—and very easily visualized—account of the structure and properties of the many-electron atom. It would appear that similar considerations might be applied to molecules (and to crystals) provided the atomic orbitals were replaced by "molecular orbitals" (or "crystal orbitals") describing the stationary states of an electron in a polycentric field.

To explore this possibility we begin by considering a single electron in the field produced by two fixed nuclei, taking for simplicity the molecular analogue—the hydrogen molecule ion H_2^+—of the system (hydrogen atom) considered in first introducing the AO's. The assumption that the nuclei may be regarded as fixed, merely determining the potential field for a pure electronic problem, requires careful justification and cannot be considered here. We remark only that the separation of electronic and nuclear variables depends on the electron-nuclear mass ratio and is rather accurately valid, except in certain exceptional cases; the total energy of the system (electrons and nuclei together) may then be taken as the sum of an electronic part, which is parametrically dependent on the nuclear coordinates, and a nuclear part containing their Coulomb repulsions. Calculation of the electronic energy for various fixed nuclear configurations (i.e. energy as a function of molecular coordinates) is thus a necessary prelude to the interpretation of molecular geometries, vibrational motion, etc., in which nuclear motion is considered. We must therefore deal first with the pure electronic problem in which the function of the nuclei is simply to provide a potential field for the electrons.

Molecular orbitals. The H_2^+ ion

The Hamiltonian for the system H_2^+ (Fig. 3.5) is, using atomic units,

$$h = -\tfrac{1}{2}\nabla^2 - \frac{1}{r_a} - \frac{1}{r_b}$$

and the corresponding Schrödinger equation may be separated (Vol. 1, App. 2) by introducing "prolate spheroidal coordinates" with the two

nuclei as foci. This course was followed by Burrau (1927), who introduced
the variables

$$\xi = (r_a + r_b)/R, \quad \eta = (r_a - r_b)/R, \quad \varphi$$

and took the wave function to be $\varphi(\xi, \eta, \varphi) = X(\xi)\,Y(\eta)\,Z(\varphi)$. Closed
form solutions of the resultant φ- and η-equations may be obtained,
while the ξ-equation may be solved numerically by standard techniques. Burrau found a lowest energy *molecular orbital* (MO) of the form
indicated in Fig. 3.6a and on repeating the calculation for a number of
internuclear distances obtained the energy curve shown in Fig. 3.6b.
This indicates an equilibrium configuration with nuclei at a distance

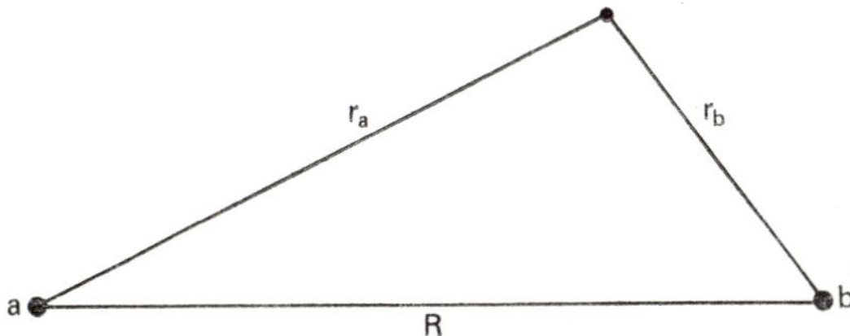

FIG. 3.5. Electronic coordinates in a diatomic molecule. The prolate
spheroidal coordinates of the electron are $\xi = (r_a + r_b)/R$, $\eta = (r_a - r_b)/R$
and the angle ϕ relative to a fixed plane containing the internuclear axis.

2·00 Bohr and an energy 2·7773 eV less than that of a single hydrogen
atom. These figures are in excellent agreement with the spectroscopically
inferred bond length and dissociation energy.[‡] The orbital in Fig. 3.6a
is thus a "bonding MO". The solution of next lowest energy is indicated
in Fig. 3.7a and gives an energy curve (Fig. 3.7b) corresponding to
repulsion at all distances; it is an "anti-bonding MO". Before discussing
the forms of other diatomic MO's, we consider briefly the physical
interpretation of these results.

The origin of the chemical bond in H_2^+ might be associated intuitively
with the enhanced probability of finding the electron of the bonding
MO (i.e. negative charge) between the (positive) nuclei. That this
"electrostatic" interpretation is essentially correct follows from a
theorem due to Hellmann (1937) and Feynman (1939). This theorem
asserts that the forces exerted by the electrons on the nuclei may be
correctly calculated by regarding their charge as being "smeared out"
into a "charge cloud" of density $P(\mathbf{r})$—the probability function
introduced in Section 3.1. The nuclei are thus attracted towards the

---

[‡] With due allowance for vibrational motion (see, for example, Pauling and Wilson,
1935, p. 337).

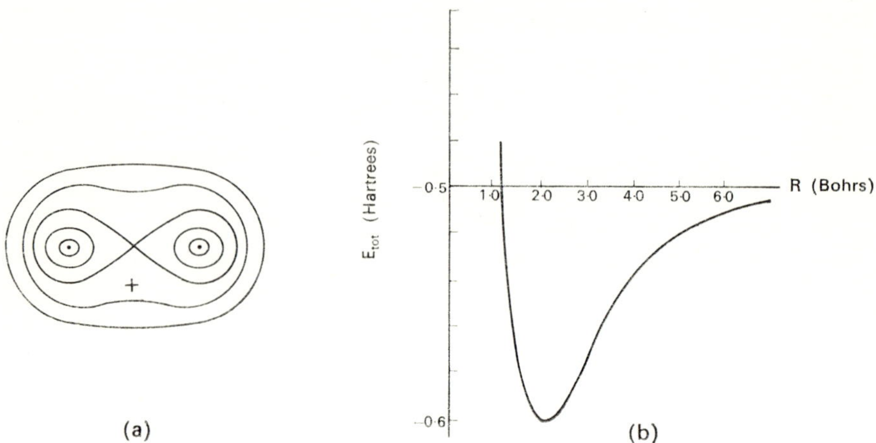

(a)   (b)

Fig. 3.6. Bonding and energy variation, H_2^+. The map shows contours corresponding to $|\psi| = 0.6,\ 0.5,\ 0.4,\ 0.3,\ 0.2$ (reading outwards from the nuclei). The energy is the total energy (electronic + nuclear repulsion) relative to that of a free hydrogen atom.

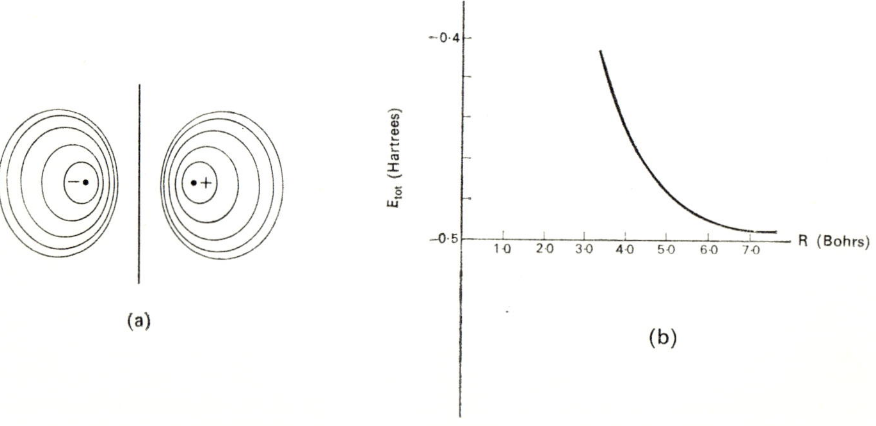

(a)   (b)

Fig. 3.7. Anti-bonding orbital and energy variation, H_2^+. The contours are drawn for $|\psi| = 0.6, 0.5, 0.4, 0.3, 0.2$. (For details see also Fig. 3.6.)

region of high charge density, in the ground state, until the attractive forces are balanced by the mutual repulsion of the nuclei.

For a one-electron system, the theorem follows very easily from perturbation theory. To calculate the electronic part of the force on a nucleus n, in say the X direction, we consider a small displacement δX_n and compute the change $\delta\varepsilon$ in the electronic energy; the force component F_{nx} is then the limit of $-(\delta\varepsilon/\delta X_n)$ for $\delta X_n \to 0$. Now *only* the potential energy term $V(\mathbf{r})$ in the Hamiltonian depends on the nuclear variables, and the perturbation due to δX_n is thus of the form $\delta\mathsf{h} = \delta V(\mathbf{r})$. Consequently, to first order, the energy change is

$$\delta\varepsilon = \int \psi^* \, \delta\mathsf{h}\psi \, d\mathbf{x} = \int \delta V(\mathbf{r})P(\mathbf{r}) \, d\mathbf{r} \qquad (3.37)$$

where the spin integration in $\psi\psi^*$ has been performed (since $\delta V(\mathbf{r})$ does not involve spin operators and is merely a factor in the integrand) to yield the density function $P(\mathbf{r})$. On dividing both sides of (3.37) by δX_n and passing to the limit $\delta X_n \to 0$ we obtain an expression for the force

$$F_{nx} = \int F_{nx}(\mathbf{r})P(\mathbf{r}) \, d\mathbf{r} \qquad (3.38)$$

where $F_{nx}(\mathbf{r}) = -(\partial V/\partial X_n)$. The electrostatic interpretation follows because $F_{nx}(\mathbf{r})$, by definition of the potential energy function $V(\mathbf{r})$, is the force on nucleus n due to unit charge (one electron) at point $\mathbf{r}$; the integrand is the force that would be exerted by an amount of charge $P(\mathbf{r}) \, d\mathbf{r}$; and the integration thus gives the total force exerted by an extended charge distribution of density $P(\mathbf{r})$.

The theorem as stated above also applies to many-electron systems, provided the wave function from which $P(\mathbf{r})$ is computed is exact. This, of course, is never the case, but the result may be shown to remain valid (as an approximation to the force) for certain types of variationally determined wave function. The theorem is perhaps most useful, however, as an interpretive tool in discussing chemical bonding in terms of orbital forms: its value for this purpose will become clear in later sections.

Molecular orbitals. Approximate forms

Since it is not normally possible to obtain molecular wave functions by direct numerical integration, it is important to develop alternative methods which can reproduce Burrau's results and can more easily be generalized. We shall do this first in the context of the H_2^+ ion already considered, later indicating how the approximate methods lead readily to a general understanding of the types of MO available in a diatomic molecule.

The approximate methods most commonly used are based on complete set expansions[‡] in which an MO is written

$$\phi = c_1\chi_1 + c_2\chi_2 + \ldots c_n\chi_n \tag{3.39}$$

in terms of a set of basis orbitals χ_1, χ_2, ... In principle, any complete set may be used; for example, a set of AO's[§] centred on a fictitious atom located anywhere in space. Although "one-centre" expansions sometimes give useful results it is normally more effective to *mix* a number of complete sets, using orbitals centred on *all* the actual nuclei in order to

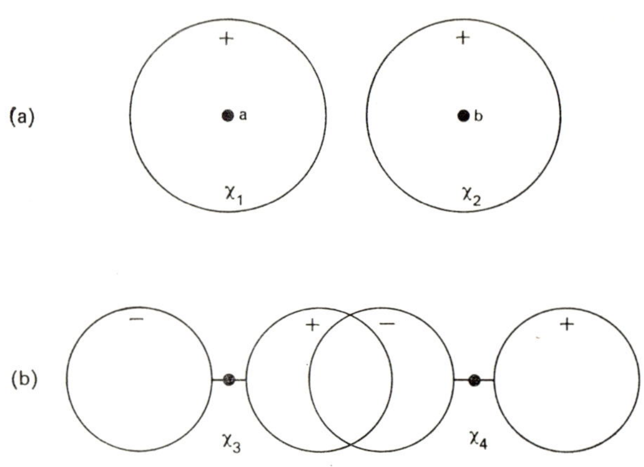

FIG. 3.8. AO's used in constructing two-centre MO's. (a) s-type AO's, χ_1 and χ_2. (b) p-type AO's, χ_3 and χ_4, allowing axial polarization.

reproduce the correct behaviour of the wave function in these important regions (e.g. the cusp of an s-orbital) without resort to slowly convergent expansions based on some other centre. This approach was developed largely by Hund, Mulliken, and Lennard–Jones during the period 1929–35. In the simplest possible application, to the system H_2^+, we might use a two-term expansion, taking χ_1 and χ_2 as $1s$-type AO's located on the nuclei a and b (Fig. 3.8a). To illustrate the general approach in a less trivial way, however, we shall include two more

---

[‡] For a discussion of basic theory see Section 3.2 of Vol. 1.

[§] If such a set is to be complete it should comprise either (i) hydrogen-like or "Slater" AO's, containing the factor $e^{-Zr/n}$, *plus* the continuum of positive energy solutions corresponding to a free scattered electron, or (ii) an alternative set which is complete without the continuum. Except in highly accurate work, however, the question of completeness is not a real issue and with only a few basis orbitals such considerations are unnecessary.

AO's, χ_3 and χ_4 (Fig. 3.8b), which will serve to "polarize" the s-like behaviour of χ_1 and χ_2 in the vicinity of the nuclei—an effect which might be expected on physical grounds.

With a four-term expansion, in terms of normalized but non-orthogonal orbitals, the mixing coefficients in (3.39) are determined as in Section 1.4 by solving the equation (symmetric across the diagonal)

$$\begin{vmatrix} (\alpha_s - \varepsilon) & (\beta_{ss} - S_{ss}\varepsilon) & \beta'_{sp} & (\beta_{sp} + \varepsilon S_{sp}) \\ & (\alpha_s - \varepsilon) & (\beta_{sp} - \varepsilon S_{sp}) & \beta'_{sp} \\ & & (\alpha_p - \varepsilon) & (\beta_{pp} + \varepsilon S_{pp}) \\ \text{etc.} & & & (\alpha_p - \varepsilon) \end{vmatrix} = 0 \qquad (3.40)$$

where the matrix elements[‡] $H_{ij} = \langle \chi_i | \mathsf{h} | \chi_j \rangle$ and $S_{ij} = \langle \chi_i | \chi_j \rangle$ have been abbreviated:

$$
\begin{aligned}
\alpha_s &= H_{11} &&= H_{22} & S_{ss} &= S_{12} = S_{21} \\
\alpha_p &= H_{33} &&= H_{44} & S_{pp} &= -S_{34} = -S_{43} \\
\beta_{ss} &= H_{12} &&= H_{21} & S_{sp} &= S_{23} = S_{32} = -S_{14} = -S_{41} \\
\beta_{pp} &= -H_{34} &&= -H_{43} \\
\beta_{sp} &= H_{23} &&= H_{32} & = -H_{14} &= -H_{41} \\
\beta'_{sp} &= H_{13} &&= H_{31} & = H_{24} &= H_{42}
\end{aligned}
\qquad (3.41)
$$

The α-, β-, and S-type quantities are referred to as "Coulomb", "resonance", and "overlap" integrals respectively. They can be evaluated in closed form, the resultant expressions depending upon the nuclear charges ($Z = 1$ for H_2^+), and internuclear distance R, and the orbital exponents in the AO's. With the present definitions, the α integrals are found to be negative, S-type integrals positive; β integrals are negative if $\chi_i \chi_j$ is positive in the overlap region, and vice versa. The fact that many integrals are the same, or differ only in sign, is a consequence of the symmetry of the system. Thus if the nuclei and orbitals in Fig. 3.8 were rotated as a whole through $180°$ about the mid-point, χ_3 would be sent into $-\chi_4$, χ_4 into $-\chi_3$, χ_1 into χ_2 and χ_2 into χ_1. The Hamiltonian, however, would be unchanged, as would be the value of any matrix element (for an observer rotated in the same way would see the same system—and the integral value cannot depend on where the observer stands): consequently, for example,

$$\langle \chi_2 | \mathsf{h} | \chi_3 \rangle = \langle \chi_1 | \mathsf{h} | -\chi_4 \rangle = -\langle \chi_1 | \mathsf{h} | \chi_4 \rangle.$$

The implications of symmetry are taken up in detail in the next chapter.

[‡] Note that the operator h, in MO secular equations, is an effective *one*-electron Hamiltonian.

Expansion of (3.40) would yield an equation of the fourth degree in ε whose lowest root would be an approximation to the ground state energy. For illustration, however, let us first ignore the $2p$ functions, considering only the 2×2 determinant in the upper left-hand corner of (3.40). The corresponding secular polynomial is then a quadratic,

$$(\alpha_s - \varepsilon)^2 = (\beta_{ss} - \varepsilon S_{ss})^2$$

with the roots

$$\varepsilon_1 = \frac{\alpha_s + \beta_{ss}}{1 + S_{ss}}, \qquad \varepsilon_2 = \frac{\alpha_s - \beta_{ss}}{1 - S_{ss}}. \tag{3.42}$$

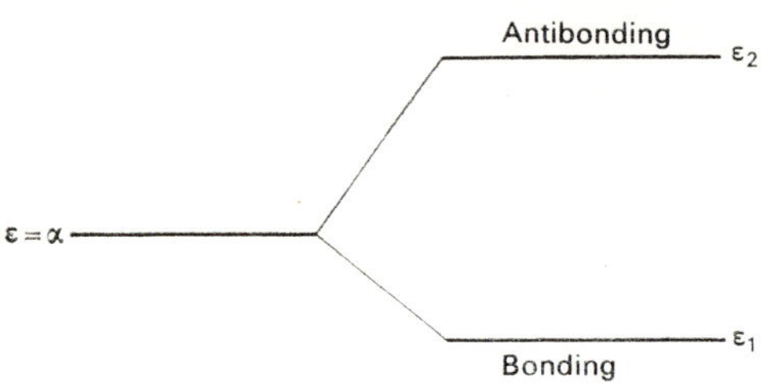

Fig. 3.8. Energy relations in formation of H_2^+. The MO energies ε_1 and ε_2 are given by equation (3.42); they lie at roughly $|\beta_{ss}|$ above and below the level $\varepsilon = \alpha_s$ corresponding to an electron on one centre only.

Since α_s and β_{ss} are negative, ε_1 is an approximate ground state energy: by substituting this value back in the first of the secular equations,

$$(\alpha_s - \varepsilon)c_1 + (\beta_{ss} - \varepsilon S_{ss})c_2 = 0$$

it follows that c_1 and c_2 are equal in the corresponding MO: $\phi = c_1(\chi_1 + \chi_2)$. The constant c_1 is eliminated by using the normalizing condition $\int \phi^2 \, d\mathbf{r} = 1$. The second root may be dealt with similarly, and in this way we find

$$\phi_1 = \frac{\chi_1 + \chi_2}{\sqrt{(2 + 2S_{ss})}}, \qquad \phi_2 = \frac{\chi_1 - \chi_2}{\sqrt{(2 - 2S_{ss})}}. \tag{3.43}$$

These MO's bear a strong resemblance to those indicated in Figs. 3.6–3.7: in particular, ϕ_1 is symmetric across a plane through the bond mid-point, and clearly approximates the bonding MO, while ϕ_2 is anti-symmetric and antibonding.

The orbital energies ε_1, ε_2 are indicated in Fig. 3.8. If S_{ss} is small these energies are about equispaced about $\varepsilon = \alpha_s$: since α_s is the expectation energy value of an electron localized in *one* of the $1s$ AO's, the bonding may be associated with the electron of a single hydrogen atom falling into the bonding MO, with a resultant drop of $\simeq |\beta_{ss}|$ in the electronic energy.

The approximate MO's are considerably improved if the AO's themselves are given a more flexible form by using the orbital exponents as parameters and making use of the variation method. Finally, when the p-orbitals are added and (3.40) is solved in a similar way the resultant electronic energy in the ground state is found to be within 0·05 eV of Burrau's value. Clearly this linear combination of atomic orbitals (LCAO) method is capable of yielding MO's of high accuracy: this is indeed fortunate, because in polycentric situations the use of numerical integration techniques (cf. those used by Hartree for atoms) is in general unmanageable and it is essential to expand over a basis, thus turning the operator form of the Schrödinger equation into a matrix eigenvalue equation which is more readily handled.

It is interesting to look at the general forms of the MO's that result from solution of (3.40): for according to general theory (p. 15) these should provide approximations to *four* states. We find these to be of the general forms

$$\phi_s = s(\chi_1 + \chi_2) + s'(\chi_3 - \chi_4),$$
$$\phi_a = a(\chi_1 - \chi_2) + a'(\chi_3 + \chi_4) \tag{3.44}$$

where ϕ_s is symmetric across the plane through the bond mid-point while ϕ_a is antisymmetric (Figs. 3.6–3.7). The bonding and anti-bonding MO's in (3.43) are of s and a type respectively and are only slightly modified by admixture of χ_3 and χ_4, i.e. the corresponding four-term approximations (above) have s' and a' small. The two MO's not mentioned so far are of s type but with $s < s'$ and of a type with $a < a'$: they are depicted in Fig. 3.10. These results conform to certain general principles, which we examine in the next chapter, and suggest also how we may build up the forms of MO's for a general homonuclear diatomic molecule.

The symmetry principles, in this simple example, may seem to state the obvious. They may be expressed as follows:

(i) Any approximate wave function, for a symmetrical system, is of a well-defined "symmetry species" (in the present case, symmetric or antisymmetric under a certain reflection, which characterizes the "mirror symmetry" of the pair of nuclei across the plane between them).

(ii) Such a wave function, of given symmetry species, is expressible in terms of "symmetry functions" *of the same species only* (in the present case, for example, $(\chi_1 + \chi_2)$ and $(\chi_3 - \chi_4)$ are both symmetric).

(iii) If the symmetry functions are regarded as new basis functions, and are used in setting up a new secular problem, then the determinant corresponding to (3.40) must break into two 2×2 blocks along the diagonal.[‡]

We add to these rules the observation that—in a simple LCAO approximation—there is little mixing between AO's of substantially different energy, on either the same or different centres: for example, valence orbitals on one centre do not mix heavily with inner-shell orbitals on another centre—the matrix elements responsible for such mixing being

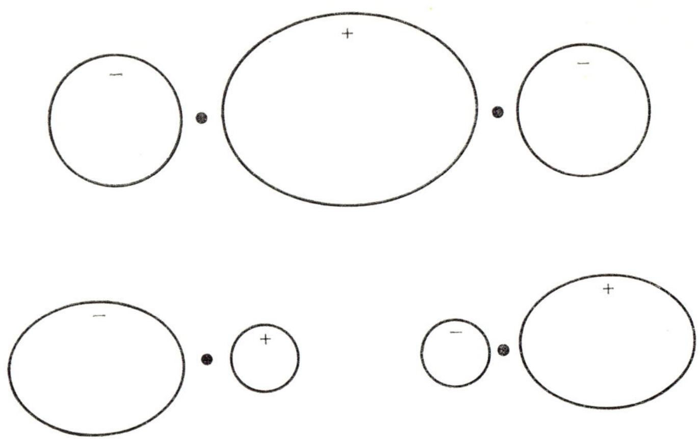

Fig. 3.10. Forms of higher energy MO's in H_2^+ (schematic). The orbitals shown are σ-type, with cylindrical symmetry.

small. These principles suggest that suitable approximate MO's can be constructed simply by combining *pairs* of AO's, one from each centre, with suitable choice of symmetry and orientation to produce an orbital which exhibits an intuitively satisfactory *overall* symmetry.

The results of combining AO's in this way, using $1s$, $2s$, and $2p$ orbitals of two identical nuclei, are indicated schematically in Fig. 3.11 along with a terminology which may be used in referring to them. Briefly, σ-type MO's have axial symmetry, π-type MO's have a nodal plane containing the axis: the AO which dominates the LCAO representation of each MO is indicated first: and a star is added to distinguish an anti-bonding MO from its bonding partner. The forms of

---

[‡] It is easily verified that matrix elements between symmetric and antisymmetric functions will vanish; e.g. $\langle (\chi_1 + \chi_2)|\mathsf{h}|(\chi_1 - \chi_2)\rangle = H_{11} - H_{22} - H_{12} + H_{21} = 0$ on using (3.41).

MO's for both homonuclear and heteronuclear molecules and the uses to which they are put are dealt with at length in all books on qualitative valence theory. In using such orbitals to describe the electronic structure of a many-electron molecule, at the level of the independent particle

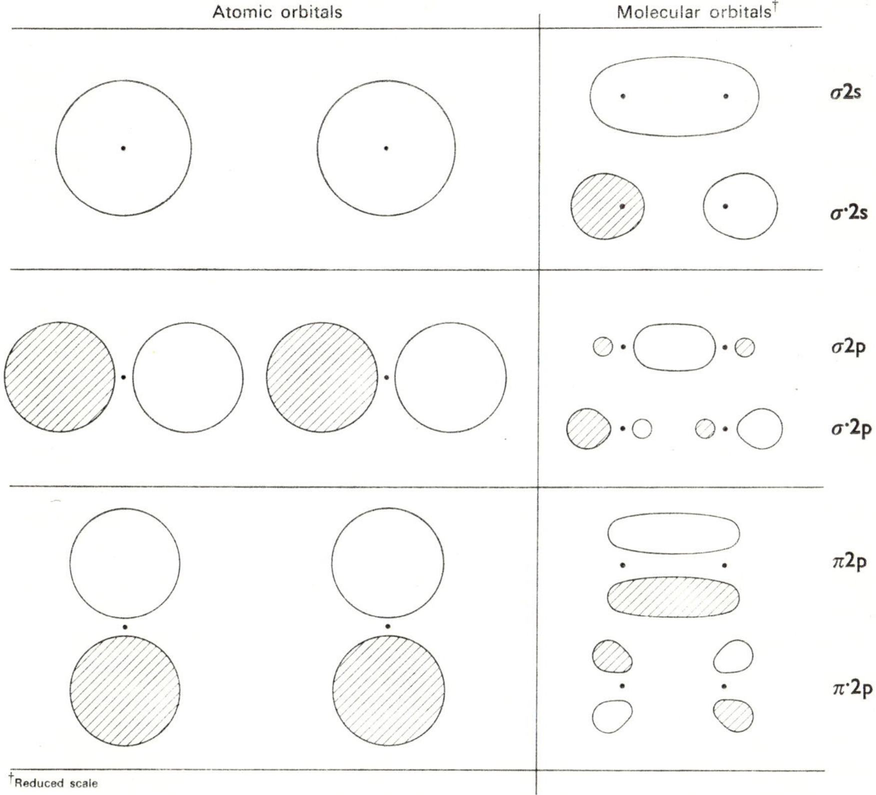

FIG. 3.11. Construction of MO's for homonuclear diatomic molecule. Line of centres is taken as the z-axis; σ, π, ... indicate type of symmetry about the axis; name of the AO is that of the principal components in LCAO approximation; unstarred and starred MO's are bonding and anti-bonding "partners". Shading indicates regions where ϕ is negative.

model, it is necessary to know the sequence in which their energies lie: for simple homonuclear diatomic molecules this is invariably

$$(\sigma 1s) < (\sigma^* 1s) < (\sigma 2s) < (\sigma^* 2s) < (\sigma 2p_z) < \begin{Bmatrix} (\pi 2p_x) \\ (\pi 2p_y) \end{Bmatrix} < \begin{Bmatrix} (\pi^* 2p_x) \\ (\pi^* 2p_y) \end{Bmatrix}$$

$$< (\sigma^* 2p_z) \dots \quad (3.45)$$

where the MO's bracketed together are degenerate owing to axial symmetry,[‡] and the energies are to be understood as orbital energies in the Hartree–Fock sense (p. 75).

3.5.   Electronic structure of some simple molecules

In Section 3.2 the electronic structure of a many-electron atom was discussed in terms of the filling of available atomic orbitals, using the general ideas of the independent particle model and assuming that each AO was an eigenfunction of a suitable *effective* Hamiltonian. Here we indicate how an exactly similar procedure can be employed in "molecule building", the orbitals now being the *molecular* orbitals whose general forms, for a diatomic molecule, have been discussed in Section 3.4.

In the many-electron case, the MO's are not those of an electron in the field of the *bare* nuclei, but refer instead to an electron in a "screened" nuclear field represented by the Hartree–Fock operator h^F. When an LCAO approximation is used, however, in setting up the MO's this causes no *formal* difficulty whatever; the α and β quantities used in the last section are simply replaced by corresponding matrix elements of the operator h^F instead of h. If we wish to actually calculate such matrix elements, expressing the Hartree–Fock operator in terms of the charge density using (3.32), then substantial computational problems arise; but much progress towards an understanding of molecular electronic structure at a qualitative or semi-quantitative level can be made simply by regarding the α's and β's as disposable parameters.

It would be out of place in this volume to give a detailed account of molecular orbital theory, and of the many approximations and semi-empirical prescriptions presently in use. Such accounts may be found in books on quantum chemistry and valence theory (see, for example, Coulson, 1961; Murrell *et al.*, 1965; Parr, 1963). Here we simply illustrate the general approach, using the most primitive approximations and considering one or two representative molecules, in order to gain some familiarity with the quantum-mechanical interpretation of the chemical bond.

Hydrogen molecule in LCAO approximation

One of the key quantities in the independent particle model is the charge density function $P(\mathbf{r})$, appearing in the form (3.11): indeed, this quantity has in some ways a deeper significance than the individual

---

[‡] As for the p-type AO's (p. 58), it is immaterial whether the degenerate MO's are chosen in real form (subscripts x, y) or complex form (subscripts ± 1, corresponding to angular momentum component $\pm \hbar$ around the axis).

MO's themselves.[‡] It is therefore of considerable interest to see what form $P(\mathbf{r})$ takes in an LCAO approximation, and to use it in developing the Hellmann–Feynman interpretation of the bond.

For the hydrogen molecule we assign two electrons to the bonding MO of lowest energy (cf. the assignment of two electrons to the $1s$ AO in the helium atom). This MO is, using the notation of Section 3.4, $\sigma 1s$ and the electronic structure is indicated by

$$H_2[\sigma 1s^2].$$

In the simplest approximation $\sigma 1s$ is represented by a linear combination of $1s$-type AO's χ_1 and χ_2, centred on the two nuclei. The normalized MO, obtained in (3.43), may be written (with S_{12} denoting the overlap integral)

$$\phi_{\sigma 1s} = \frac{\chi_1 + \chi_2}{\sqrt{(2 + 2S_{12})}}.$$

The charge density is then $P(\mathbf{r}) = 2|\phi(\mathbf{r})|^2$, which becomes

$$P(\mathbf{r}) = (1 + S_{12})^{-1}[\chi_1{}^2(\mathbf{r}) + \chi_2{}^2(\mathbf{r}) + 2\chi_1(\mathbf{r})\chi_2(\mathbf{r}),$$

the AO's being real functions. This may be written

$$P(\mathbf{r}) = q_1\, d_1(\mathbf{r}) + q_2\, d_2(\mathbf{r}) + q_{12}\, d_{12}(\mathbf{r}) \tag{3.46}$$

where the d's are normalized density functions

$$d_1(\mathbf{r}) = \chi_1{}^2(\mathbf{r}), \quad d_2(\mathbf{r}) = \chi_2{}^2(\mathbf{r}), \quad d_{12}(\mathbf{r}) = \chi_1(\mathbf{r})\chi_2(\mathbf{r})/S_{12} \tag{3.47}$$

and the coefficients are

$$q_1 = q_2 = \frac{1}{1 + S_{12}}, \qquad q_{12} = \frac{2S_{12}}{1 + S_{12}}. \tag{3.48}$$

The three terms in (3.46) evidently represent (i) a $1s$-like charge density on centre 1, with weight factor q_1, (ii) a similar density on centre 2, with an equal weight factor, and (iii) a normalized "overlap density" with weight factor q_{12}. The overlap density is so called because it is large only where χ_1 and χ_2 simultaneously take substantial values, i.e. where they "overlap" appreciably.

It is clear that the coefficients defined in (3.48) represent amounts of charge, or more accurately "populations" (in numbers of electrons), in

---

‡ A one-determinant wave function is unchanged if the orbitals are replaced by new orbitals obtained by a unitary transformation; this corresponds to replacing the columns of the determinant (3.26) by new linear combinations, an operation which does not affect the value of the determinant. The MO's are therefore not unique (except in a certain conventional sense): the charge density, however, is an *invariant* and is not affected by changing from one set of orbitals to another.

the regions to which they refer. If $P(\mathbf{r})$ is integrated over all space we must obtain 2, the total number of electrons smeared out with density $P(\mathbf{r})$, and hence

$$\int P(\mathbf{r})\, d\mathbf{r} = q_1 + q_2 + q_{12} = 2. \tag{3.49}$$

This is a "conservation equation" stating simply that the sum of the populations of all the orbital and overlap regions must equal the total number of electrons. The populations themselves give a very easily visualized picture of the form of the charge density. In particular, the fact that q_{12} is positive means that there is an accumulation of electron density in the overlap region, as required for bond formation. An energy calculation confirms this result, giving a curve of E_{tot} against inter-nuclear distance (R) with features similar to those in Fig. 3.5b. A simple MO calculation of this kind predicts a bond length 1·38 Bohrs and a binding energy (relative to two H atoms) of 3·47 eV, in fair agreement with the experimental values of 1·40 Bohrs and 4·722 eV, respectively.

Diatomic molecules of first row elements

We now examine briefly one or two of the main features of the MO interpretation of bonding in diatomic molecules, using three illustrative examples involving atoms in the first row of the Periodic Table. Perhaps the most characteristic feature of a chemical bond is a "saturation property". Most bonds involve *pairs* of electrons, and on adding further electrons the bond is not strengthened but weakened. The MO approximation provides a simple interpretation of this result in terms of the electron density: it is therefore useful to define orbital and overlap populations more generally.

There is no difficulty in obtaining simple expressions for the electron populations in a many-electron system, at the level of the independent particle model. Let us call the occupied MO's ϕ_A, ϕ_B, ..., with the LCAO forms

$$\phi_A = \sum_i c_{Ai}\, \chi_i, \quad \phi_B = \sum_i c_{Bi}\, \chi_i, \dots . \tag{3.50}$$

The charge density expression (3.11) then becomes, with n_R electrons in ϕ_R,

$$P(\mathbf{r}) = \sum_{R(\text{occ})} n_R \sum_{i,j} c_{Ri}\, \chi_i(\mathbf{r}) c_{Rj}{}^* \chi_j{}^*(\mathbf{r}).$$

It is convenient to introduce a matrix $\mathbf{P}$, whose general element is

$$P_{ij} = \sum_{R(\text{occ})} n_R\, c_{Ri}\, c_{Rj}{}^* \tag{3.51}$$

for we then obtain simply

$$P(\mathbf{r}) = \sum_{ij} P_{ij}\chi_i(\mathbf{r})\chi_j{}^*(\mathbf{r}). \tag{3.52}$$

By comparison with (3.46) and (3.47) it is clear that the populations of orbital and overlap regions (assuming for simplicity that all orbitals are real) are given by

$$q_{ii} = P_{ii}, \quad q_{ij} = 2P_{ij}S_{ij}. \tag{3.53}$$

The populations may be used in various ways (e.g. McWeeny, 1952, 1954) and "population analysis" has been developed extensively by Mulliken (1955). We now discuss some simple chemical bonds in terms of populations.

The helium molecule

If two helium atoms come together, *four* electrons will be available for bonding and the usual procedure suggests the electron configuration

$$\text{He}_2[\sigma 1s^2\sigma^*1s^2].$$

On using the LCAO approximations defined in (3.43) the definitions (3.51) and (3.53) give populations

$$q_1 = q_2 = \frac{2}{1 - S_{12}{}^2}, \qquad q_{12} = \frac{-4S_{12}{}^2}{1 - S_{12}{}^2},$$

referring, respectively, to the two $1s$ orbitals and their region of overlap. Consequently, since $S_{12}{}^2$ is normally small compared with unity, the charge distribution does not differ significantly from that for two independent helium atoms—as if each kept two electrons in its own $1s$ orbital—and there is *no* building up of charge density in the overlap region (in fact there is a small flow of electron density out of the "bond"). In view of the Hellmann–Feynman theorem (p. 89), we therefore expect no resultant bonding. To verify quantitatively that there is no bond, we must of course calculate the energy but here we are concerned only with the qualitative relationship between electron configuration and chemical bonding, and the "saturation property" thus receives at least a highly plausible interpretation in terms of the populations. This conclusion may be stated as follows. When two electrons occupy a bonding MO they give rise to a normal "electron-pair" bond; if two more electrons are added to the anti-bonding partner the bond is destroyed, and the electron density becomes roughly equivalent to that due to the separate doubly occupied AO's. The principle is equally useful in more complicated situations, as the following examples indicate.

The nitrogen molecule

Reference to the energy level sequence (3.45) suggests that the fourteen electrons can be accommodated in diatomic MO's to give the electron configuration

$$N_2[(\sigma 1s)^2(\sigma^*1s)^2(\sigma 2s)^2(\sigma^*2s)^2(\sigma 2p)^2(\pi 2p_x)^2(\pi 2p_y)^2].$$

The doubly degenerate π-type MO's are here denoted $\pi 2p_x$ and $\pi 2p_y$ (built up from the *real* $2p_x$ and $2p_y$ AO's pointing normal to the internuclear axis); but it is easily verified that the complex forms, $\pi 2p_{+1}$ and $\pi 2p_{-1}$, would give exactly the same charge density (see footnote on p. 99).

The inner shells overlap very little and $(\sigma 1s)^2(\sigma^*1s)^2$ then gives (as for two helium atoms) a density essentially identical with $(1s_a)^2(1s_b)^2$; this part of the electron configuration is frequently denoted by KK, indicating two atomic K-shells. The bonding and anti-bonding MO's formed from the $2s$ AO's are also *both* filled and should therefore again give no resultant bonding—or even a little repulsion. The strongest bond comes from the two electrons in $\sigma 2p$, represented by an axially symmetric electron density concentrated along the bond axis: such a bond is a typical "σ-bond". The remaining MO's give somewhat weaker bonds, each with a concentration of charge on either side of the bond axis, in the x and y directions respectively: such bonds are "π bonds". When the two π-bond densities are superposed the resultant electron density becomes axially symmetric, though it is zero on the axis itself and so forms a "sheath" of charge around the σ bond. This is the electronic interpretation of the nitrogen "triple bond" in which three electron pairs participate.

The oxygen molecule

As a final example we consider the oxygen molecule, which contains two further electrons. The next available orbitals are π^*2p_x and π^*2p_y which are degenerate, and (anticipating the effects of electron repulsion) we assign one electron to each. The electron configuration is then

$$O_2[KK(\sigma 2s)^2(\sigma^*2s)^2(\sigma 2p)^2(\pi 2p_x)^2(\pi 2p_y)^2(\pi^*2p_x)(\pi^*2p_y)].$$

As in the nitrogen molecule, there is a strong σ-bond from the two electrons in $(\sigma 2p)$. But the π bonds are now weakened, since each doubly occupied bonding MO is accompanied by a singly occupied anti-bonding MO: the enhancement of electron density in the π-bond region is roughly half that due to the bonding electrons alone, and the situation corresponds to two "half" π-bonds. In other words, diatomic oxygen may be regarded as containing a "double bond". This description is supported by experiment: thus, for example, ionization of the molecule leads to a strengthening of the bond (reflected in bond shortening) due to removal of an *anti*-bonding electron; and the presence of singly occupied MO's provides a natural interpretation of the observed paramagnetism of the system in terms of "unpaired" spins.

3.6. Crystal orbitals. Electronic structure of solids

The LCAO approximation used in discussing the electronic structure of molecules is equally applicable, with trivial extensions, to crystals. Although for practical reasons its implementation is difficult, and other methods are commonly employed, it is instructive to derive the so-called "tight-binding" approximation of solid state physics as an exercise in molecular orbital theory.

For simplicity let us consider a linear array of N atoms each with just one valence electron (i.e. a one-dimensional "crystal" of a monovalent metal) and set up MO's in LCAO approximation to describe the valence electrons *in the crystal*. Each MO takes the form

$$\phi = c_1\chi_1 + c_2\chi_2 + \ldots + c_N\chi_N \tag{3.54}$$

where χ_1, χ_2, ..., χ_N are identical AO's on the N centres, and the coefficients are determined by solving the usual secular equations (Section 1.4c), namely

$$(H_{11} - \varepsilon)c_1 + (H_{12} - \varepsilon S_{12})c_2 + \ldots = 0$$
$$(H_{21} - \varepsilon)c_1 + (H_{22} - \varepsilon)c_2 \quad + \ldots = 0$$
$$\cdot \quad \cdot \quad \cdot \quad \cdot \quad \cdot \quad \cdot \quad \cdot \quad \cdot \quad \cdot \quad \cdot$$

The matrix elements are of two types

$$H_{rs} = \langle \chi_r | \mathsf{h} | \chi_s \rangle, \qquad S_{rs} = \langle \chi_r | \chi_s \rangle$$

where in general h may be regarded as a Hartree–Fock operator, taking account of the main electron interaction effects. A simple and widely used approximation is to ignore non-orthogonality of the AO's and to neglect H_{rs} unless r and s are coincident or refer to nearest neighbours: this amounts to setting

$$S_{rs} = \delta_{rs}, \quad H_{rr} = \alpha \text{ (all } r), \quad H_{rs} = \beta \text{ (}r, s \text{ nearest neighbours)}.$$

With this "nearest-neighbour approximation" the secular equations take the form

$$(\alpha - \varepsilon)c_1 + \beta c_2 \qquad = 0$$
$$\beta c_1 + (\alpha - \varepsilon)c_2 + \beta c_3 \qquad = 0$$
$$\beta c_2 + (\alpha - \varepsilon)c_3 + \beta c_4 \qquad = 0 \tag{3.55}$$
$$\cdots$$
$$\beta c_{N-1} + (\alpha - \varepsilon)c_N = 0.$$

Every equation is of the same type, expressing the periodicity of the crystal, namely

$$\beta c_{n-1} + (\alpha - \varepsilon)c_n + \beta c_{n+1} = 0, \tag{3.56}$$

and even those for the end atoms, 1 and N, can be thrown into this form by adding terms βc_0 and βc_{N+1} and then imposing "boundary conditions" $c_0 = c_{N+1} = 0$. We now consider two types of solution.

Standing wave solutions

It might be expected that for N finite the MO's will resemble the standing waves for a particle in a box (Vol. 1, p. 22) and this suggests a "trial" solution exhibiting periodicity in the variable n. Let us choose

$$c_n = e^{in\theta}$$

where θ is to be determined, if possible, to satisfy (3.56). Substitution yields

$$e^{in\theta}[\beta\, e^{-i\theta} + (\alpha - \varepsilon) + \beta\, e^{i\theta}] = 0$$

and, since this must hold for all value of n, the factor in square brackets must vanish: θ, which determines the "wave length" with which the coefficients vary, must therefore be related to the energy. In fact, since $e^{i\theta} + e^{-i\theta} = 2\cos\theta$

$$\varepsilon = \alpha + 2\beta\cos\theta. \tag{3.57}$$

For a given α, however, the sign of θ is arbitrary and a general solution is thus

$$c_n = A\, e^{in\theta} + B\, e^{-in\theta} \tag{3.58}$$

where the constants are dependent on boundary conditions.

In the system considered, $c_0 = 0$ for $n = 0$ and it follows that $A + B = 0$. The variation is thus sinusoidal:

$$c_n = C\sin n\theta.$$

The second boundary condition, $c_{N+1} = 0$ for $n = N+1$, then determines the acceptable values of the parameter θ:

$$\sin(N+1)\theta = 0, \qquad \theta = \frac{\kappa\pi}{N+1} \qquad (\kappa \text{ integral}).$$

Each choice of θ yields the coefficients defining an MO, together with its energy from (3.57): thus

$$\varepsilon_\kappa = \alpha + 2\beta \cos \frac{\kappa\pi}{N+1},$$

$$c_n^{(\kappa)} = C_\kappa \sin \frac{n\kappa\pi}{N+1} \qquad (3.59)$$

where κ is a positive integer.

Some solutions for the case $N = 4$ are indicated schematically in Fig. 3.12. They arise for $\kappa = 1, 2, 3, 4$ and evidently on taking $\kappa = 5$

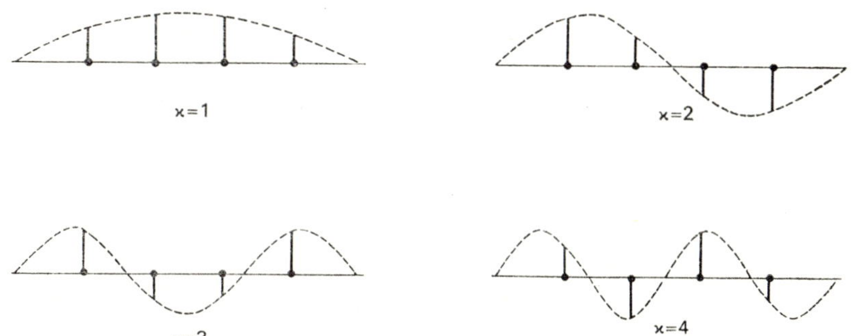

FIG. 3.12. MO's for a linear chain. Vertical lines indicate values of the mixing coefficients $c_n^{(\kappa)}$ ($n = 1, \ldots 4$) in each of the MO's $\kappa = 1, \ldots 4$.

the solution for $\kappa = 1$ is merely reproduced: in general there are just N solutions, for $\kappa = 1, \ldots N$. The normalizing factor C_κ follows easily (with normalized AO's and overlap neglected) from the condition

$$\sum_n \left| c_n^{(\kappa)} \right|^2 = 1.$$

This example is of interest in indicating the similarity between, on the one hand, the behaviour of a free electron, and, on the other hand, that of a "mobile" electron in the field of a periodic array of charges (Fig. 3.13). Near a nucleus the plane wave form is obviously inappropriate: but nevertheless the sinusoidal factors establish a one-to-one correspondence between the "crystal orbitals" and the "free-electron" functions. It should be noted that for $\beta = 0$ there would be N degenerate levels $\varepsilon = \alpha$ (energy of an electron in *one* AO, with no interaction) and that interaction ($\beta \neq 0$) splits the level into a "band" of N distinct levels, or width 4β. When N is large, as in a crystal, the band of levels become virtually a continuum.

Travelling wave solutions

Let us now consider the case $N \to \infty$. In this case (cf. Vol. 1, Section 2.4) there will be a normalization difficulty which may be overcome in various ways. The one most common in solid state theory depends on the introduction of "periodic boundary conditions". Instead of considering the whole infinite crystal, we mentally divide it into identical parts ("microcrystals") which are repeated indefinitely along the three crystal axes, and assume the wave function to be repeated in a similar way within each part (Fig. 3.14). Attention may then be confined to

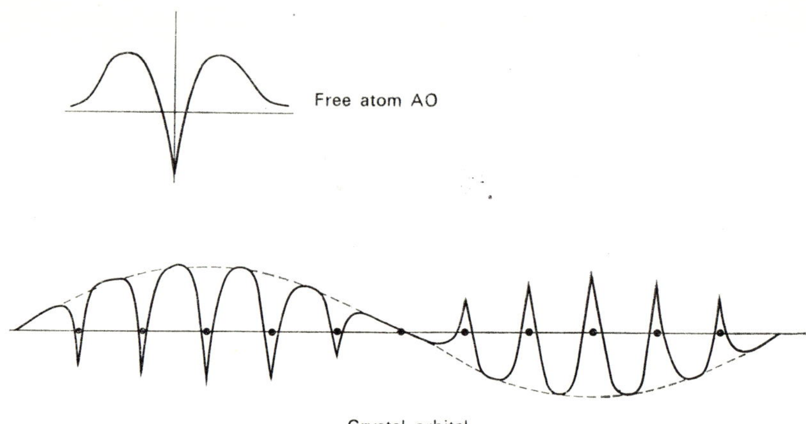

Fig. 3.13. Relationship between typical crystal orbital and a corresponding free-electron wave function. Near each nucleus the function varies rapidly, as it would in a single atom; away from the nuclei it may closely approximate a plane wave.

one "fundamental volume", within which normalization is straightforward, endless repetition of the wave function being imposed by applying the periodic boundary conditions.

To confirm that this procedure leads to acceptable solutions, we return to the secular equations (3.56) for the one-dimensional case. The substitution $c_n = e^{in\theta}$ again satisfies all equations, subject to the condition (3.57). Now, however, we regard N as the number of atoms in the fundamental "micro-crystal", and require repetition of the values of the coefficients in accordance with the periodicity condition $c_{n+N} = c_n$, or

$$e^{i(n+N)\theta} = e^{in\theta} \qquad \text{(all } n\text{).}$$

This implies

$$N\theta = \kappa \times 2\pi \qquad (\kappa = 0, \pm 1, \pm 2, \ldots)$$

and instead of (3.59) we therefore obtain

$$\varepsilon_\kappa = \alpha + 2\beta \cos \frac{2\pi\kappa}{N} \qquad (\kappa = 0,\ \pm 1,\ \pm 2,\ \ldots),$$

$$c_n^{(\kappa)} = C_\kappa \exp \frac{2\pi i\kappa n}{N}.$$

(3.60)

The normalization condition shows that $C_\kappa = 1/\sqrt{N}$: and it is readily verified that κ values differing by an integral multiple of N lead to exactly the same ϕ_κ, so that any N consecutive values from the set $\{-\infty \ldots -3,\ -2,\ -1,\ 0,\ +1,\ +2,\ +3 \ldots +\infty\}$ will define N independent solutions or "crystal orbitals".

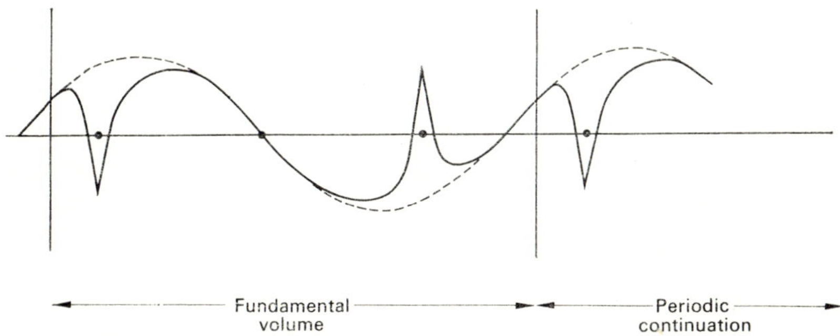

FIG. 3.14. Illustrating periodic boundary conditions. The fundamental volume here contains three atoms; periodic boundary conditions ensure that the behaviour of the wave function within this volume is repeated in all translationally equivalent regions.

Linear momentum in a crystal orbital

In view of the close correspondence between plane wave solutions for a free electron and crystal orbitals for an electron in a periodic field, it would be expected that the standing wave solutions (3.59) describe states in which the linear momentum has zero expectation value, while the travelling wave solutions (3.60) can describe an electron moving through the crystal with a non-zero linear momentum.

To verify this possibility in the one-dimensional case we simply compute the expectation momentum in a travelling wave state. From (3.54) we find for state ϕ_κ,

$$\langle \phi_\kappa | \mathsf{P}x | \phi_\kappa \rangle = \sum_{n,m} c_m^{(\kappa)} c_n^{(\kappa)*} \langle \chi_n | \mathsf{P}x | \chi_m \rangle. \tag{3.61}$$

It is easily seen that, with x along the crystal direction, the matrix elements are $\langle\chi_n|\mathsf{p}_x|\chi_n\rangle = 0$ (zero *linear* momentum for an electron in AO χ_n), while

$$\langle\chi_n|\mathsf{p}_x|\chi_m\rangle = (\hbar/i)\,d_{nm}, \qquad d_{nm} = -d_{mn}.$$

The antisymmetry of $d_{nm} = \langle\chi_n|\partial/\partial x|\chi_m\rangle$ becomes apparent on turning the x-axis and the system through $180°$ about the origin: this cannot change the matrix element but interchanges χ_m and χ_n and gives $(\partial/\partial x)_{\text{rotated frame}} = -\partial/\partial x$. The expression (3.61) for the linear momentum, in nearest-neighbour approximation and with the co-efficients given in (3.60), then reduces to

$$\langle p_x\rangle = \sum_n \frac{1}{N}\frac{\hbar}{i}\,\{-e^{2\pi i\kappa/N} + e^{-2\pi i\kappa/N}\}\,d = -2\hbar\,d\,\sin\,(2\pi\kappa/N). \qquad (3.62)$$

Here $d = d_{mn}$ $(m = n-1)$ is the nearest-neighbour matrix element and the two terms in the sum come from $m = n+1$ and $m = n-1$ respectively (nothing more arising from the m-summation (3.61)).

This result has an interesting interpretation. For orbitals of the kind shown in Fig. 3.13, d is a positive quantity, small for weakly overlapping AO's but increasing with overlap. The total linear momentum along the positive x direction thus increases as κ takes the values 1, 2, 3, ... and is clearly analogous to the momentum in the free electron case (Vol. 1, p. 37). The momentum does not continue to increase, however; it reaches a maximum for $\kappa = \frac{1}{4}N$ and falls to zero as κ approaches $\frac{1}{2}N$. The behaviour for states of negative κ is similar except for direction.

The results are summarized in Fig. 3.15. When N is large the energy levels form a band, of width 4β, the energy as a function of quantum number κ increasing from $\kappa = 0$ to $\kappa = \pm\frac{1}{2}N$. The magnitude of the momentum in these states, however, begins to fall off as the energy increases above the midpoint of the bands. At the bottom of the band the electron behaves essentially like a *free* electron (energy $\propto \kappa^2$, momentum $\propto \kappa$) but for higher energies the scattering of the electron by the lattice becomes important and in the uppermost states the momentum expectation value falls to zero, the scattering being so complete that the electron is equally likely to be travelling to right or to left.

Electronic structure of crystals

The preceding considerations are readily extended to three-dimensional crystals. A simple lattice is defined by three basis vectors $\mathbf{a}_1$, $\mathbf{a}_2$, $\mathbf{a}_3$ that define the edges of a unit cell. The fundamental volume to which periodic boundary conditions are applied is then a prism with edges $G_1\mathbf{a}_1$, $G_2\mathbf{a}_2$, $G_3\mathbf{a}_3$, say, corresponding to G_1 unit cells in the $\mathbf{a}_1$ direction, etc. If there is one atom per unit cell, at the point $\mathbf{R}_n = n_1\mathbf{a}_1 + n_2\mathbf{a}_2 + n_3\mathbf{a}_3$, the argument used in the one-dimensional case now leads to a solution (nearest-neighbour approximation)

$$\varepsilon_{\kappa_1\kappa_2\kappa_3} = \alpha + 2\beta_1 \cos\left(2\pi\kappa_1/G_1\right) + 2\beta_2 \cos\left(2\pi\kappa_2/G_2\right)$$
$$+ 2\beta_3 \cos\left(2\pi\kappa_3/G_3\right),$$

$$c_{n_1n_2n_3}^{(\kappa_1\kappa_2\kappa_3)} = C \exp 2\pi i\left(\frac{\kappa_1 n_1}{G_1} + \frac{\kappa_2 n_2}{G_2} + \frac{\kappa_3 n_3}{G_3}\right). \tag{3.63}$$

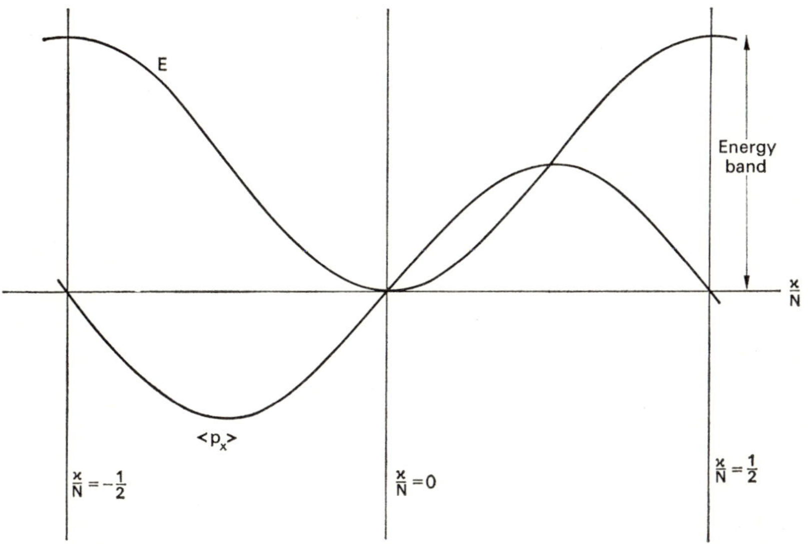

FIG. 3.15. Behaviour of energy and momentum for states in an energy band.

It is customary in solid state physics to classify these states in terms of vectors in a "reciprocal lattice" of the type used in crystallography; this lattice has basis vectors $\mathbf{b}_1$, $\mathbf{b}_2$, $\mathbf{b}_3$ such that, for instance, $\mathbf{b}_1$ is orthogonal to $\mathbf{a}_2$ and $\mathbf{a}_3$ with a "length" such that $\mathbf{a}_1\cdot\mathbf{b}_1 = 1$—and similarly for $\mathbf{b}_2$, $\mathbf{b}_3$. Thus

$$\mathbf{a}_i\cdot\mathbf{b}_i = 1, \qquad \mathbf{a}_i\cdot\mathbf{b}_j = 0 \quad (i \neq j) \qquad (i, j = 1, 2, 3). \tag{3.64}$$

The solutions (3.63) can then be put in clear formal correspondence with the free-electron functions $\phi_\mathbf{k}(\mathbf{r}) = C\,e^{i\mathbf{k}\cdot\mathbf{r}}$. We introduce a

"$\mathbf{k}$-space" whose basis vectors are $2\pi\mathbf{b}_1$, $2\pi\mathbf{b}_2$ and $2\pi\mathbf{b}_3$: a general vector in k-space is then

$$\mathbf{k} = k_1(2\pi\mathbf{b}_1) + k_2(2\pi\mathbf{b}_2) + k_3(2\pi\mathbf{b}_3). \tag{3.65}$$

The exponential factor in (3.63) can then be neatly expressed, bearing in mind (3.64), in the form $\exp i\mathbf{k}\cdot\mathbf{R}_n$ where $\mathbf{k}$ is a vector whose numerical components are

$$k_1 = \kappa_1/G_1, \qquad k_2 = \kappa_2/G_2, \quad k_3 = \kappa_3/G_3. \tag{3.66}$$

As κ_1, κ_2, κ_3 each run through G_1, G_2 or G_3 successive integral values we obtain all $G_1G_2G_3$ distinct solutions of the type (3.63). The corresponding wave functions are, using (3.63) in the form $c_n{}^k = C \exp i\mathbf{k}\cdot\mathbf{R}_n$ for the AO coefficients,

$$\phi_\mathbf{k}(\mathbf{r}) = C \sum_{\mathbf{R}_n} \exp (i\mathbf{k}\cdot\mathbf{R}_n)\chi_n(\mathbf{r}) \tag{3.67}$$

where $\chi_n(\mathbf{r})$ is the atomic orbital located at the lattice point $\mathbf{R}_n$. Such functions were first used by Bloch, and play a fundamental role in solid state theory. The fact that the expansion coefficients in the LCAO approximation to a crystal orbital are completely determined in analytical form is a consequence of the symmetry of the lattice.

Once the forms of the crystal orbitals have been established, the complete electronic structure may be discussed using the independent particle model, essentially as in the case of a many-electron atom or molecule. From each orbital of a free atom there arises, in the crystal, an energy band whose corresponding orbitals are characterized by a set of k vectors, the energy level distribution becoming quasi-continuous as the number of unit cells increases. The properties of actual crystals may then be discussed in terms of the widths, spacings, degree of filling, etc., of the available bands of levels. For developments of this kind the reader is referred to standard textbooks on solid state theory (see, for example, Kittel, 1966; Donovan, 1967).

REFERENCES

Burrau, O. (1927) *Kgl. Danske Videnskab. Selskab.* **7**, 1.

Condon, E. U. and Shortley, E. (1935) *The Theory of Atomic Spectra*, Cambridge University Press, London.

Coulson, C. A. (1961) *Valence*, 2nd ed., Oxford University Press, London.

Donovan, B. (1967) *Elementary Theory of Metals*, Pergamon Press.

Eyring, H., Walter, J. and Kimball, G. E. (1944) *Quantum Chemistry*, John Wiley & Sons, London and New York.

Feynman, R. P. (1939) *Phys. Rev.* **41**, 721.

Fock, V. (1930) *Z. Physik* **61**, 126.

Hartree, D. R. (1928) *Proc. Camb. Phil. Soc.* **24**, 89, 111 (see also *Calculation of Atomic Structures*, Wiley, New York).

HELLMANN, H. (1937) *Einführung in die Quantchemie*, Franz Deuticke, Leipzig, Germany.

KITTEL, C. (1966) *Introduction to Solid State Physics* Wiley, London and New York.

LINDGREN, I. (1966) *Ark. Fysik* **31**, 59.

McWEENY, R. (1952) *Acta Crystallogr.* **5**, 463.

McWEENY, R. (1954) *Proc. Roy. Soc.* A **223**, 63.

MULLIKEN, R. S. (1955) *J. Chem. Phys.* **23**, 1833, 2343.

MURRELL, J. N., KETTLE, S. F. A. and TEDDER, J. M. (1965) *Valence Theory*, Wiley & Sons, London and New York.

PARR, R. G. (1963) *The Quantum Theory of Molecular Electronic Structure*, Benjamin, New York.

PAULING, L. and WILSON, E. B. (1935) *Introduction to Quantum Mechanics*, McGraw-Hill, New York and London.

SLATER, J. C. (1929) *Phys. Rev.* **34**, 1293.

SLATER, J. C. (1930a) *Phys. Rev.* **35**, 210; (1930b) **36**, 57.

SLATER, J. C. (1951) *Phys. Rev.* **81**, 385.

CHAPTER 4

IMPLICATIONS OF SYMMETRY

4.1. Symmetry and group theory

From time to time we have referred to the "symmetry properties" of
wave functions. Whenever a system exhibits any kind of symmetry it is
possible to draw important conclusions about the allowed forms of the
wave functions, the nature of degeneracies in the energy spectrum (and
their possible resolution by perturbation of the system) and the
"selection rules" which are of fundamental importance in discussing
spectroscopic transitions between different states. Sometimes the
implications of symmetry can be discussed using intuitive or pictorial
arguments (cf. p. 93) but more generally it is necessary to employ the
mathematical machinery of group theory. It might be objected that
highly symmetrical systems are the exception rather than the rule,
and that the applicability of group theoretical ideas might therefore be
too restricted to justify their inclusion in a book on quantum mechanics:
but in fact symmetry in a wider sense[‡] is almost always present and
group theoretical arguments are now widely used not only in the
applications of quantum mechanics but also in obtaining a deeper
interpretation of fundamental principles.

In this chapter we shall be concerned only with spatial symmetry,
but this will allow us to define all the necessary concepts, and to review
and illustrate basic theorems and methods in a quantum-mechanical
context. Further applications will be made in Chap. 5 and in the
theory of many-electron systems (Vol. 3). For a fuller discussion of
group theory itself the reader is referred elsewhere (e.g. McWeeny,
1963, for a simple account; Hamermesh, 1962, for a more comprehensive treatment).

Example of a group

For the purposes of exposition, we shall consider a system comprising
a positive point charge at the centre of a cube, with six negative charges

---

[‡] For example, the interchange of any two identical particles is a symmetry operation,
as will become clear presently.

112

at the centre of the cube faces (Fig. 4.1a). This is a typical model for the discussion of a "crystal field" problem in which the positive charge may represent a transition metal ion, while the negative charges may represent attached "ligands". The array of charges determines a potential field of high symmetry, characterized by that of the cube, and the quantum-mechanical problem we shall consider is the determination of wave functions for an electron in this field.

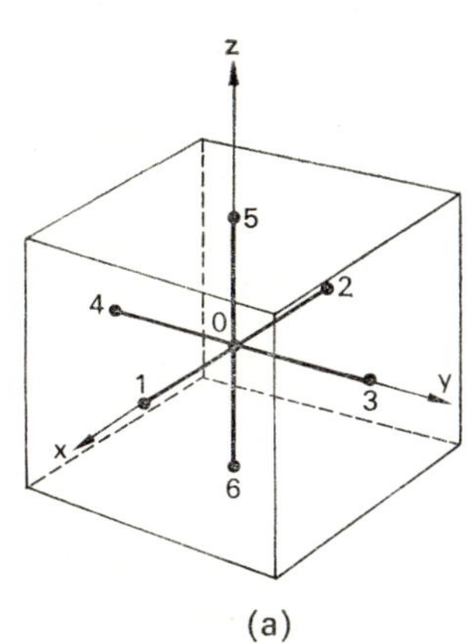

(a)

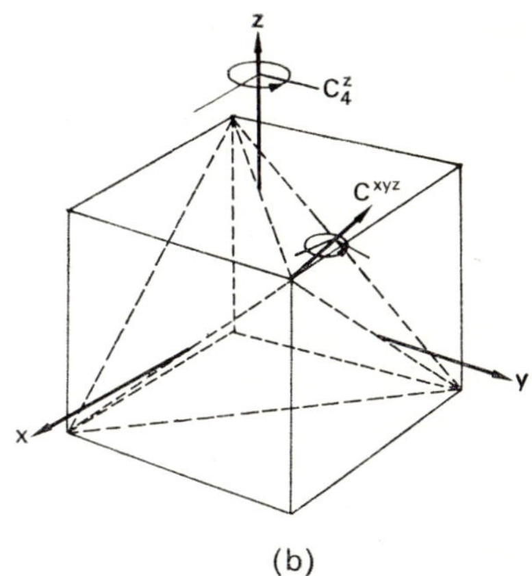

(b)

FIG. 4.1. Cubic symmetry. (a) Model of a crystal field situation. Six equal negative charges (1, 2, ... 6), at the centres of the cube faces, surround a central positive charge (0). The field has the symmetry of the cube; so has the octahedron whose edges are formed by joining the six vertices. (b) Typical symmetry operations. C_4^z (quarter of a turn about the z-axis) and C_4^{xyz} (third of a turn about the body diagonal) each bring the cube into self-coincidence. Repeated combination of such operations (*generators*) produces twenty-four symmetry operations comprising the *octahedral group* (O). Note that the inscribed tetrahedron (broken line) has a lower symmetry: C_4^z is replaced by C_2^z and there follow twelve symmetry operations comprising the *tetrahedral group* ($\mathscr{T}$) which is a *subgroup* of O.

First we define the *symmetry group* of the system. The whole array of charges, and the cube which contains them, will be brought into "self-coincidence" by certain operations: the two basic operations are (Fig. 4.1b):

C_4^z: positive rotation through $\pi/2$ about the z-axis,

C_3^{xyz}: positive rotation through $2\pi/3$ about the body-diagonal on which $x = y = z$.

The symbols C_3 and C_4 denote rotations (through $2\pi/4$ and $2\pi/3$ respectively) in the Schoenflies notation,[‡] while the superscripts indicate the rotation axes. Operations of this kind are called *symmetry operations*. It is clear that the cube is characterized by a whole set of symmetry operations, and that these may be specified individually by defining an axis and a rotation angle. Thus rotations through $2\pi/n$ about the various symmetry axes may all be denoted by

$$\mathsf{C}_n{}^{\alpha}, \quad \mathsf{C}_n{}^{\alpha\beta}, \quad \mathsf{C}_n{}^{\alpha\beta\gamma} \qquad (\alpha,\ \beta,\ \gamma = x,\ y,\ z,\ \bar{x},\ \bar{y},\ \bar{z})$$

where, for example, $\bar{x}$ denotes the negative x-axis and $\bar{x}y$ the axis midway between the $\bar{x}$- and y-axes. The cube evidently has two-fold axes of symmetry (a typical operation being $\mathsf{C}_2{}^{xy}$), threefold axes (a typical operation being $\mathsf{C}_3{}^{xyz}$) and four-fold axes (a typical operation being $\mathsf{C}_4{}^{z}$). The cube is also brought into self-coincidence by reflections in which every point on one side of a plane is sent into its mirror image on the other side; but for present purposes it will be sufficient to consider the rotations alone.

Two symmetry operations may be combined by *sequential perfor-mance*: if S sends an arbitrary point P into P' and R sends P' into P'' the operation "S followed by R" sends P into P''. We call this composite operation the "product" of R and S, and denote it by RS where the order of the operations is from right to left (i.e. the standard convention employed throughout this book). If we can find an operation T which leads to exactly the same final configuration as does S followed by R we regard RS and T as completely equivalent and write

$$\mathsf{RS} = \mathsf{T}.$$

In the case of the cube, for example, $\mathsf{C}_4{}^{z}\mathsf{C}_4{}^{z}$ (which we abbreviate to $(\mathsf{C}_4{}^{z})^2$) has the same effect as a single operation $\mathsf{C}_2{}^{z}$ and we write

$$(\mathsf{C}_4{}^{z})^2 = \mathsf{C}_2{}^{z}.$$

For every symmetry operation, R say, we can find an *inverse*, generally denoted by R^{-1}, such that combination of the two operations is equivalent to doing nothing: we therefore need to include an *identity* operation ("leave the cube alone"), generally denoted by E. With this convention

$$\mathsf{RR}^{-1} = \mathsf{E}.$$

In the C notation for the symmetry operations of the cube, it is convenient to indicate the inverse of a given operation by adding a bar: thus

$$\overline{\mathsf{C}}_4{}^{z} = \text{negative rotation through } \pi/4 \text{ about the } z\text{-axis}$$

[‡] A detailed account of the notation, as used in molecular spectroscopy, is given by Herzberg (1945).

and since this is equivalent in final effect to a positive rotation through $3\pi/4$ or three successive rotations through $\pi/4$ we can write

$$(\mathsf{C}_4{}^z)^3 = \overline{\mathsf{C}}_4{}^z.$$

With this terminology, the full set of (rotational) symmetry operations for a cube is easily found. The results are collected in Table 4.1. There are therefore twenty-four operations, of five geometrically distinct types. The set is complete in the sense that (associative) combination of any of its members yields a member already listed: it contains the identity E, and for every operation R there is an inverse R^{-1} (e.g. $\overline{\mathsf{C}}_3{}^{xyz}$ is the inverse of $\mathsf{C}_3{}^{xyz}$, while $\mathsf{C}_2{}^x$ is *its own* inverse, for $\mathsf{C}_2{}^x\overline{\mathsf{C}}_2{}^x = \mathsf{E}$).

TABLE 4.1

Symmetry operations of the Group $\mathcal{O}$

E							
$\mathsf{C}_2{}^x$	$\mathsf{C}_2{}^y$	$\mathsf{C}_2{}^z$					
$\mathsf{C}_3{}^{xyz}$	$\mathsf{C}_3{}^{x\bar{y}\bar{z}}$	$\mathsf{C}_3{}^{\bar{x}y\bar{z}}$	$\mathsf{C}_3{}^{\bar{x}\bar{y}z}$	$\overline{\mathsf{C}}_3{}^{xyz}$	$\overline{\mathsf{C}}_3{}^{x\bar{y}\bar{z}}$	$\overline{\mathsf{C}}_3{}^{\bar{x}y\bar{z}}$	$\overline{\mathsf{C}}_3{}^{\bar{x}\bar{y}z}$
$\mathsf{C}_2{}^{xy}$	$\mathsf{C}_2{}^{x\bar{y}}$	$\mathsf{C}_2{}^{yz}$	$\mathsf{C}_2{}^{y\bar{z}}$	$\mathsf{C}_2{}^{zx}$	$\mathsf{C}_2{}^{z\bar{x}}$		
$\mathsf{C}_4{}^x$	$\overline{\mathsf{C}}_4{}^x$	$\mathsf{C}_4{}^y$	$\overline{\mathsf{C}}_4{}^y$	$\mathsf{C}_4{}^z$	$\overline{\mathsf{C}}_4{}^z$		

Note. Each row contains a different *class* of geometrically similar operations (rotations through a given angle about a given type of symmetry axis). The first three rows comprise a *subgroup*, the group $\mathcal{T}$ of the regular tetrahedron.

A set of entities, $\mathcal{G}$ say, with the above properties is called a *group*, the members of the set being *group elements*. The properties are usually summarized formally in the following way:

(i) There is an associative law of combination, such that any two elements of the group may be combined to yield a third.

(ii) The group contains an *identity* element (E) which combines with any element to leave it unchanged ($\mathsf{ER} = \mathsf{R}$).

(iii) For every element R, the group contains an *inverse* R^{-1} such that $\mathsf{RR}^{-1} = \mathsf{E}$.

The number of elements is the *order* of the group. The symmetry group just discussed is a *finite* group of order 24: it is also called a "point group" since it comprises operations which leave one point of a symmetrical figure fixed in space, namely the cube centre. It will be referred to as "the octahedral group $\mathcal{O}$". We must now list one or two basic definitions and properties relating to finite groups.

Subgroups

It is possible that a restricted set of elements within a group may themselves possess the properties which define a group; such a set is a "subgroup". Thus, the elements in the first three lines of Table 4.1 are found to form a subgroup of $\mathcal{O}$; combination of these operations always leads to an operation included in the first three lines only, and the subgroup so defined is the symmetry group $\mathcal{T}$ of the tetrahedron indicated by the broken lines in Fig. 4.1b. We say that $\mathcal{T}$ is contained in $\mathcal{O}$, and that the tetrahedron has a "lower" symmetry than the cube.

Generators

A group is in general completely determined by stating the properties of a relatively small number of elements which, by successive combination among themselves, yield all the elements of the group: such elements constitute a set of *generators*. Thus, the group $\mathcal{O}$ is completely determined by *two* generators, which may conveniently be taken as the elements C_4^z and C_3^{xyz}. By suitably combining the two elements we may generate any of the 24 elements of the group: for example,

$$C_2^x = C_3^{xyz}(C_4^z)^2(C_3^{xyz})^2.$$

The same is true of the subgroup $\mathcal{T}$ if we replace the generators C_4^z and C_3^{xyz} by C_2^z and C_3^{xyz}.

Maximum economy and simplicity can be achieved by giving special attention to the generators. In Table 4.2 the structures of both groups are exhibited in terms of their generators.

Classes

Each line in Table 4.1 contains a certain *class* of elements of the group. It is readily verified that any two elements A, A′ which belong to the same class are related by

$$A' = RAR^{-1} \tag{4.1}$$

where R is some element of the group: such elements are said to be "conjugate with respect to R", and a class is in fact defined as a set of elements which are all conjugate to each other (with respect to some element of the group). To illustrate the geometrical interpretation we note that $C_3^{xyz}C_2^z\bar{C}_3^{xyz} = C_2^x$, which is an operation "like" C_2^z except that the *axis* of rotation has itself been subjected to the symmetry operation C_3^{xyz}—a rotation which sends the z-axis over into the place of the x-axis. Elements in the same class, in this context, are therefore symmetry operations "of the same kind". For example, if C_n is a rotation through $(2\pi/n)$ and $C_n' = RC_nR^{-1}$ then

$$C_n'^{\,n} = (RC_nR^{-1})(RC_nR^{-1}) \ldots (n \text{ factors}) = C_n{}^n = E.$$

Hence C_n' is also a rotation through $(2\pi/n)$.

TABLE 4.2

Structure of the groups $\mathcal{T}$ and $\mathcal{O}$ in terms of generators

E							
A	B	C					
S	ASA	BSB	CSC	$\bar{S}$	$A\bar{S}A$	$B\bar{S}B$	$C\bar{S}C$
RA	RB	RS	RSCS	$RB\bar{S}B$	$RA\bar{S}A$		
R	R^3	RASA	RBSB	$RC\bar{S}C$	$R\bar{S}$		

Note. The generators of $\mathcal{O}$ are taken as $R = C_4{}^z$, $S = C_3{}^{xyz}$. Those of $\mathcal{T}$ are thus $R^2\,(=C_2{}^z)$ and S. In the table we have introduced, for convenience,

$$E = R^4, \qquad \bar{S} = S^{-1} = S^2,$$
$$A = C_2{}^x = SR^2\bar{S}, \quad B = C_2{}^y = \bar{S}R^2S, \quad C = C_2{}^z = R^2.$$

The elements in the table are displayed in the same order as in Table 4.1: every one can be written in terms of R and Δ. The first three rows comprise the subgroup $\mathcal{T}$.

4.2. Group representations

The general formulation of quantum mechanics depends heavily on geometrical ideas relating to vector spaces.[‡] In dealing with symmetry groups (and indeed groups in general) the same language is again appropriate; for a point group operation which sends every point P of ordinary 3-space into an *image* P' is simply a rotation[§] sending a position vector r (point P) into position vector r' (point P'). If the symmetry operation is R we write

$$r' = Rr, \qquad (4.2)$$

noting that R plays the part of an *operator* in vector space. On introduc-

---

[‡] A more leisurely development of the vector space concepts used in this section may be found in Chap. 3 of Vol. 1.

[§] Normally we use the term to include also "improper" rotations (e.g. reflection or inversion).

E

ing a row matrix of basis vectors, $\mathbf{e} = (e_1\, e_2\, e_3)$, it appears that the set of images $\mathbf{e}' = (e_1'\, e_2'\, e_3')$ may be written

$$\mathbf{e}' = \mathsf{R}\mathbf{e} = \mathbf{e}\mathbf{R} \tag{4.3}$$

where $\mathbf{R}$ is a square matrix which characterizes the rotation R. It is customary to use a "functional" notation for the matrix associated in this way with the group element R, writing[‡]

$$\mathbf{R} = \mathbf{D}(\mathsf{R}). \tag{4.4}$$

If we write an arbitrary vector r in terms of its Cartesian components $r_1, r_2, r_3\,(=x, y, z)$, collected into a column $\mathbf{r}$, then

$$\mathsf{r} = \mathbf{e}\mathbf{r}. \tag{4.5}$$

It follows at once that a rotation of the whole reference frame, and all vectors defined therein, sends r into a new vector

$$\mathsf{r}' = \mathbf{e}'\mathbf{r} = \mathbf{e}\mathbf{R}\mathbf{r} = \mathbf{e}\mathbf{r}' \tag{4.6}$$

whose components, relative to the unrotated basis, are thus

$$\mathbf{r}' = \mathbf{R}\mathbf{r}. \tag{4.7}$$

It is then clear, by considering either rotation of the basis vectors according to (4.3) or the rotation of a vector with arbitrary components according to (4.7) that a set of *matrices*

$$\mathbf{D} = \{\mathbf{D}(\mathsf{A}), \mathbf{D}(\mathsf{B}), \ldots \mathbf{D}(\mathsf{R}), \ldots\}.$$

may be associated with the group

$$\mathscr{G} = \{\mathsf{A}, \mathsf{B}, \ldots \mathsf{R}, \ldots\}.$$

It is also readily shown that any relationship

$$\mathsf{R}\mathsf{S} = \mathsf{T} \tag{4.8a}$$

among three group elements implies a corresponding relationship among the associated matrices, namely

$$\mathbf{D}(\mathsf{R})\mathbf{D}(\mathsf{S}) = \mathbf{D}(\mathsf{T}). \tag{4.8b}$$

The set of matrices, referred to briefly as D, is then a *matrix* group (with matrix multiplication as the law of combination) which exhibits the

---

‡ $\mathbf{D}(\mathsf{R})$ and $\Gamma(\mathsf{R})$ are the most common notations—D from the German *Darstellung*. Other points concerning notation and interpretation are fully discussed in Chap. 3 of Vol. 1.

same structure as $\mathscr{G}$: it is a *representation* of $\mathscr{G}$ and the association of matrices with the group elements is often indicated in symbols by

$$R \rightarrow \mathbf{D}(R)$$

or in words by "with R is associated the matrix $\mathbf{D}(R)$".

The representation just introduced is said to be "faithful" or "one-to-one", because each matrix is associated only with *one* symmetry operation and (4.8b) implies unambiguously which elements are related by (4.8a). The structure of the group (Table 4.2) may therefore be verified unambiguously by multiplication of the representative matrices. It is always possible, however, to find also "many-to-one" representations in which the same matrix is associated with *several* group elements and in this case the equality (4.8b) does *not* imply that the group elements R, S, T are related by (4.8a). One very trivial case is the "identity representation" in which the number 1 (1×1 matrix) is associated with *every* element: this example emphasizes the fact that the "representation space", in which matrix transformations are associated with the group elements, is not necessarily ordinary 3-space—it is set up for formal purposes (whose importance will soon become clear) and the number of dimensions is unrestricted.

An example

We consider the group $\mathscr{O}$ and obtain (cf. Vol. 1, p, 62) by inspection of Fig. 4.2a

$$C_4{}^z \rightarrow \begin{pmatrix} 0 & -1 & 0 \\ 1 & 0 & 0 \\ 0 & 0 & 1 \end{pmatrix}, \quad C_3{}^{xyz} \rightarrow \begin{pmatrix} 0 & 0 & 1 \\ 1 & 0 & 0 \\ 0 & 1 & 0 \end{pmatrix} \tag{4.9}$$

From (4.8) it follows that the matrix associated with any product of $C_4{}^z$ and $C_3{}^{xyz}$, and hence with any group element, may be obtained by simple matrix multiplication. The set of matrices obtained in this way provides the "T_1 representation of the group $\mathscr{O}$", T being the standard name for a three-dimensional representation; this is shown in Table 4.3. We already have the identity representation, called A_1 in Table 4.3, and note that another unidimensional representation can be obtained by making the association (cf. (4.9))

$$C_4{}^z \rightarrow -1 \qquad C_3{}^{xyz} \rightarrow +1. \tag{4.10}$$

In this way we obtain the A_2 representation in Table 4.3.

From the A_1, A_2 and T_1 representations it is possible to generate others. Let us take the matrices of T_1 and multiply each one by the corresponding element of the A_2 representation (i.e. by the number ± 1): then if we denote the resultant matrices by $\mathbf{D}(R) = D_{A_2}(R)\mathbf{D}_{T_1}(R)$ it is

TABLE 4.3

Irreducible representations of the group $\mathcal{O}$

	E	C_2^x	C_2^y	C_2^z	C_3^{xyz}	$C_3^{x\bar y\bar z}$	$C_3^{\bar x y\bar z}$	$C_3^{\bar x\bar y z}$	$\bar C_3^{xyz}$	$\bar C_3^{x\bar y\bar z}$	$\bar C_3^{\bar x y\bar z}$	$\bar C_3^{\bar x\bar y z}$
A_1, A_2	1	1	1	1	1	1	1	1	1	1	1	1
E	$\begin{bmatrix}1&0\\0&1\end{bmatrix}$	$\begin{bmatrix}1&0\\0&1\end{bmatrix}$	$\begin{bmatrix}1&0\\0&1\end{bmatrix}$	$\begin{bmatrix}1&0\\0&1\end{bmatrix}$	$\begin{bmatrix}\bar c&\bar s\\s&\bar c\end{bmatrix}$	$\begin{bmatrix}\bar c&\bar s\\s&\bar c\end{bmatrix}$	$\begin{bmatrix}\bar c&\bar s\\s&\bar c\end{bmatrix}$	$\begin{bmatrix}\bar c&\bar s\\s&\bar c\end{bmatrix}$	$\begin{bmatrix}\bar c&s\\\bar s&\bar c\end{bmatrix}$	$\begin{bmatrix}\bar c&s\\\bar s&\bar c\end{bmatrix}$	$\begin{bmatrix}\bar c&s\\\bar s&\bar c\end{bmatrix}$	$\begin{bmatrix}\bar c&s\\\bar s&\bar c\end{bmatrix}$
T_1, T_2	$\begin{bmatrix}1&0&0\\0&1&0\\0&0&1\end{bmatrix}$	$\begin{bmatrix}1&0&0\\0&\bar1&0\\0&0&\bar1\end{bmatrix}$	$\begin{bmatrix}\bar1&0&0\\0&1&0\\0&0&\bar1\end{bmatrix}$	$\begin{bmatrix}\bar1&0&0\\0&\bar1&0\\0&0&1\end{bmatrix}$	$\begin{bmatrix}0&0&1\\1&0&0\\0&1&0\end{bmatrix}$	$\begin{bmatrix}0&0&\bar1\\\bar1&0&0\\0&1&0\end{bmatrix}$	$\begin{bmatrix}0&0&1\\\bar1&0&0\\0&\bar1&0\end{bmatrix}$	$\begin{bmatrix}0&0&\bar1\\1&0&0\\0&\bar1&0\end{bmatrix}$	$\begin{bmatrix}0&1&0\\0&0&1\\1&0&0\end{bmatrix}$	$\begin{bmatrix}0&\bar1&0\\0&0&1\\\bar1&0&0\end{bmatrix}$	$\begin{bmatrix}0&\bar1&0\\0&0&\bar1\\1&0&0\end{bmatrix}$	$\begin{bmatrix}0&1&0\\0&0&\bar1\\\bar1&0&0\end{bmatrix}$

	C_2^{xy}	$C_2^{x\bar y}$	C_2^{yz}	$C_2^{y\bar z}$	C_2^{zx}	$C_2^{z\bar x}$	C_4^x	$\bar C_4^x$	C_4^y	$\bar C_4^y$	C_4^z	$\bar C_4^z$
A_1	1	1	1	1	1	1	1	1	1	1	1	1
A_2	-1	-1	-1	-1	-1	-1	-1	-1	-1	-1	-1	-1
E	$\begin{bmatrix}1&0\\0&\bar1\end{bmatrix}$	$\begin{bmatrix}1&0\\0&\bar1\end{bmatrix}$	$\begin{bmatrix}\bar c&\bar s\\\bar s&c\end{bmatrix}$	$\begin{bmatrix}\bar c&\bar s\\\bar s&c\end{bmatrix}$	$\begin{bmatrix}\bar c&s\\s&c\end{bmatrix}$	$\begin{bmatrix}\bar c&s\\s&c\end{bmatrix}$	$\begin{bmatrix}\bar c&\bar s\\\bar s&c\end{bmatrix}$	$\begin{bmatrix}\bar c&\bar s\\\bar s&c\end{bmatrix}$	$\begin{bmatrix}\bar c&s\\s&c\end{bmatrix}$	$\begin{bmatrix}\bar c&s\\s&c\end{bmatrix}$	$\begin{bmatrix}1&0\\0&\bar1\end{bmatrix}$	$\begin{bmatrix}1&0\\0&\bar1\end{bmatrix}$
T_1	$\begin{bmatrix}0&1&0\\1&0&0\\0&0&\bar1\end{bmatrix}$	$\begin{bmatrix}0&\bar1&0\\\bar1&0&0\\0&0&\bar1\end{bmatrix}$	$\begin{bmatrix}\bar1&0&0\\0&0&1\\0&1&0\end{bmatrix}$	$\begin{bmatrix}\bar1&0&0\\0&0&\bar1\\0&\bar1&0\end{bmatrix}$	$\begin{bmatrix}0&0&1\\0&\bar1&0\\1&0&0\end{bmatrix}$	$\begin{bmatrix}0&0&\bar1\\0&\bar1&0\\\bar1&0&0\end{bmatrix}$	$\begin{bmatrix}1&0&0\\0&0&\bar1\\0&1&0\end{bmatrix}$	$\begin{bmatrix}1&0&0\\0&0&1\\0&\bar1&0\end{bmatrix}$	$\begin{bmatrix}0&0&1\\0&1&0\\\bar1&0&0\end{bmatrix}$	$\begin{bmatrix}0&0&\bar1\\0&1&0\\1&0&0\end{bmatrix}$	$\begin{bmatrix}0&\bar1&0\\1&0&0\\0&0&1\end{bmatrix}$	$\begin{bmatrix}0&1&0\\\bar1&0&0\\0&0&1\end{bmatrix}$
T_2	$\begin{bmatrix}0&\bar1&0\\\bar1&0&0\\0&0&1\end{bmatrix}$	$\begin{bmatrix}0&1&0\\1&0&0\\0&0&1\end{bmatrix}$	$\begin{bmatrix}1&0&0\\0&0&\bar1\\0&\bar1&0\end{bmatrix}$	$\begin{bmatrix}1&0&0\\0&0&1\\0&1&0\end{bmatrix}$	$\begin{bmatrix}0&0&\bar1\\0&1&0\\\bar1&0&0\end{bmatrix}$	$\begin{bmatrix}0&0&1\\0&1&0\\1&0&0\end{bmatrix}$	$\begin{bmatrix}\bar1&0&0\\0&0&1\\0&\bar1&0\end{bmatrix}$	$\begin{bmatrix}\bar1&0&0\\0&0&\bar1\\0&1&0\end{bmatrix}$	$\begin{bmatrix}0&0&\bar1\\0&\bar1&0\\1&0&0\end{bmatrix}$	$\begin{bmatrix}0&0&1\\0&\bar1&0\\\bar1&0&0\end{bmatrix}$	$\begin{bmatrix}0&1&0\\\bar1&0&0\\0&0&\bar1\end{bmatrix}$	$\begin{bmatrix}0&\bar1&0\\1&0&0\\0&0&\bar1\end{bmatrix}$

Note. The upper part of the table refers to the subgroup $\mathcal{T}$, in which the A_1, A_2 representations coincide (and similarly for T_1, T_2); the distinction, by a sign difference, arises for the elements in the second part of the table. Negative signs have been replaced by bars, to make the table compact, and $s = \sin(2\pi/6) = \frac{1}{2}\sqrt 3$, $c = \cos(2\pi/6) = \frac{1}{2}$.

at once clear that $RS = T$ implies $D(R)D(S) = D(T)$. The representation
which arises in this way is a special case of the "direct product of two
representations"; it is the "T_2 representation" of the group $\mathcal{O}$.

The last representation we shall construct is the so-called "E
representation". To this end, we look for a simple *two*-dimensional
representation of the subgroup consisting of E, $C_3{}^{xyz}$ and $\overline{C}_3{}^{xyz}$, choosing
another coordinate system in which (Fig. 4.2b) the new z-axis ($\bar{e}_3$ say)
is along the three-fold xyz-axis and the new x-axis bisects a cube edge.

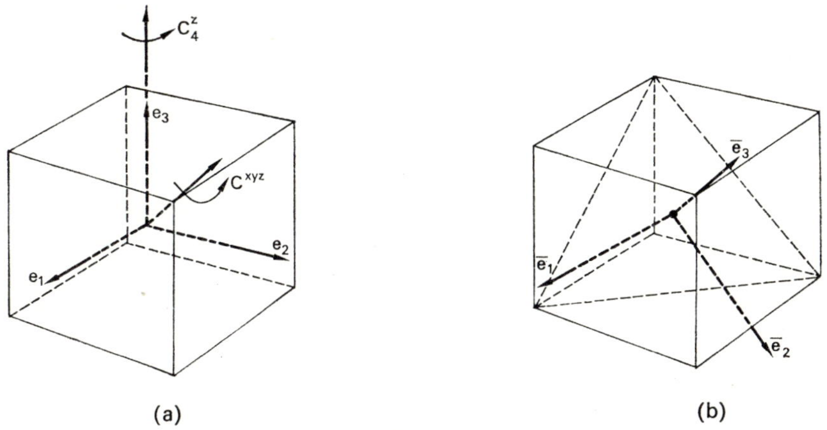

FIG. 4.2. Alternative bases for representations of the group $\mathcal{O}$. (a) Unit
vectors normal to cube faces (defining x-, y-, z-axes). (b) Alternative choice
with body diagonal as new z-axis ($\bar{e}_3$) and $\bar{e}_1$, $\bar{e}_2$ chosen perpendicular and
parallel to a side of the inscribed triangle (broken lines).

Rotations about $\bar{e}_3$ then affect only $\bar{e}_2$ and $\bar{e}_1$ and the matrix describing
$C_3{}^{xyz}$ according to (4.3) takes the form

$$D(C_3{}^{xyz}) = \begin{pmatrix} -\frac{1}{2} & -\frac{1}{2}\sqrt{3} \\ \frac{1}{2}\sqrt{3} & -\frac{1}{2} \end{pmatrix}. \tag{4.11}$$

This matrix properly reflects the properties of rotations about the
three-fold axis—or *any* three-fold axis: we adopt this association of a
matrix with $C_3{}^{xyz}$, and now abandon the geometrical picture, simply
looking for a suitable 2×2 matrix to associate with the other generator
$C_4{}^z$, so as to satisfy (4.8 a, b). Is it possible to choose this matrix so that
every subgroup of rotations about a three-fold axis will have the same
set of matrices associated with it? For this to be so, we should have to
choose $D(C_4{}^z)$ so that, for example,

$$D(C_3{}^{\overline{xyz}}) = D(C_2{}^z)D(C_3{}^{xyz})D(C_2{}^z) = D(C_3{}^{xyz}).$$

This is satisfied (i) by associating the *unit* matrix with $C_4{}^z$, or (ii) by using

$$D(C_4{}^z) = \begin{pmatrix} 1 & 0 \\ 0 & -1 \end{pmatrix}. \tag{4.12}$$

It is readily verified that the second choice conforms to (4.8) for all elements of the group, the first choice failing because it would imply $D(C_2{}^{yz}) = D(C_4{}^z)D(C_3{}^{xyz}) = D(C_3{}^{xyz})$, which leads to incompatibilities. By making the associations (4.11) and (4.12) we obtain the E representation shown in Table 4.3.

The conclusions reached may be summarized as follows: we have associated with a certain group of symmetry operations five sets of matrices which exhibit similar properties in the sense of (4.8b). Each set of matrices may be regarded as defining a set of rotations in a vector space, a "representation space", and we have obtained representations in spaces of one, two and three dimensions. Although geometrical ideas were used in setting up these particular representations, a representation space may in general be a pure abstraction introduced merely so that the matrices may be regarded as describing rotations of basis vector sets. The interesting fact is that we shall find entities of various kinds (e.g. wave functions, operators) which behave under rotations of a physical system (e.g. a molecule) exactly like these basis vectors, and in this way provide "ready-made" representation spaces.

4.3.   Equivalence and irreducibility

Any representation (in particular those in Table 4.3) is clearly arbitrary in the sense that any two independent sets of basis vectors will serve equally well to describe a set of rotations in a given n-dimensional vector space. The actual form of the matrix associated with a rotation R then depends on the basis (see Vol. 1, Section 3.6) and if two bases are related by

$$\bar{e} = eU \tag{4.13}$$

where (assuming each basis orthonormal) U is a unitary matrix, then the matrices associated with R in the two cases are related by

$$\bar{R} = U^\dagger R U \tag{4.14}$$

or, using the notation (4.4) of representation theory,

$$\bar{D}(R) = U^\dagger D(R) U. \tag{4.15}$$

It is of course clear, from the property $U^\dagger U = U U^\dagger = 1$, that when

RS = T and $D(R)D(S) = D(T)$ then $\overline{D}(R)\overline{D}(S) = \overline{D}(T)$—so that $\overline{D}$ is indeed a representation. Two representations, D and $\overline{D}$, which are connected in this way are not regarded as distinct for group theoretical purposes; they are referred to as *equivalent representations*. Equivalence is usually indicated by an equality sign between the representation symbols. Thus

$$\overline{D} = D \tag{4.16}$$

means simply that the *matrices* of the two representations are related by (4.15). When such a relationship does not exist, the representations are *inequivalent*. The representations in Table 4.3 are all inequivalent.

The representations in Table 4.3 illustrate a second important property; that of "irreducibility". The meaning of this term is illustrated by the matrices associated with $C_3{}^{xyz}$ in the representations based on e_1, e_2, e_3 and $\overline{e}_1$, $\overline{e}_2$, $\overline{e}_3$, respectively, in Fig. 4.2. The matrix describing the rotation of $\overline{e}_1$ and $\overline{e}_2$ appears in (4.11), while $\overline{e}_3$ is left unchanged by the rotation. Consequently, abbreviating sine and cosine of the rotation angle by s and c,

$$C_3{}^{xyz}(e_1\ e_2\ e_3) = (e_1\ e_2\ e_3)\begin{pmatrix} c & \overline{s} & 0 \\ s & c & 0 \\ 0 & 0 & 1 \end{pmatrix}$$

and the matrix on the right is $\overline{D}(C_3{}^{xyz})$. The corresponding matrix in the representation D, based on e_1, e_2, e_3, appears in (4.9). Thus

$$D(C_3{}^{xyz}) = \begin{pmatrix} 0 & 0 & 1 \\ 1 & 0 & 0 \\ 0 & 1 & 0 \end{pmatrix} \qquad \overline{D}(C_3{}^{xyz}) = \left(\begin{array}{cc|c} c & \overline{s} & 0 \\ s & c & 0 \\ \hline 0 & 0 & 1 \end{array}\right) \tag{4.17}$$

The partitioning, in the second case, emphasizes that $\overline{D}(C_3{}^{xyz})$ takes "block form", the only non-zero blocks lying along the diagonal. If attention is confined to the group E, $C_3{}^{xyz}$, $\overline{C}_3{}^{xyz}$ (i.e. the subgroup of O obtained from the single generator $C_3{}^{xyz}$), all three matrices of the corresponding representation take the same block form. It is of course easy to verify, expressing the barred basis vectors in Fig. 4.2b in terms of the unbarred to obtain the matrix U, that the matrices in (4.17) are related by the unitary transformation (4.15)—as must be the matrices associated with *any* of the three group elements.

The property of *reducibility* may now be defined. When, by means of a basis change (4.13) and the corresponding unitary transformation

(4.15), all the matrices $\mathbf{D}(R)$ of a representation D can be brought to a common block form

$$\mathbf{U}^{\dagger}\mathbf{D}(R)\mathbf{U} = \left[\begin{array}{c|c} \mathbf{D}_1(R) & \mathbf{0} \\ \hline \mathbf{0} & \mathbf{D}_2(R) \end{array}\right] = \overline{\mathbf{D}}(R) \tag{4.18}$$

the representation $\mathbf{D}(R)$ is said to be *reducible*.[‡] It is at once clear that if D is a representation, then so are the collections of matrices $\mathbf{D}_1(R)$ and $\mathbf{D}_2(R)$ respectively; in this sense the original representation D is equivalent to a pair of representations D_1 and D_2.

Just as a representation of n dimensions can be *reduced* to two representations D_1 and D_2, of n_1 and n_2 dimensions respectively $(n_1 + n_2 = n)$, so it is possible to *build up* a representation from two given representations simply by constructing the "block-form" matrices of (4.18). We then write, using the equality to imply equivalence and introducing a purely formal $+$ between the representation symbols[§]

$$D = D_1 + D_2. \tag{4.19}$$

The representation D is said to be the *direct sum* of D_1 and D_2, and the shorthand statement (4.19) simply indicates that the matrices of the three representations can be connected according to (4.18) by choosing $\mathbf{U}$ suitably. Direct summation is simply the converse of reduction.

The process of reduction may clearly be continued, though not indefinitely, by looking for further unitary transformations which will express D_1 and D_2 in terms of smaller blocks. When no further reduction can be achieved the matrices of the reduced representation all have the common form

$$\mathbf{U}^{\dagger}\mathbf{D}(R)\mathbf{U} = \begin{bmatrix} \mathbf{D}_1(R) & & & \\ & \mathbf{D}_2(R) & & \\ & & \mathbf{D}_3(R) & \\ & & & \ddots \end{bmatrix} \tag{4.20}$$

[‡] Strictly the term "*completely* reducible" should be used, "reducible" implying that only one off-diagonal block need vanish. For all the groups with which we deal, reducibility implies complete reducibility and the terms may be used interchangeably.

[§] Other symbols, such as $\dotplus$ and $\oplus$, are also frequently used for this purpose.

in which non-zero elements appear only in the diagonal blocks. The representations D_1, D_2, ... are then *irreducible representations*. If two or more of them are equivalent, we know that their matrices may be made *identical* by suitably choosing the unitary transformation and we shall usually assume this done. If D_α occurs n_α times we then write

$$D = n_1 D_1 + n_2 D_2 + \ldots = \sum_\alpha n_\alpha D_\alpha \qquad (4.21)$$

where the interpretation of the $=$ and $+$ signs is exactly as in (4.19) and, for example, $2D_1$ simply means $D_1 + D_1$.

The representations in Table 4.3 are all both irreducible and inequivalent; there are five of them and no other inequivalent representations can be found. This limitation arises from certain fundamental properties of irreducible representations, on which nearly all applications in quantum mechanics rest.

4.4. Fundamental properties

We introduce the basic properties with reference to the octahedral group, using the unitary representations already constructed, and then state them generally.

The set of g ($=24$) numbers in each of the first two rows of Table 4.3 may be regarded as the components of a vector in a g-dimensional space. We introduce this space for purely formal reasons, assuming the components refer to an orthonormal basis. The two vectors set up in this way ($\mathbf{u}$ and $\mathbf{v}$, say) are seen to be orthogonal in the sense

$$\sum_i u_i v_i = 0.$$

Now let us take corresponding elements from all g matrices of any representation (D) in Table 4.3 to get the vector with components

$$D(\mathsf{E})_{ij} \quad D(\mathsf{C}_2{}^x)_{ij} \quad D(\mathsf{C}_2{}^y)_{ij} \ldots D(\mathsf{C}_4{}^y)_{ij}.$$

We can obtain one such vector for a one-dimensional representation, four for a two-dimensional and n^2 for an n-dimensional—to obtain, altogether, for the group $\mathcal{O}$, twenty-four different vectors in a twenty-four-dimensional space (since $g = 24$). The remarkable fact is that *all* these vectors are mutually orthogonal. This statement may be translated into symbols, and generalized so as to admit complex components, by introducing the Hermitian scalar product

$$\langle \mathsf{u} | \mathsf{v} \rangle = \sum_i u_i{}^* v_i$$

E*

and writing the orthogonality condition

$$\sum_R D_\alpha(R)_{ij}{}^* D_\beta(R)_{kl} = 0 \quad (\text{unless } \alpha = \beta, \ i = j \text{ and } k = l)$$

in which D_α and D_β are any two irreducible representations.

On turning to the scalar products of each vector with itself, we find similarly that the square of the length of the vector is the same for all vectors formed from a given representation; for a representation D_α, of dimension g_α, it turns out to be g/g_α. Thus the vectors formed from the A_1 and A_2 representations have square modulus 24, while for those from E it is 12, and for those from T_1 and T_2 it is 8.

The orthogonality relations

All the above properties are summarized by writing

$$\sum_R D_\alpha(R)_{ij}{}^* D_\beta(R)_{kl} = \left(\frac{g}{g_\alpha}\right) \delta_{\alpha\beta} \, \delta_{ik} \, \delta_{jl}. \tag{4.22}$$

This fundamental equation expresses the *orthogonality relations* for the (unitary) irreducible representations of *any* finite group. It must be noted that $\alpha \neq \beta$ means the representations are *inequivalent*, and that in case of equivalence ($\alpha = \beta$) *the representations are taken to be identical.*[‡]

The orthogonality relations (4.22), whose proof is given in thestandard books on group theory, are very widely used in quantum mechanics. Sometimes, however, the explicit representations may not be available and then a weaker form of the result may be used: this involves only the "characters ' of the elements.

Group characters

The character of an element R in representation D is simply the trace of its representative matrix

$$\chi(R) = \text{tr} \, \mathbf{D}(R). \tag{4.23}$$

The "character system" in the given representation is the full set of characters as R runs through the group; and this characterizes the representation in a manner independent of choice of basis—for the trace of a matrix is invariant under a unitary transformation. Character tables, showing the character systems for all irreducible representations, are available in many textbooks.

[‡] If the representations are merely equivalent, the result does not apply.

It is readily shown, from (4.22), that the character systems exhibit similar orthogonality properties

$$\sum_{R} \chi_\alpha(R)^* \chi_\beta(R) = g\, \delta_{\alpha\beta} \qquad (4.24)$$

for any two irreducible representations D_α and D_β. An important consequence of orthogonality is that identity of character systems is a test (necessary and sufficient) for the equivalence of two representations. Thus, it may be verified at a glance that all the irreducible representations in Table 4.3 are inequivalent.

Reduction of representations

In considering the reduction of representations, it is often necessary to know which irreducible representations will appear. Reference to (4.20) shows that (cf. (4.21)), using $\chi(R)$ for the characters in the given representation,

$$\chi(R) = n_1 \chi_1(R) + n_2 \chi_2(R) + \ldots \sum_{\alpha} n_\alpha \chi_\alpha(R) \qquad (4.25)$$

and (4.24) then leads to a formula for the number of times each irreducible representation will be found, namely

$$n_\alpha = \frac{1}{g} \sum_{R} \chi_\alpha(R)^* \chi(R). \qquad (4.26)$$

As a further consequence, we obtain a test for irreducibility of a representation: a representation is irreducible if its character system satisfies

$$\sum_{R} \chi(R)^* \chi(R) = g. \qquad (4.27)$$

Thus, if say $\chi = \chi_\alpha$, use of (4.24) in (4.26) shows that $n_\alpha = 1$, $n_\beta = 0$ $(\beta \neq \alpha)$.

Naturally, in setting up irreducible representations, it is necessary to know when to stop: for, when the group is finite, the number of inequivalent irreducible representations that can be found is also finite. The required criteria are as follows:

> The sum of the squares of the dimensions of the distinct irreducible representations of a finite group is equal to the order of the group. $\qquad (4.28)$

—a result which, for example, indicates the possible dimensions of any missing representations—and secondly,

The number of distinct irreducible representations of a finite group is equal to the number of classes.

(4.29)

which fixes the total number of irreducible representations obtainable in terms only of the *structure* of the group.

It is easy to confirm, using these results, that Table 4.3 contains all distinct irreducible representations of the group $\mathcal{O}$.

4.5.　Symmetry properties of operators and functions

Let us return to the quantum-mechanical problem stated in Section 4.1: we wish to determine the stationary states of an electron in the field produced by an array of charges whose symmetry is, we shall suppose,[‡] described by the group $\mathcal{O}$. So far we have used an *active* interpretation of the symmetry operations, in which the system of nuclei is actually moved by each operation; sometimes a "passive" description is used in which the system is left unchanged but the observer (or the reference frame which he defines) is rotated in the opposite direction. We shall continue to use the active interpretation, but at the same time must distinguish carefully the physical meaning of terms such as "rotated system", "rotated wave functions" and "rotated operators".

Rotation of operators. Invariants

First we note that the array of nuclei, which provides the potential field for the electronic Hamiltonian, may be defined by giving the positions of the nuclei in some reference frame.

It is convenient to adopt exactly the same convention in defining rotated wave functions and operators: for instance, if L_x is an angular momentum operator the "rotated" angular momentum operator $\mathsf{L}_x{}'$ is the operator "constructed like L_x" but in the rotated reference frame— it is thus associated with angular momentum about the x'-axis. It must be stressed that this is a convention—we can rotate a framework of nuclei, since this is a physical object, but we cannot rotate a mathe-

---

‡ Actually, the symmetry is somewhat higher than we assume here, because reflection operations are also admissible. To keep the group small we confine attention to the rotations.

matical quantity except by setting up some suitable rule. The convention adopted is a natural one because the orientation of a physical system in space is immaterial and therefore quantities defined for the rotated system must satisfy equations of exactly the same form as before rotation. In particular, the Schrödinger equation[‡]

$$H\Psi = E\Psi$$

for any system will, after rotation, be replaced by

$$H'\Psi' = E\Psi' \tag{4.30}$$

where H', Ψ' are set up in the rotated form by recipes exactly similar to those defining H and Ψ. The quantity E is of course simply a numerical parameter making no reference to orientation of the system; the eigenvalues of E are numbers to be determined by solving the equation.

It is customary to regard Ψ' as the result of applying a rotation operator R to Ψ and to write[§]

$$\Psi' = R\Psi. \tag{4.31a}$$

It is then clear that R must be a *unitary* operator (see, for example, Vol. 1, p. 114) since preservation of normalization requires that $\langle\Psi'|\Psi'\rangle = \langle R\Psi|R\Psi\rangle = \langle\Psi|R^\dagger R\Psi\rangle = 1$ (any Ψ) and this means the adjoint of the operator coincides with its inverse

$$R^\dagger R = I, \qquad R^\dagger = R^{-1}.$$

It is also possible to express the effect of rotation upon an arbitrary operator A formally in terms of R. Thus, subjecting all quantities to a common rotation must leave any matrix element invariant:

$$\langle\Psi'|A'|\Psi'\rangle = \langle R\Psi|A'|R\Psi\rangle = \langle\Psi|R^\dagger A'R|\Psi\rangle = \langle\Psi|A|\Psi\rangle$$

(all Ψ). This requires $R^\dagger A'R = A$ or, multiplying from left and right by R and $R^\dagger$ respectively,

$$A' = RAR^\dagger. \tag{4.31b}$$

Equations such as (4.31a, b) are essentially formal: examples of the actual construction of rotated functions and operators occur later.

The notion of symmetry in quantum mechanics arises when we compare the Hamiltonian for the rotated system with that for the original system: to do this, we must of course refer both operators to the same reference frame, which we shall take to be the original or

---

[‡] Although the discussion and examples are based on one-electron systems, the conclusions are generally applicable: H and Ψ may refer to a many-electron system.

[§] When no misunderstanding is likely, we use the same symbol R for the rotation in real space and for the corresponding rotation in the function space to which Ψ belongs.

"laboratory fixed" frame. Thus, the potential energy term in H for the system considered so far (Fig. 4.1) would be, using atomic units,

$$V = \frac{-Z_0}{r_0} - Z\left(\frac{1}{r_1} + \frac{1}{r_2} + \dots \frac{1}{r_6}\right)$$

where Z_0 is the central charge (nucleus at the origin) while Z is an effective charge on each of the ligands. Here r_1, for example, is the distance from nucleus 1 to the "field point" to which the electronic coordinates refer. In the rotated system the nuclei may occupy new positions $1', 2', \dots$ and the corresponding potential term is

$$V' = -\frac{Z_0}{r_0} - Z\left(\frac{1}{r_1'} + \frac{1}{r_2'} + \dots \frac{1}{r_6'}\right)$$

where r_1' is the distance of the field point from nucleus 1 in its new position.

We note that, in general, V and V' are *different functions* of the electronic coordinates x, y, z. In a *symmetry operation*, however, the points $1', 2', \dots$ come into coincidence with $1, 2, \dots$ (merely occurring in some different order)[‡] and V' is exactly the same as V apart from a trivial rearrangement of terms in the parentheses. This is a general result: a symmetry operator leaves the form of the potential energy function invariant. In the same way, the Laplacian ∇^2 is replaced after rotation by ∇'^2 which differs only in the axial directions along which the derivatives are taken; and it is well known that rotation of the axes leaves the Laplacian invariant. These observations suggest a somewhat wider definition of a symmetry operation: in the context of the Schrödinger equation a symmetry operation is any operation or "transformation" *under which the Hamiltonian operator is invariant*. In symbols, we require from (4.31b)

$$H' = RHR^\dagger = H. \tag{4.32}$$

It is thus clear, for example, that any *permutation* of the electrons in a many-electron system will also be a symmetry operation. For, since the particles are indistinguishable, their space and spin variables must all enter the Hamiltonian in exactly the same way and a permutation can produce only a trivial rearrangement of the terms in H.

Vector and tensor operators

Many operators, of course, will *not* be invariant under a symmetry operation characterized by (4.32). For instance, the position vector of

---

‡ For example (Fig. 4.1), the rotation C_2^z gives $1 \to 1' = 2$ (image of "1" coincides with original 2), etc., or in full $(1, 2, 3, 4, 5, 6) \to (2, 1, 4, 3, 5, 6)$.

a field point (r) defines the electronic variables x, y, z, after choosing the unit vectors (e_1, e_2, e_3) that specify the orientation of the system: if the system is rotated, the axes now being specified by e_1', e_2', e_3' with

$$\mathbf{e}' = \mathbf{eR}, \tag{4.33}$$

then the position variables are replaced by (cf. Section 3.6 of Vol. 1)

$$\mathbf{r}' = \mathbf{R}^{-1}\mathbf{r}. \tag{4.34}$$

For a Cartesian reference frame $\mathbf{R}$ is a real orthogonal matrix and $\mathbf{R}^{-1} = \tilde{\mathbf{R}}$ (transpose of $\mathbf{R}$): thus, by transposition of both sides of the last equation, $\tilde{\mathbf{r}}' = \tilde{\mathbf{r}}\mathbf{R}$ (since double transposition gives the original matrix) or, in full,

$$(x' \, y' \, z') = (x \, y \, z) \begin{bmatrix} R_{11} & R_{12} & R_{13} \\ R_{21} & R_{22} & R_{23} \\ R_{31} & R_{32} & R_{33} \end{bmatrix}. \tag{4.35}$$

Thus, *for a Cartesian reference frame*[‡] the coordinates of a field point follow the same transformation law (4.33) as the unit vectors. The operators x, y, z (i.e. multiplication by the coordinates) are replaced by x', y', z', given by (4.35) when the system is rotated; for example, the electric moment along the axis of a diatomic molecule (z, say) is essentially an expectation value of z; while after rotation it is an expectation value of z'. Sets of three operators transforming in this way are referred to as components of a *vector operator*. The transformation law

$$(V_x' \, V_y' \, V_z') = (V_x V_y V_z)\mathbf{R} \tag{4.36}$$

may in fact be taken as the definitive property of a vector. The momentum components (p_x, p_y, p_z) follow the same law, and so also must the associated operators (p_x, p_y, p_z).[§]

Here it must be noted that the angular momentum operators $\mathsf{L}_x = y\mathsf{p}_z - z\mathsf{p}_y$, etc., are not strictly vector operators because they depend on *two* vectors and are therefore components of a *second-rank tensor*.[‖] However, they fortuitously follow the transformation law (4.36) so long as $\mathbf{R}$ is a real orthogonal matrix with det $\mathbf{R} = +1$: this means

---

‡ The argument is valid only for a Cartesian system, where there is no distinction between covariant and contravariant tensor components. For more detailed discussions of vectors, tensors, pseudovectors, etc., see other volumes in this series—Hall (1963) or McWeeny (1963).

§ This may be verified directly in the Schrödinger language by considering $\mathsf{p}_x = (\hbar/i) \, \partial/\partial x$, etc., and transforming the derivatives to the rotated axes.

‖ See, for example, Hall (1963), McWeeny (1963).

that for proper rotations L_x, L_y, L_z behave like vector components but for "improper" rotations (which incorporate a reflection operation, and have det $\mathbf{R} = -1$) they behave like vector components apart from an additional change of sign. The angular momentum is therefore a pseudo-vector or "axial vector"; its behaviour is often represented pictorially (Fig. 4.3) by means of vectors with a "sense" indicated by curved arrows. In considering only proper rotations, as in the group $\mathcal{O}$,

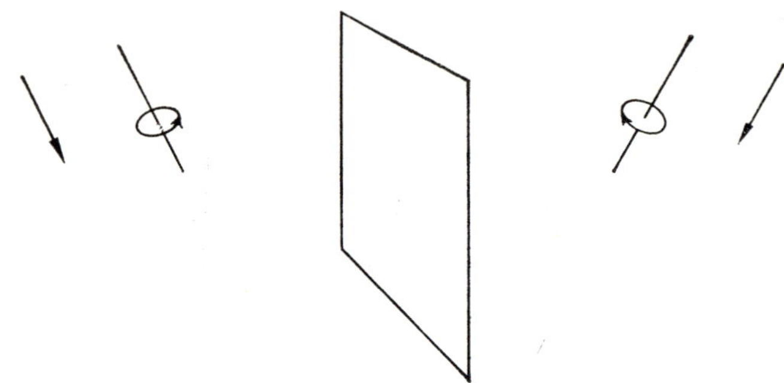

Fig. 4.3. Pictorial representation of a pseudovector. The pseudovector represented by the curved arrow (left) has an image under reflection (right) in which the sense of rotation, corresponding to corkscrew motion in the upward direction, has been reversed. If the pseudovector is represented by a straight arrow (axial vector), the result of any reflection (improper rotation) must be "corrected" by adding a further factor of -1. The axial vector behaves like a true (polar) vector only under proper rotations.

it is not necessary to distinguish between vectors and pseudo-vectors. Operators with more complicated transformation properties may be dealt with using similar principles.

Transformation properties of functions

Even when the Hamiltonian of a system is invariant under some group of symmetry operations, its eigenfunctions do not, in general, share this property. Thus, although the Hamiltonian of the hydrogen molecule ion (p. 90) is invariant under a rotation which interchanges the two nuclei, this is not true of the wave function of the first excited state—which is multiplied by -1 under this operation. Similarly, in the case of a crystal (p. 110), the wave function is multiplied by a phase factor $e^{-i\mathbf{k}\cdot\mathbf{d}}$ in a displacement $\mathbf{d}$ which carries the lattice into self-coincidence. And under an arbitrary rotation about the nucleus, the atomic orbitals in Fig. 2.5 are mixed among themselves in a more complicated way.

Again, we may define a "rotated function" conventionally as the function which "looks like" the original function but is referred to the rotated reference frame (Fig. 4.4). A function $f(x, y, z)$ is a prescription for associating a certain number with a field point whose coordinates, in a given frame, are x, y, z. If we rotate the frame, the coordinates become x', y', z' and $f(x', y', z')$ defines the function constructed by the same prescription as $f(x, y, z)$ but in the rotated frame: to find its form

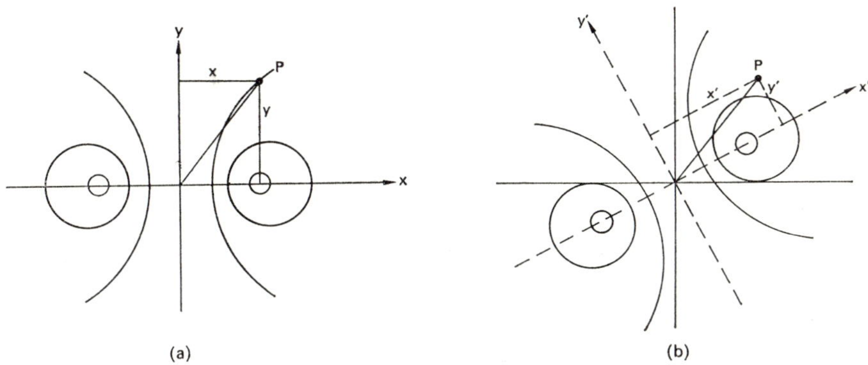

FIG. 4.4. Convention defining rotation of a function. The "rotated function" (b) looks exactly like the original function (a), but relative to the rotated frame.

in "fixed-frame" language we must express the variables in terms of x, y, z, using (4.35), and the result will be a new function $f'(x, y, z)$. We call $f'(x, y, z)$ the "rotated function" and the definition is thus

$$\mathsf{R}f(x, y, z) = f(x', y', z') \qquad (4.37\mathrm{a})$$

or, using $\mathbf{r}$ to stand for the set of coordinates and noting from (4.34) that $\mathbf{r}' = \mathbf{R}^{-1}\mathbf{r}$,

$$\mathsf{R}f(\mathbf{r}) = f(\mathbf{R}^{-1}\mathbf{r}). \qquad (4.37\mathrm{b})$$

We note, on replacing $\mathbf{r}$ by $\mathbf{Rr}$, that $\mathsf{R}f$ evaluated at the rotated point $\mathbf{Rr}$ will take the same numerical value as f at the unrotated point $\mathbf{r}$; $\mathsf{R}f$ therefore "looks like" f in the sense that its contour map (Fig. 4.4) is exactly like that of f, but rotated in space.

The rotated function may be set up by various rules, but basically one definition is sufficient: the rotated quantity is constructed by the same rules as the original quantity but in the rotated frame—and is then expressed in fixed-frame language. This definition is sufficient to cover all situations,[‡] though in many cases (4.37) coupled with (4.35) is adequate.

EXAMPLE. *Rotation properties of atomic orbitals.* Let us consider the d functions (p. 58). These functions have Cartesian forms

$$\tfrac{1}{2}(x^2-y^2)f(r), \quad \frac{1}{2\sqrt{3}}(3z^2-r^2)f(r), \quad yzf(r), \quad zxf(r), \quad xyf(r)$$

and we require their transformation properties under all rotations R about the origin. It is at once clear that each function, on rotation, turns into some linear combination of the full set: for example, the angle-dependent factor xy in d_{xy} is replaced by

$$x'y' = (R_{11}x + R_{21}y + R_{31}z)(R_{21}x + R_{22}y + R_{23}z),$$

which is a linear combination of the six products x^2, y^2, ..., zx. Only five products are linearly independent, however, since $x^2+y^2+z^2 = r^2$ and it is readily verified that d_{xy}' $(= R\, d_{xy})$ may be expressed as a linear combination of the five original d-functions (e.g. by writing down the equality and comparing coefficients of x^2, y^2, ..., zx): the same is true for each of the d-functions and *the set therefore carries a five-dimensional representation of the group of rotations.*

The mixing of the d-orbital basis functions describes a rotation in a five-dimensional *function space,* which is said to be *induced* by the rotation of the physical system in ordinary three-dimensional space. Although we have used the same symbol R for both types of rotation it is important to appreciate the difference. In exactly the same way, a set of p-functions carries a three-dimensional representation. From (4.33) and (4.35) this is clearly *identical* with that carried by the Cartesian unit vectors; and p-functions are sometimes said to exhibit "vector properties". For example, a p-function pointing along an axis with direction cosines l, m, n, is $p' = lp_x + mp_y + np_z$, just as if each function were a unit vector. The f-functions, on the other hand, carry a seven-dimensional representation and have a more complicated behaviour. We know that every group has a trivial one-dimensional representation, the identity, carried by an invariant basis vector: each s function carries such a representation.

4.6.   Symmetry and degeneracy

It is no accident that the eigenfunctions of any set that carries a representation of the symmetry group of a system are *degenerate*. In general, all the eigenfunctions that are mixed among themselves under the symmetry operations (which leave H invariant) are essentially degenerate.

To see why this is so, let us consider any eigenfunction Ψ_1 of H, with eigenvalue E. There may be other eigenfunctions with the same energy E, in which case we call them Ψ_2, ... Ψ_g, where g is the number of linearly independent solutions. From (4.30) and (4.32) it follows that

---

[‡] Applying both to wave functions and operators, with or without spin dependence.

the rotated function $\Psi_1' = R\Psi_1$, where R is any symmetry operation of the group under which H is invariant, must also satisfy the eigenvalue equation

$$H(R\Psi_1) = E(R\Psi_1).$$

But the most general solution with eigenvalue E is a linear combination of $\Psi_1, \ldots, \Psi_g$. Hence $R\Psi_1 = \Psi_1 R_{11} + \Psi_2 R_{21} + \ldots \Psi_g R_{g1}$, where the coefficients have been labelled to conform to the usual matrix notation as in (4.35). Since each one of the g-eigenfunctions must behave in a similar way we may write

$$R(\Psi_1 \Psi_2 \ldots \Psi_g) = (\Psi_1 \Psi_2 \ldots \Psi_g)\mathbf{R} \qquad (4.38)$$

where $\mathbf{R}$ is the array of coefficients. In this way we make the standard association of a matrix with an operator and, by considering the whole group of symmetry operations (or, more simply, the generators), determine a representation. We also note that since

$$HR\Psi_i = ER\Psi_i = RH\Psi_i$$

for all eigenfunctions, and hence for an arbitrary linear combination, the operators HR and RH are identical:

$$HR = RH. \qquad (4.39)$$

Another way of stating the invariance property (4.32) is therefore to say that *symmetry operators commute with the Hamiltonian.*

The significance of the symmetry properties of the central field functions, studied in the last Example, is now clear: if they were collected into a row matrix, they would carry a representation of the block form (4.20) with a 3×3 matrix for each set of p-functions, a 5×5 for each set of d-functions, etc., and each set, by construction, would comprise eigenfunctions with a common eigenvalue. In a very special case (Coulombic central field) we have seen (Section 2.5) that *different* sets may have the same eigenvalue; for example, $E_{3s} = E_{3p} = E_{3d}$. But, *in general* the degeneracy of different sets is not *necessary*, because the functions of one set would never be mixed, as in (4.38), with those of another and would therefore not be required by symmetry to have the same E value. Thus, for any other potential, the eigenfunctions providing different representations would belong to different eigenfunctions, as in Fig. 2.4. A "chance" equality of eigenvalues is referred to as an "accidental degeneracy"; and the ns, np, nd degeneracy for central field functions with a strictly Coulombic potential is the best-known example.[‡]

---

‡ The corresponding Hamiltonian actually possesses a higher invariance group: "accidental" degeneracy is nearly always a result of either a "hidden" symmetry, or of approximations.

If accidental degeneracies are excluded, we may now make the following assertion:

Degenerate groups of eigenfunctions of a Hamiltonian H normally carry irreducible representations of the symmetry group under which H is invariant.   (4.40)

The reason the representations may be assumed irreducible in general (they were clearly irreducible in the actual central field example) is that reducibility would imply an accidental degeneracy—which we have agreed to exclude. For reducibility would require that mixing of the degenerate functions could yield two or more *subsets* of functions that would mix only among themselves under these symmetry operations (to give block-form matrices) and the fact that different subsets possessed the same eigenvalue would then constitute an accidental degeneracy.

The consequences of (4.40) are particularly important in allowing us to predict how a degeneracy will be resolved or "lifted" when the symmetry of a Hamiltonian is disturbed (e.g. by the inclusion of a perturbation). The central field functions are again useful in providing an example:

EXAMPLE. *Resolution of degeneracies by reduction of symmetry.* Let us suppose that the spherical symmetry of a central potential is reduced by applying a field of *cubic* form (e.g. the crystal field discussed in Section 4.1). The question is whether the s, p, d, ... levels will retain their one-, three-, five-fold degeneracies or will be resolved—and if so, in what way.

The perturbed Hamiltonian is no longer invariant under *all* rotations but only under those belonging to the cubic group $\mathcal{O}$—of which we need consider only the generators C_3^{xyz} and C_4^z (p. 116). The symmetry properties of the degenerate eigenfunctions, which carry irreducible representations of $\mathcal{O}$, will be characterized by their behaviour under these two rotations. Now we know from Table 4.3 that there are no irreducible representations of dimension greater than 3, and the implication is therefore that $d, f,$ ... levels, at least, should be resolved by a cubic perturbation of the central field.

To find how this resolution occurs we suppose the perturbation becomes very weak: the actual eigenfunctions must then go over smoothly into central field eigenfunctions and the appropriate "zero-order" functions, carrying the different irreducible representations of $\mathcal{O}$, will be certain mixtures of the central field functions of each degenerate set. For example, if a d-level splits into two sublevels the zero-order functions of both sublevels must be linear combinations of $d_{+2}, d_{+1}, \ldots d_{-2}$: and to determine the nature of the splitting we therefore need to know what linear combinations of the d-functions carry irreducible representations of $\mathcal{O}$.

We now consider again the real d-functions (p. 58) with x, y, z directions defined by the cube axes, and show that these particular linear combinations of d_m already carry irreducible representations of $\mathcal{O}$. It is only necessary to use the particular rotation matrices given in (4.9) and to set up rotated functions by making the substitutions, indicated in (4.35), namely:

$$\mathsf{C_4}^z: \quad x \to x' = y \qquad\qquad \mathsf{C_3}^{xyz}: \quad x \to x' = y$$
$$y \to y' = -x \qquad\qquad\qquad y \to y' = z$$
$$z \to z' = z \qquad\qquad\qquad z \to z' = z.$$

Substitution in (4.37a) then leads to the rotated functions. For example, $d_{xy}' = \mathsf{C_4}^z d_{xy} = x'y'f(r) = -d_{xy}$. The two sets $\{d_{z^2}, d_{x^2-y^2}\}$ and $\{d_{yz}, d_{zx}, d_{xy}\}$ are called the d_λ and d_ε orbitals. Clearly they transform quite separately, as follows:

$$\mathsf{C_4}^z(d_{z^2}\, d_{x^2-y^2}) = (d_{z^2}\, d_{x^2-y^2}) \begin{pmatrix} 1 & 0 \\ 0 & -1 \end{pmatrix},$$

$$\mathsf{C_3}^{xyz}(d_{z^2}\, d_{x^2-y^2}) = (d_{z^2}\, d_{x^2-y^2}) \begin{pmatrix} -\tfrac{1}{2} & -\tfrac{1}{2}\sqrt{3} \\ \tfrac{1}{2}\sqrt{3} & -\tfrac{1}{2} \end{pmatrix},$$

while

$$\mathsf{C_4}^z(d_{yz}\, d_{zx}\, d_{xy}) = (d_{yz}\, d_{zx}\, d_{xy}) \begin{pmatrix} 0 & 1 & 0 \\ -1 & 0 & 0 \\ 0 & 0 & -1 \end{pmatrix},$$

$$\mathsf{C_3}^{xyz}(d_{yz}\, d_{zx}\, d_{xy}) = (d_{yz}\, d_{zx}\, d_{xy}) \begin{pmatrix} 0 & 0 & 1 \\ 1 & 0 & 0 \\ 0 & 1 & 0 \end{pmatrix}.$$

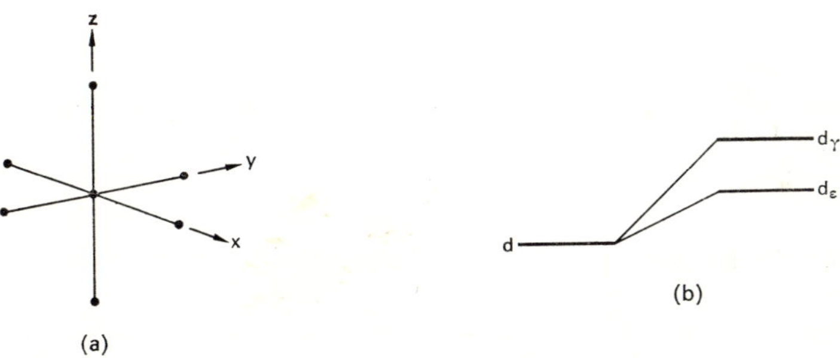

FIG. 4.5. Effect of crystal field on d-orbital energies. (a) Octahedral array of negative ligands. The energies of all five states are raised by the surrounding negative charges; but (cf. Fig. 2.5) the d_γ-orbitals, lying directly over the charges, are more affected than the d_ε, which lie between the charges, suggesting the splitting indicated in (b).

The real d-functions thus carry separate representations of the point group $\mathcal{O}$ and by comparing the matrices with those in Table 4.3 it is clear that d_{z^2}, $d_{x^2-y^2}$ carry the E representation while d_{yz}, d_{zx}, d_{xy} carry the T_1 representation. The full set of d-functions carries a five-dimensional representation of $\mathcal{R}(3)$ but on restriction to the subgroup $\mathcal{O}$ this representation breaks into block form, with a 2×2 block and a 3×3 block. The real d-functions are said to be "adapted" to a reduction of the representation; the complex functions d_{+2}, d_{+1}, ... d_{-2} are not.

The two sets of functions are limiting forms (zero-order approximations for vanishing perturbation) of two sets of cubic field eigenfunctions which are not required by symmetry to have the same eigenvalue. The cubic field is therefore expected to resolve a five-fold degenerate d level into doubly and triply degenerate components, the eigenfunctions of the two components having the symmetry properties of the $\{d_{z^2}, d_{x^2-y^2}\}$ and $\{d_{yz}, d_{zx}, d_{xy}\}$ sets respectively. In the presence of the cubic field, the two sets of d functions could be used as *approximate* eigenfunctions, and would in general yield different energy approximations. Thus, according to first-order perturbation theory (Section 1.1), the approximate energies could be found by averaging the cubic perturbation over the unperturbed d-orbitals: and for a field of the type indicated in Fig. 4.1 such considerations suggest the sequence of levels in Fig. 4.5.

The above example shows that the resolution of degeneracies by modification of symmetry may be predicted by knowing which irreducible representations of the "lower" symmetry group are contained in a given irreducible representation of the symmetry group of the unperturbed system. To obtain this kind of information, however, it is not necessary to actually perform the reduction (i.e. to set up the appropriate linear combinations, as in the last example); and to conclude this section we consider a powerful alternative procedure.

Suppose that an n-fold degenerate set of unperturbed eigenfunctions carries a representation D, and that the symmetry of the system is reduced to that of a group $\mathcal{G}$ with irreducible representations D_1, D_2, ... (of dimension g_1, g_2, ... $< n$). The set of eigenfunctions thus carries a *reducible* representation of $\mathcal{G}$. If we were to adapt this set to the reduction as in the last Example, we should find appropriate mixtures carrying the different irreducible representations and these would *not* in general all correspond to the same energy: in fact there would be n_1 g_1-fold degenerate levels, n_2 g_2-fold degenerate levels, etc., instead of the original n-fold degenerate level. But n_1, n_2, ... are simply the numbers of times the various irreducible representations occur when we write $D = n_1 D_1 + n_2 D_2 + ...$ and this information follows at once from (4.26) together with standard character tables. The following Example shows the power of the method.

Example. *Resolution of degeneracy. Use of character tables.* Consider seven central field f-functions, carrying a representation D_f. The characters of all rotations[‡] through an angle ω, in a representation provided by functions of angular momentum l, are given by

$$\chi_l(\mathsf{R}_\omega) = \frac{\sin (l+\tfrac{1}{2})\tfrac{1}{2}\omega}{\sin \tfrac{1}{2}\omega} \qquad (\omega \neq 0).$$

The characters in this way are compared below with those of the irreducible representations of the group $\mathscr{O}$:

	E	$3C_2$	$8C_3$	$6C_4$	$6C_2$
A_1	1	1	1	1	1
A_2	1	1	1	-1	-1
E	2	2	-1	0	0
T_1	3	-1	0	1	-1
T_2	3	-1	0	-1	1
D_f	7	-1	1	-1	-1

Here the abbreviated notation $8C_3$, for example, indicates the class of eight C_3 operations with a common character. Application of (4.26) then shows at once that

$$D = A_2 + T_1 + T_2.$$

An f-level is thus split by a cubic perturbation into a singlet and two triplets. The appropriate mixtures of free-atom f-functions which provide bases for the A_2, T_1 and T_2 representations must also have transformation properties as indicated in Table 4.3; the A_2 functions, for example, must be invariant under rotations such as $C_2{}^z$ and $C_3{}^{xyz}$ but must change sign under $C_4{}^z$ and $C_2{}^{xy}$.

It is clear from the preceding Examples that group theory is a powerful tool in obtaining information about the symmetry properties of wave functions, about their degeneracies, and about the effects of change of symmetry. But it is also clear that to obtain further information (e.g. about the energy sequence of the levels as in Fig. 4.5) it is necessary to make energy calculations using explicit forms of the wave functions. We must therefore turn, at some stage, to the actual construction of functions with given symmetry properties.

4.7. Symmetry functions

In constructing approximate wave functions, the importance of a classification according to symmetry has already been noted. Thus in Section 3.4 (p. 95) ϕ_s and ϕ_a were recognized as MO's of two different

---

‡ Change of rotation axis corresponds to an orthogonal transformation under which the trace, and hence the character, is invariant; choosing the axis in the z direction, and noting that the eigenfunctions contain a factor $e^{im\varphi}$, we easily obtain the rotation matrix and hence the formula given.

"symmetry species": and in LCAO approximation, ϕ_s was constructed from symmetric functions alone, ϕ_a from antisymmetric functions alone. These results embody two important group theoretical ideas:

(i) an eigenfunction of given symmetry species may be expanded in terms of basis functions of the same species only;

(ii) *arbitrary* functions (e.g. AO's) can usefully be combined into "symmetry functions", using symmetry considerations alone, suitable for use in such an expansion.

The advantages of distinguishing functions according to symmetry species are clear; in general, a function of given species becomes a linear combination of a relatively small number of predetermined *symmetry functions*. As a consequence the secular problem is replaced by a set of smaller secular problems—one for each symmetry species.

To formalize such ideas, we must first define the term "symmetry species". If a set of functions provides an irreducible representation D_α, we may label its members $\Phi_1^{(\alpha)}$, $\Phi_2^{(\alpha)}$, ..., $\Phi_{g\alpha}^{(\alpha)}$. On expressing (4.38) in component form and using the standard notation (4.4), we then obtain

$$R\Phi_i^{(\alpha)} = \sum_j \Phi_j^{(\alpha)}\, D_\alpha(R)_{ji}. \qquad (4.41)$$

A function which transforms under rotations according to (4.41)—or *behaves like the i-th basis vector of representation* D_α—is said to be of species (α, i).[‡] There will generally be many sets of functions carrying any given representation (e.g. the $2p$, $3p$, ... np functions all carry the same D_p) and to characterize a symmetry function fully a third index is therefore necessary. We shall generally use $\Phi_i^{(\alpha,\mu)}$ to denote the function of species (α, i) which comes from the μth set of functions carrying D_α. The statement that an eigenfunction of symmetry (α, i) can be expanded in terms of symmetry functions of the same species then becomes

$$\Psi_i^{(\alpha)} = \sum_\mu C_i^{(\alpha,\mu)}\Phi_i^{(\alpha,\mu)} \qquad (4.42)$$

where the summation is over μ *only*. Since functions of different species do not mix, it is often possible to focus attention on one species at a time and to suppress the labels (α, i): Ψ is then written simply as a linear combination of basis functions, $\Psi = \sum c_\mu \Phi_\mu$, all understood to be of the given species.

---

‡ Often the term "species" is applied to α alone, addition of i giving the "sub-species", but usually the distinction is unnecessary.

A second basic result concerns the construction of symmetry functions from an arbitrary function Φ: it states that a set of functions $\Phi_i^{(\alpha)}$, forming a basis for D_α according to (4.41), may be constructed as

$$\Phi_i^{(\alpha)} = \sum_R D_\alpha(R)_{ij}{}^* R\Phi. \tag{4.43}$$

For given α, it may be possible to obtain several distinct sets of g_α functions by making different choices of j, and a third label μ, as in (4.42), would then be added to distinguish the different sets. In other cases, it may be impossible to find a non-vanishing result by application of (4.43): this merely indicates that the function Φ, from which $\Phi_i^{(\alpha)}$ is "extracted", contains no component of the required species. For some purposes it is useful to write (4.43) as

$$\Phi_i^{(\alpha)} = \rho_{ij}^{(\alpha)}\Phi \tag{4.44}$$

where

$$\rho_{ij}^{(\alpha)} = (g_\alpha/g) \sum_R D_\alpha(R)_{ij}{}^* R \tag{4.45}$$

is a linear combination of symmetry operators, whose effect on any operand Φ is given, apart from normalization, by (4.43). The operators $\rho_{ij}^{(\alpha)}$ may then be handled independently of the operand and have the following basic properties:

$$\rho_{ij}^{(\alpha)}\rho_{kl}^{(\beta)} = \delta_{\alpha\beta}\,\delta_{jk}\rho_{il}^{(\alpha)}. \tag{4.46}$$

When $i = j$, $\rho_{ij}^{(\alpha)}$ becomes a *projection operator* (see Vol. 1, App. 4); otherwise it is a "shift operator" (similar to the step-up and step-down operators in Section 2.2) which turns the jth basis function of a representation into the ith. The following example illustrates the use of the operators $\rho_{ij}^{(\alpha)}$.

EXAMPLE. *Octohedral symmetry orbitals.* Let us consider an octohedral complex, in which there is a metal atom at the origin in Fig. 4.1 and six identical ligand atoms at the centres of the cube faces. Suppose we wish to construct symmetry functions, carrying the various irreducible representations of the group $\mathcal{O}$ shown in Table 4.3, using s-, p- and d-orbitals on the central atom and s- and p-functions on the surrounding ligand atoms.

First we note that the functions on the central atom, taken in real form, already carry irreducible representations as follows:[‡]

Representation	Functions
A_1	s
E	$\{d_{z^2}, d_{x^2-y^2}\}$
T_1	$\{p_x, p_y, p_z\}$
T_2	$\{d_{yz}, d_{zx}, d_{xy}\}$

---

‡ It should be noted that the order of the functions is important, as well as their definition (p. 137): thus d_{yz}, d_{zx}, d_{xy} transform like the basis vectors e_1, e_2, e_3, respectively, of the T_2 representation.

It is therefore only necessary to construct, from the twenty-four ligand orbitals linear combinations of the various species. To do this we take a typical orbital and note how it behaves under all the symmetry operations: it is then possible to set up the linear combinations (4.43) which correspond to application of the operators (4.45). The ligand s-orbitals are merely permuted under symmetry operations, but the p-orbitals are *rotated* in addition to being transferred from one centre to another. When the orbitals are numbered as in Fig. 4.1a, typical transformation properties may be collected as in Table 4.4 (notation as in Table 4.3). From this table it is easy to "project" from s_1, p_{1x} and p_{1y} functions of the various species: we simply take the results $(\mathrm{R}\,\phi)$ in the table and attach the appropriate numerical coefficients $\mathrm{D}_\alpha(\mathrm{R})_{ij}{}^*$ to obtain the symmetry combinations (4.43).

TABLE 4.4

R	E	$C_2{}^x$	$C_2{}^y$	$C_2{}^z$	$C_3{}^{xyz}$	$C_3{}^{x\overline{yz}}$	$C_3{}^{\overline{x}y\overline{z}}$	$C_3{}^{\overline{xy}z}$	$\overline{C}_3{}^{xyz}$	$C_3{}^{x\overline{yz}}$	$\overline{C}_3{}^{\overline{x}\overline{y}z}$	$\overline{C}_3{}^{\overline{xy}z}$
$\mathrm{R}s_1$	s_1	s_1	s_2	s_2	s_3	s_4	s_4	s_3	s_5	s_6	s_5	s_6
$\mathrm{R}p_{1x}$	p_{1x}	p_{1x}	$-p_{2x}$	$-p_{2x}$	p_{3y}	$-p_{4y}$	$-p_{4y}$	p_{3y}	p_{5z}	$-p_{6z}$	p_{5z}	$-p_{6z}$
$\mathrm{R}p_{1y}$	p_{1y}	$-p_{1y}$	p_{2y}	$-p_{2y}$	p_{3z}	p_{4z}	$-p_{4z}$	$-p_{3z}$	p_{5x}	$-p_{6x}$	$-p_{5x}$	p_{6x}

R	$C_4{}^z$	$C_4{}^{\overline{z}}$	$C_4{}^x$	$\overline{C}_4{}^x$	$C_4{}^y$	$\overline{C}_4{}^y$	$C_2{}^{xy}$	$C_2{}^{x\overline{y}}$	$C_2{}^{yz}$	$C_2{}^{y\overline{z}}$	$C_2{}^{zx}$	$C_2{}^{\overline{z}x}$
$\mathrm{R}s_1$	s_3	s_4	s_1	s_1	s_6	s_5	s_3	s_4	s_2	s_2	s_5	s_6
$\mathrm{R}p_{1x}$	p_{3y}	$-p_{4y}$	p_{1x}	p_{1x}	$-p_{6z}$	p_{5z}	p_{3y}	$-p_{4y}$	$-p_{2x}$	$-p_{2x}$	p_{5z}	$-p_{6z}$
$\mathrm{R}p_{1y}$	$-p_{3x}$	p_{4x}	p_{1z}	$-p_{1z}$	p_{6y}	p_{5y}	p_{3x}	$-p_{4x}$	p_2	$-p_{2z}$	$-p_{5y}$	$-p_{6y}$

It will be sufficient to consider two or three examples. If we want to construct an A_1 function from the s-orbitals the coefficients are all unity and we obtain (disregarding normalization)

$$\phi^{A_1} = (s_1 + s_2 + s_3 + s_4 + s_5 + s_6)$$

as would be expected. On the other hand, no A_2-type function can be constructed since each rotated orbital occurs once with a coefficient $+1$ and once with a -1. For the E representation we can construct two functions, one transforming like each basis vector: the first is

$$\sum_R D_E(\mathrm{R})_{11}\mathrm{R}s_1 = (s_1 + s_1 + s_2 + s_2) + c(s_3 + s_4 + s_4 + s_3 + s_5 + s_6 + s_5 + s_6)$$
$$+ (s_3 + s_4) + c(s_1 + s_1 + s_6 + s_5) + (s_3 + s_4) + c(s_2 + s_2 + s_5 + s_6)$$

where $c = -\frac{1}{2}$, while the second function follows on using the $D_E(\mathrm{R})_{21}$ coefficients. The results are (unnormalized)

$$\phi_1{}^E(s) = (s_1 + s_3 - 2s_5 + s_2 + s_4 - 2s_6),$$
$$\phi_2{}^E(s) = (-s_1 + s_3 - s_2 + s_4).$$

These functions are of the same species as d_{z^2} and $d_{x^2-y^2}$ (on the central atom), respectively: thus, for example, a molecular orbital of species (E, 1) will contain d_{z^2} and $\phi_1{}^E$ but will have no component of $d_{x^2-y^2}$ or $\phi_2{}^E$. The reader should show that three more linearly independent mixtures of s-functions can be set up, these providing a T_1 representation.

The p-functions may be handled in a similar way. Thus, for example, linear combinations transforming like the two basis vectors of the E representation are

$$\phi_1{}^{\mathrm{E}}(p) = (p_{1x} + p_{3y} - 2_5 p_z - p_{2x} - p_{4y} + 2p_{6z}),$$

$$\phi_2{}^{\mathrm{E}}(p) = (-p_{1x} + p_{3y} + p_{2x} - p_{4y}).$$

The reader should find all the remaining symmetry orbitals in the same way. The implications of equation (4.42) *et seq.* are then clear: the most general function of (E, 1) symmetry that can be constructed from the thirty-three given AO's is a mixture of d_{z^2}, $\phi_1{}^{\mathrm{E}}(s)$, $\phi_1{}^{\mathrm{E}}(p)$ and any other combinations *of the same species only*.

4.8. Matrix element theorems

In Section 4.4 various types of operator were distinguished according to their transformation properties: the Hamiltonian for a rotated molecule preserved its form, provided the rotation belonged to the group of symmetry operations, and was thus an *invariant* for the given group; the operators V_x, V_y, V_z (vector components referred to molecule-fixed axes) followed the same transformation law (4.36) as the orthogonal unit vectors in (4.33), and were called *vector operators*; and more generally we could distinguish *tensor operators* with more complicated transformation properties, vectors and invariants being the special cases of tensors of "rank" 1 and 0.

It is convenient now to classify tensor operators generally according to their transformation properties under the symmetry group of the system in question, defining the operators in such a way that they carry standard irreducible representations of the group. Thus, in the last Example (p. 141), it is clear from (4.36) *et seq.* that the operators $(\mathsf{V}_x, \mathsf{V}_y, \mathsf{V}_z)$ associated with the three components of a vector, transform like the basis vectors carrying the T_1 representation of the group $\mathcal{O}$. In general, the operators of a system may be grouped into sets carrying irreducible representations of the symmetry group; such sets are described as "irreducible tensorial sets" (Racah, 1942) and a large literature has built up around their properties (e.g. Fano and Racah, 1959; Griffith, 1961). The operators carrying a representation D_α may be denoted by $\mathsf{T}_i{}^{(\alpha)}$, the implication being that, under a symmetry operation, $\mathsf{R}\mathsf{T}_i{}^{(\alpha)} \to \mathsf{T}_i{}^{(\alpha)\prime}$ where

$$\mathsf{T}_i{}^{(\alpha)\prime} = \sum_j \mathsf{T}_j{}^{(\alpha)} \, \mathrm{D}_\alpha(\mathsf{R})_{ji}. \tag{4.47}$$

The sets of operators must of course be determined with reference to the particular group considered. The three angular momentum operators, for example, form a set irreducible under the operations of $\mathcal{O}$: but if the symmetry is reduced to D_{4h} (corresponding to flattening of the cube), the pair (L_x, L_y) carries one irreducible representation (E_g) while L_z carries another (A_{2g}). Also it may be necessary to make linear combinations of the operators in order to obtain sets carrying any given "standard" irreducible representations.

The possibility of classifying operators according to irreducible representations has important implications in the determination of matrix elements. Any matrix element involves three parts (function|operator|function) and if each part is classified according to symmetry species then it turns out that large numbers of matrix elements are either numerically related or identically zero.

We shall in fact require only one basic theorem, due to Wigner (1959) and Eckart (1930), in the slightly extended form (see, for example, McWeeny (1963), Section 8.6) applicable to finite groups. The theorem may be stated as follows:

The matrix elements $\langle \Phi_i^{(\alpha)} | T_j^{(\beta)} | \Phi_k^{(\gamma)} \rangle$, where (α, i), (β, j), (γ, k) indicate symmetry species, vanish identically unless $D_\beta \times D_\gamma$ contains D_α at least once. Non-zero elements are then related by

$$\langle \Phi_i^{(\alpha)} | T_j^{(\beta)} | \Phi_k^{(\gamma)} \rangle = \begin{pmatrix} \beta & \gamma & | & \alpha \\ j & k & | & i \end{pmatrix}_1^* A_1 + \begin{pmatrix} \beta & \gamma & | & \alpha \\ j & k & | & i \end{pmatrix}_2^* A_2 \ldots \quad (4.48)$$

where A_1, A_2, ... are independent of i, j, k and their coefficients are "coupling constants" *depending only on symmetry*.

This result involves the idea of the "direct product" (cf. Vol. 1, Section 3.8), to which we return in Chap. 5. If two sets of quantities $\phi_1^{(\beta)}$, $\phi_2^{(\beta)}$ and $\phi_1^{(\gamma)}$, $\phi_2^{(\gamma)}$, ... provide irreducible representations D_β and D_γ, then the set of $g_\beta \times g_\gamma$ products $\phi_j^{(\beta)} \phi_k^{(\gamma)}$ carries another representation which is called the "direct-product representation" and is denoted purely formally by $D_\beta \times D_\gamma$. This is often reducible, which means the products can be put together with suitable coefficients to yield basis functions for various irreducible representations: these coefficients are the "coupling constants" appearing in (4.48). Special methods must be developed for determining the coupling constants (e.g. McWeeny, *loc. cit.*, Section 8.7).

For many groups, in particular for the three-dimensional rotation group, no irreducible representation occurs more than once in the direct product $D_\beta \times D_\gamma$, and the right-hand side of (4.48) then contains only one term. This is the case considered by Wigner and Eckart, to which we return in Chaps. 5 and 6, and (4.47) then becomes

$$\langle \Phi_i{}^{(\alpha)}|\mathsf{T}_j{}^{(\beta)}|\Phi_k{}^{(\gamma)}\rangle = \begin{pmatrix} \beta\ \gamma \ \Big|\ \alpha \\ j\ k\ \Big|\ i \end{pmatrix}^{*} A. \tag{4.49}$$

When the coupling coefficients are known (these are tabulated by various authors) the application of (4.49) is simple: the constant A is evaluated by calculating *one* matrix element (the most convenient), and the matrix elements for all other choices of i, j, k follow on multiplying by the appropriate coefficient. For some point groups it may be necessary to use (4.48) and to calculate two matrix elements.

In the context of the rotation group, (4.49) amounts to the Wigner–Eckart theorem in its original form, and the coupling constants are the "Clebsch–Gordan coefficients". The corresponding results are of fundamental importance in the theory of atomic spectra. The full power of group theoretical methods only becomes apparent in dealing with complicated many-electron systems, but the principles discussed in this chapter are sufficient for the majority of applications; their importance will become clear in later chapters.

<h2 style="text-align:center">REFERENCES</h2>

ECKART, C. (1930) *Rev. Mod. Phys.* **2**, 395.

FANO, U. and RACAH, G. (1959) *Irreducible Tensorial Sets*, Academic Press, New York.

GRIFFITH, J. S. (1961) *The Theory of Transition Metal Ions*, Cambridge University Press, London.

HALL, G. G. (1963) *Vectors, Matrices and Tensors*, Pergamon Press.

HAMERMESH, M. (1962) *Group Theory and Its Application to Physical Problems.* Addison–Wesley, Reading, Mass.

HERZBERG, G. (1945) *Infrared and Raman Spectra of Polyatomic Molecules*, Van Nostrand, New York.

McWEENY, R. (1963) *Symmetry—An Introduction to Group Theory*, Pergamon Press.

RACAH, G. (1942) *Phys. Rev.* **62**, 438.

WIGNER, E. P. (1959) *Group Theory and Its Application to the Quantum Mechanics of Atomic Spectra*, Academic Press, New York.

CHAPTER 5

ANGULAR MOMENTUM AND THE ROTATION GROUP

5.1. Spin representations of the rotation group

One of the most important groups in (non-relativistic) physics is the three-dimensional rotational group $\mathscr{R}(3)$. Its properties provide a mathematical expression of the principle that all directions in space are equivalent—that space is invariant under all three-dimensional rotations. We have already seen how this principle leads to the commutation properties of the Pauli operators (Vol. 1, Section 4.9). On introducing basis vectors α and β, for the spin eigenstates, the associated matrices become

$$\mathbf{S}_x = \begin{pmatrix} 0 & \tfrac{1}{2} \\ \tfrac{1}{2} & 0 \end{pmatrix}, \quad \mathbf{S}_y = \begin{pmatrix} 0 & -\tfrac{1}{2}i \\ \tfrac{1}{2}i & 0 \end{pmatrix}, \quad \mathbf{S}_z = \begin{pmatrix} \tfrac{1}{2} & 0 \\ 0 & -\tfrac{1}{2} \end{pmatrix}. \tag{5.1}$$

These are essentially the "Pauli spin matrices". It is now time to lay the foundations of the general theory of angular momentum and to examine its connection with symmetry principles: to this end we try to construct the basic irreducible representations of $\mathscr{R}(3)$ by studying the transformation properties of the spin states α and β.

First we note that $\mathscr{R}(3)$ is not a *finite* group; it contains an infinite number of elements, each labelled by three continuously variable parameters (rotation angles), and is thus a three-parameter continuous group. Most of the ideas and principles in Chapter 4 remain valid, but summation over the elements of the group is replaced by "integration over the group manifold". Fortunately, however, the technical considerations peculiar to continuous groups can be almost completely avoided by specializing to the group $\mathscr{R}(3)$ and constructing the irreducible representations directly. This is the procedure to be followed.

We start by considering some system with an axis of quantization and two spin-states α, β. A rotation R then produces a rotated system with spin states $\alpha' = \mathbf{R}\alpha$, $\beta' = \mathbf{R}\beta$ corresponding to spin z-components $\pm\tfrac{1}{2}$ along the *new* z-axis.[‡] Let us describe these in the usual way by

$$(\alpha'\ \beta') = \mathbf{R}(\alpha\ \beta) = (\alpha\ \beta)\mathbf{U} \tag{5.2}$$

---

‡ As usual, we employ the same symbol for the rotation in real space and for the rotation it induces in a representation space, remembering that $\mathbf{R}^\dagger\mathbf{R} = \mathbf{R}\mathbf{R}^\dagger = \mathbf{I}$ (the unit operator) for both real and induced rotations.

where $\mathbf{U}$ is a 2×2 matrix describing the rotation induced in spin space[‡]. Now α' and β' are eigenfunctions of the spin operator S_z' for the rotated axis (see Vol. 1, Section 4.9), and to obtain a corresponding transformation property of the operators we remember that simultaneous rotation of states *and* operators must leave expectation values unchanged: thus (cf. p. 129) if θ is an *arbitrary* spin state

$$\langle \theta' | \mathsf{S}_\mu' | \theta' \rangle = \langle \theta | \mathsf{S}_\mu | \theta \rangle \quad (\mu = x, y, z).$$

Thus, since $\theta' = \mathsf{R}\theta$ where R is a unitary operator the rule $\langle \mathsf{R}\theta | \theta \rangle = \langle \theta | \mathsf{R}^\dagger \theta \rangle$ gives $\langle \theta | \mathsf{R}^\dagger \mathsf{S}_\mu' \mathsf{R} | \theta \rangle = \langle \theta | \mathsf{S}_\mu | \theta \rangle$ and since θ is arbitrary this implies equality of the operators

$$\mathsf{S}_\mu' = \mathsf{R}\mathsf{S}_\mu\mathsf{R}^\dagger \quad (\mu = x, y, z).$$

Since we are constructing a representation, in which the matrix $\mathbf{U}$ is associated with R by (5.2), the corresponding relationship between the matrices is

$$\mathbf{S}_\mu' = \mathbf{U}\mathbf{S}_\mu\mathbf{U}^\dagger \quad (\mu = x, y, z). \tag{5.3}$$

We must now determine the forms of $\mathbf{U}$ corresponding to typical rotations of the system.

Generally, being associated with a rotation, $\mathbf{U}$ is a unitary matrix with unit determinant. Its most general form is

$$\mathbf{U} = \begin{pmatrix} a & b \\ -b^* & a^* \end{pmatrix} \quad (aa^* + bb^* = 1). \tag{5.4}$$

To identify the rotations described by $\mathbf{U}$, we now consider special choices of a and b. For $b = 0$, the $\mathbf{S}_z$ transformation in (5.3) is

$$\mathbf{S}_z' = \begin{pmatrix} a & 0 \\ 0 & a^* \end{pmatrix}\begin{pmatrix} \tfrac{1}{2} & 0 \\ 0 & -\tfrac{1}{2} \end{pmatrix}\begin{pmatrix} a^* & 0 \\ 0 & a \end{pmatrix} = aa^*\mathbf{S}_z = \mathbf{S}_z$$

where $\mathbf{S}_z$ is the Pauli matrix in (5.1). Thus, unitary transformations with $b = 0$ are induced by rotations about the z-axis. To obtain the rotation angle we put $a = e^{i\theta}$ and consider the $\mathbf{S}_x$ transformation in (5.3), obtaining easily

$$\mathbf{S}_x' = \cos 2\theta \; \mathbf{S}_x - \sin 2\theta \; \mathbf{S}_y.$$

Now since the spin operators transform according to (4.36), and hence, like the basis vectors $\mathbf{e}_1, \mathbf{e}_2, \mathbf{e}_3$, a rotation through φ about the z-axis *in real space*, described by the second matrix in (4.17), must give

$$\mathbf{S}_x' = \cos \varphi \; \mathbf{S}_x + \sin \varphi \; \mathbf{S}_y$$

---

‡ The matrix is here denoted by $\mathbf{U}$ (a general unitary matrix) to avoid confusion with $\mathbf{R}$ used elsewhere for transformations of Cartesian coordinates.

and by comparison we see $\theta = -\frac{1}{2}\varphi$. It may be verified similarly that under the same unitary transformation $\mathbf{S}_y \to \mathbf{S}_y' = \sin 2\theta\, \mathbf{S}_x - \cos 2\theta\, \mathbf{S}_y$ and hence that $\mathbf{S}_y$ behaves like $\mathbf{e}_2$. Thus the rotation which we shall denote by $\mathsf{R}_z(\varphi)$ induces a rotation in spin space with matrix given by

$$\mathsf{R}_z(\varphi) \to \mathbf{U}_z(\varphi) = \begin{pmatrix} e^{-i\varphi/2} & 0 \\ 0 & e^{i\varphi/2} \end{pmatrix}. \tag{5.5}$$

The eigenstates α', β' of the rotated system, corresponding to the original α, β, are

$$\alpha' = e^{-i\varphi/2}\alpha, \quad \beta' = e^{i\varphi/2}\beta$$

and are thus still eigenstates of $\mathbf{S}_z$, only the phases being changed.

Similar considerations may be applied to rotations about the y-axis, characterized by putting $\mathbf{S}_y' = \mathbf{S}_y$ in (5.3). This restricts the $\mathbf{U}$-matrix parameters to *real* values, and since $a^2 + b^2 = 1$ we may put $a = \cos\theta$, $b = \sin\theta$. It then follows that

$$\mathbf{S}_x' = \cos 2\theta\, \mathbf{S}_x + \sin 2\theta\, \mathbf{S}_z,$$

which corresponds to a positive rotation $\mathsf{R}_y(\varphi)$ about the y-axis with $\varphi = -2\theta$. The behaviour of $\mathbf{S}_z$ also conforms to such a rotation, for we find in a similar way

$$\mathbf{S}_z' = -\sin 2\theta\, \mathbf{S}_x + \cos 2\theta\, \mathbf{S}_z$$

and we therefore obtain

$$\mathsf{R}_y(\varphi) \to \mathbf{U}_y(\varphi) = \begin{pmatrix} \cos\frac{1}{2}\varphi & -\sin\frac{1}{2}\varphi \\ \sin\frac{1}{2}\varphi & \cos\frac{1}{2}\varphi \end{pmatrix}. \tag{5.6}$$

Such relations are useful in defining states with definite spin components along an axis other than the z-axis. Thus with $\varphi = \pi/2$, $2\theta = -\pi/2$ and $\mathbf{S}_z' = \mathbf{S}_x$: the α' spin state, which corresponds to spin component $+\frac{1}{2}$ *along the original x-axis*, is accordingly $\alpha' = (\alpha + \beta)/\sqrt{2}$ —as follows at once from (5.2) and (5.6).

The rotations R_z and R_y are of basic importance, for it is well known that any one point on the surface of a sphere may be sent into any other by the following sequence of operations:[‡]

    (i) rotation through some angle γ about the z-axis,

    (ii)   ,,   ,,   ,,   ,, β ,, ,, y-axis,

    (iii)   ,,   ,,   ,,   ,, α ,, ,, z-axis.

The general rotation described by this sequence may be indicated by $\mathsf{R}(\alpha\beta\gamma)$, the operations being performed in the above order *about axes*

[‡] α and β denoting rotation angles should not be confused with the α and β spin states.

fixed in space.[‡] Exactly the same result is in fact achieved by performing α first about the z-axis, then β about the *rotated* y-axis, and finally γ about the *rotated* z-axis. These three continuously variable parameters, which determine all elements of the rotation group $\mathscr{R}(3)$, are called the "Euler angles" of the rotation. α and γ lie in the range 0 to 2π, β in the range 0 to π. We now write, for a general rotation $R = R(\alpha\beta\gamma)$,

$$R = R_z(\alpha)R_y(\beta)R_z(\gamma) \tag{5.7}$$

and observe that a 2×2 matrix may be associated with this rotation in terms of those determined in (5.5) and (5.6): thus

$$\mathbf{U} = \mathbf{U}_z(\alpha)\mathbf{U}_y(\beta)\mathbf{U}_z(\gamma). \tag{5.8}$$

It has been noted already in Section 4.5 that $\mathscr{R}(3)$ has irreducible representations of dimension $2l+1$, carried by the spherical harmonics of order l ($l = 0, 1, 2, \ldots$): it is customary to denote such representations by $\mathbf{D}_l$ and we therefore denote by $\mathbf{D}_{1/2}$ the two-dimensional representation with matrices $\mathbf{U}(\alpha\beta\gamma)$ defined in (5.8). The spin eigenvectors α, β are thus found to carry a two-dimensional representation of the rotation group:

$$R \rightarrow \mathbf{D}_{1/2}(R) = \begin{pmatrix} a & b \\ -b^* & a^* \end{pmatrix}$$

$$= \begin{pmatrix} e^{-i(\alpha+\gamma)/2} \cos \tfrac{1}{2}\beta & -e^{-i(\alpha-\gamma)/2} \sin \tfrac{1}{2}\beta \\ e^{i(\alpha-\gamma)/2} \sin \tfrac{1}{2}\beta & e^{i(\alpha+\gamma)/2} \cos \tfrac{1}{2}\beta \end{pmatrix}. \tag{5.9}$$

The parameters a and b are called the "Cayley–Klein parameters" of the rotation,

$$a = e^{-i(\alpha+\gamma)/2} \cos \tfrac{1}{2}\beta, \quad b = -e^{-i(\alpha-\gamma)/2} \sin \tfrac{1}{2}\beta \tag{5.10}$$

and are determined by its Euler angles. With the notation of (5.9) the basis vector rotation (5.2) becomes

$$R(\alpha\ \beta) = (\alpha\ \beta)\mathbf{D}_{1/2}(R). \tag{5.11}$$

From this basic two-dimensional representation all the irreducible representations of $\mathscr{R}(3)$ may be constructed.

5.2. The standard irreducible representations

The rotation of an *arbitrary* vector in spin space, $\theta = \xi\alpha + \eta\beta$ say, induced by rotation R, is described in terms of components by

$$\begin{pmatrix} \xi \\ \eta \end{pmatrix} \rightarrow \begin{pmatrix} \xi' \\ \eta' \end{pmatrix} = \mathbf{D}_{1/2}(R)\begin{pmatrix} \xi \\ \eta \end{pmatrix} \tag{5.12}$$

[‡] It should be noted that many notations and conventions are in use: these must be carefully observed.

F

where $\mathbf{D}_{1/2}(\mathsf{R})$ is given in (5.9). When the components ξ, η are changed in this way, their distinct products ξ^2, $\xi\eta$, η^2 will suffer a corresponding change: for example, when $\xi \to \xi' = a\xi + b\eta$ and $\eta \to \eta' = -b^*\xi + a^*\eta$, we obtain

$$\xi\eta \to (a\xi + b\eta)(-b^*\xi + a^*\eta) = -ab^*\xi^2 + (aa^* - bb^*)\xi\eta + ba^*\eta^2.$$

It is clear that the "second degree monomials" ξ^2, $\xi\eta$, η^2 turn into new linear combinations of themselves as a result of rotation and hence, interpreted as components of a vector in a new space, define a *three-dimensional* representation of $\mathscr{R}(3)$. This representation is not unitary, but may be made so by renormalizing the basis, each monomial then being multiplied by a suitable factor.

The construction of new representations in this way may clearly be generalized and it may be verified that the set of monomials of degree $2j$, defined by

$$f_\mu^{(j)} = \frac{\xi^{(j+\mu)}\eta^{(j-\mu)}}{\sqrt{\{(j+\mu)!(j-\mu)!\}}} \qquad (\mu = j, j-1, \ldots, -j) \tag{5.13}$$

provides a unitary representation of dimension $(2j+1)$. The $f_\mu^{(j)}$, regarded as a vector component in conformity with (5.12), may be collected into a column $\mathbf{f}^{(j)}$; and the rotation induced by R then yields a new vector with components

$$\mathbf{f}^{(j)\prime} = \mathbf{D}_j(\mathsf{R})\mathbf{f}^{(j)}. \tag{5.14}$$

The $(2j+1)$ *basis vectors* of the corresponding representation space, $\mathbf{e}^{(j)} = \{\mathsf{e}_j^{(j)}\ \mathsf{e}_{j-1}^{(j)} \ldots \mathsf{e}_{-j}^{(j)}\}$ are therefore rotated according to[‡]

$$\mathbf{e}^{(j)} \to \mathsf{R}\,\mathbf{e}^{(j)} = \mathbf{e}^{(j)}\,\mathbf{D}_j(\mathsf{R}). \tag{5.15}$$

It is evident that the last two equations are generalizations of (5.12) and (5.11) respectively. When the matrices $\mathbf{D}_j(\mathsf{R})$ are constructed by the prescription just outlined, equation (5.15) provides a natural definition of the *standard basis* for a $(2j+1)$-dimensional representation of the rotation group. To find the matrices explicitly, in terms of the Euler angles of the rotation, we need only replace ξ and η in (5.13) by $(a\xi + b\eta)$ and $(-b^*\xi + a^*\eta)$ respectively, expand by the binomial theorem, and pick out the coefficient of $f_\nu^{(j)}$. The coefficient of $f_\nu^{(j)}$ in the expansion of $f_\mu^{(j)}$ will be $D_j(\mathsf{R})_{\mu\nu}$, where R is specified by the Euler angles and hence by the parameters (5.10). The result is

$$D_j(\mathsf{R})_{\mu\nu} = \sum_t{}' (-1)^t \frac{\sqrt{\{(j+\mu)!(j-\mu)!(j+\nu)!(j-\nu)!\}}}{(j-\mu-t)!(j+\nu-t)!(t+\mu-\nu)!\,t!}$$

$$\times a^{j+\nu-t}a^{*j-\mu-t}b^{t+\mu-\nu}b^{*t} \tag{5.16}$$

[‡] If $\mathbf{e} \to \mathbf{e}' = \mathbf{e}\mathsf{R}$, the corresponding image $\mathbf{v}'$ of any vector $\mathbf{v}$ (with components $\mathbf{v}$) has components $\mathbf{v}' = \mathbf{R}\mathbf{v}$ (cf. equations (4.3) and (4.7), p. 118).

where t takes only those values for which the factorials are meaningful (i.e. non-negative integers).

It is easy to obtain rotation matrices from (5.16) for any particular value of j: thus for $j = 1$ we find a three-dimensional representation with

$$\mathbf{D}_1(\mathsf{R}) = \begin{bmatrix} \tfrac{1}{2}e^{-i(\alpha+\gamma)}(1+\cos\beta) & -\tfrac{1}{2}\sqrt{2}\,e^{-i\alpha}\sin\beta & \tfrac{1}{2}e^{-i(\alpha-\gamma)}(1-\cos\beta) \\[4pt] \tfrac{1}{2}\sqrt{2}\,e^{-i\gamma}\sin\beta & \cos\beta & -\tfrac{1}{2}\sqrt{2}\,e^{i\gamma}\sin\beta \\[4pt] \tfrac{1}{2}e^{i(\alpha-\gamma)}(1-\cos\beta) & \tfrac{1}{2}\sqrt{2}\,e^{i\alpha}\sin\beta & \tfrac{1}{2}e^{i(\alpha+\gamma)}(1+\cos\beta) \end{bmatrix}$$

$$(5.17)$$

in which the rows and columns are ordered according to decreasing μ-values $(+1, 0, -1)$.

The matrices defined above are of fundamental importance because they provide a complete system of *inequivalent irreducible* representations of the rotation group. They are also of great practical value since they describe the effect of rotation upon the sets of spherical harmonics which characterize the angular behaviour of the central field eigenfunctions in Chap. 2. Thus $\mathbf{D}_1(\mathsf{R})$ describes an arbitrary rotation of p-type atomic orbitals containing the factor $Y_{l,m}$ (with $l = 1$ and $m = +1, 0, -1$), *provided* the phase convention of Section 2.4 is adopted. Similarly $\mathbf{D}_2(\mathsf{R})$, $\mathbf{D}_3(\mathsf{R})$, ... describe the rotation of d, f, ... atomic orbitals. It must be emphasized that, although any linear combinations of the $Y_{l,m}$ ($m = l, l-1, ..., -l$) (in particular the real functions, such as p_x, p_y, p_z) will carry an irreducible representation, it is the complex harmonics with Condon–Shortley phase conventions which carry the *standard* representations defined in (5.16).

It is interesting, in the three-dimensional case, to compare the standard representation D_1 with that carried by the ordinary Cartesian basis vectors $(\mathbf{e}_1, \mathbf{e}_2, \mathbf{e}_3)$, or correspondingly by the real p-functions. Rotations about the z-axis ($\alpha = \beta = 0$) change the standard basis according to

$$\mathbf{e}_{+1}{}^{(1)} \to e^{-i\gamma}\,\mathbf{e}_{+1}{}^{(1)}, \quad \mathbf{e}_0{}^{(1)} \to \mathbf{e}_0{}^{(1)}, \quad \mathbf{e}_{-1}{}^{(1)} \to e^{i\gamma}\,\mathbf{e}_{-1}{}^{(1)}$$

whereas for the Cartesian basis vectors (putting $\cos\gamma = c$, $\sin\gamma = s$),

$$\mathbf{e}_1 \to c\mathbf{e}_1 + s\mathbf{e}_2, \quad \mathbf{e}_2 \to -s\mathbf{e}_1 + c\mathbf{e}_2, \quad \mathbf{e}_3 \to \mathbf{e}_3.$$

The linear combinations which provide the standard basis for D_1 are evidently, on normalizing to unity,

$$\mathbf{e}_{+1}{}^{(1)} = -(\mathbf{e}_1 + i\mathbf{e}_2)/\sqrt{2}, \quad \mathbf{e}_0{}^{(1)} = \mathbf{e}_3, \quad \mathbf{e}_{-1}{}^{(1)} = (\mathbf{e}_1 - i\mathbf{e}_2)/\sqrt{2} \qquad (5.18)$$

which are the so-called *spherical basis vectors*. Since the real p-functions defined in Section 2.5 have the same properties under rotation as e_1, e_2, e_3, it follows that functions of the form

$$f_{+1}(x, y, z) = -\tfrac{1}{2}\sqrt{2}(x+iy)f(r), \quad f_0(x, y, z) = zf(r),$$
$$f_{-1}(x, y, z) = \tfrac{1}{2}\sqrt{}(x-iy)f(r),$$

where $f(r)$ is any function of r ($= (x^2+y^2+z^2)^{1/2}$), provide the representation D_1. The Cartesian factors are harmonic polynomials (in this case of degree 1): if they are divided by r they become functions of the polar angles θ, φ. Apart from a normalizing factor they are the spherical harmonics $Y_{1,1}(\theta, \varphi)$, $Y_{1,0}(\theta, \varphi)$, $Y_{1,-1}(\theta, \varphi)$ defined in (2.36). We shall show presently that similar results hold for all *integral j* values. First, however, a peculiarity of the representations with non-integral j must be cleared up.

5.3.　Double groups

The matrices $\mathbf{D}_{1/2}(R)$ for the basic two-dimensional representation have been associated with the three-dimensional rotations $R(\alpha\beta\gamma)$ through (5.11). Closer inspection shows, however, that increasing the value of any rotation angle by 2π changes the sign of the matrix; but such a change corresponds physically to the identity operation, and two matrices which differ in sign are therefore associated *with the same rotation*. In other words, two distinct matrices may be associated with every single element of the rotation group. As a result of this ambiguity, the matrices do not, strictly speaking, form a representation of $\mathscr{R}(3)$. The ambiguity arises because the special unitary group in two dimensions, comprising all matrices of the form (5.4), is larger than $\mathscr{R}(3)$; it contains *two* elements (matrices differing in sign) for each element of $\mathscr{R}(3)$ and is said to be a "covering group" of $\mathscr{R}(3)$. *Two* revolutions (i.e. a rotation of 4π) in three-dimensional space is equivalent to the identity in the unitary group $D_{1/2}$. It can be verified by inspection that for all representations with j half an odd integer a similar ambiguity occurs, but that when j is integral (as for example in (5.17)) there is no such ambiguity and we obtain a representation in the usual sense.

Since *all* D_j are important in quantum mechanics owing to the need to introduce spin, it is desirable to shift attention to the covering group itself—in this case called the "double group". The D_j provide true representations of this group and the usual theorems may be applied to them directly. Similar considerations apply to every *sub*-group of $\mathscr{R}(3)$ and hence to the finite point groups; with every point group we can associate a double group and the double groups will be required whenever spin properties must be taken into account.

The double groups may be dealt with algebraically by a method due to Bethe (1929). Since rotation through 2π is distinguished from zero rotation in the matrix group $D_{1/2}$ we "invent" an operation Q (commuting with all other rotations) corresponding to any rotation through 2π, and associate with Q the negative unit matrix in the group $D_{1/2}$. The double group then consists of all the elements of $\mathscr{R}(3)$ and all the elements $Q\mathscr{R}(3)$; and these elements are now in *one-to-one* correspondence with the matrices of $D_{1/2}$, which therefore provides a representation of the double group. The requirement that the two sets of quantities have the same multiplication table implies that $Q^2 = E$ is the basic property of the new element (a new generator). The properties of the original elements need formal modification so that, for example, $(C_n)^n = E$ is replaced by $(C_n)^n = Q$, and consequently $(C_n)^{2n} = E$. With this device, the matrices of D_j provide true representations of the double group, the new element Q having associated with it the matrix ± 1 (the $(2j+1)$-dimensional unit matrix), the lower sign applying when $j = \frac{1}{2}, \frac{3}{2}, \ldots$. When j is integral, and there is no ambiguity in sign of the matrices, the upper sign applies, giving identical matrices for elements R and QR.

The theory of the double groups is more interesting in the case of the point groups, which are finite subgroups of $\mathscr{R}(3)$. The matrices $\mathbf{D}_j(R)$ may then be reduced to block form by suitable basis change and there is a non-trivial problem of determining the irreducible representations of the *double* point groups. Usually, it is sufficient to know the character system of the group and this is easily determined as the following Example shows.

EXAMPLE. *The octahedral double group.* The classes of the group $\mathscr{O}$ are $\{E\}$, $\{8C_3\}$, $\{3C_2\}$, $\{6C_2\}$, $\{6C_4\}$, where, for example, $\{3C_2\}$ indicates the set of three rotations $C_2^{(x)}$, $C_2^{(y)}$, $C_2^{(z)}$. The double group $\mathscr{O}'$ will now contain elements such as $\{8QC_3\}$. To see whether these form a new class we note that an operation R with matrix $\mathbf{D}(R)$ will be accompanied by an operation QR with matrix $\pm \mathbf{D}(R)$; the upper sign corresponds to j integral in the full rotation group, showing the set of matrices (and characters) is simply duplicated in some representations; the lower sign corresponds to the non-trivial new representations in which R and QR have different characters and therefore cannot be elements in the same class (within a class characters are always identical). Thus $\{8C_3\}$ and $\{8QC_3\}$ comprise distinct classes of $\mathscr{O}'$, their characters being identical in some representations but the sign of the latter being changed in the new representations. This argument breaks down for C_2 operators, for which the character vanishes, but it may then be shown (Opechowski, 1940) that C_2 and QC_2 are in the same class if and only if there is another two-fold axis, perpendicular to that of C_2: this is the case in the group $\mathscr{O}$ and hence $\{3C_2, 3QC_2\}$ forms one class and $\{6C_2, 6QC_2\}$ another. The class structure and the character table of $\mathscr{O}'$ is set out in Table 5.1 (with the nomenclature of Bethe, 1929). The character systems for the first five representations are identical, apart from duplication for the new operations, with those of the group $\mathscr{O}$. The representation Γ_6 is the one carried by spin functions α, β (a subgroup of the $D_{1/2}$ representation of $(\mathscr{R}3)$); Γ_1 is obtained by multiplying the matrices of Γ_6 by the elements of Γ_2 (forming a new representation, by definition, with sign changes producing an orthogonal character vector); and Γ_8 corresponds to the $D_{3/2}$ representation of the octahedral subgroup of $\mathscr{R}(3)$ with characters obtained from the usual character formula (p. 139). It is easily verified, using the basic results of Section 4.3 that the representations so determined are irreducible and exhaust all possibilities.

TABLE 5.1

$\mathscr{O}'$	E	R	$8C_3$	$8QC_3$	$3C_2+3QC_2$	$6C_2+6QC_2$	$6C_4$	$6QC_4$
Γ_1	1	1	1	1	1	1	1	1
Γ_2	1	1	1	1	1	-1	-1	-1
Γ_3	2	2	-1	-1	2	0	0	0
Γ_4	3	3	0	0	-1	-1	1	1
Γ_5	3	3	0	0	-1	1	-1	-1
Γ_2	2	-2	1	-1	0	0	2	$-\sqrt{2}$
Γ_1	2	-2	1	-1	0	0	$-\sqrt{2}$	2
Γ_2	4	-4	-1	1	0	0	0	0

5.4. The infinitesimal operators

Any rotation can be described as the result of a sequence of *infinitesimal* rotations. It should, therefore, be possible to study the rotation group via the "infinitesimal operators" which describe small rotations about the axes.

Let us consider a small rotation $\delta\varphi$, in real space, about the z-axis. This changes the basis vectors according to

$$R_z(\delta\varphi)(e_1\,e_2\,e_3) = (e_1\,e_2\,e_3)\left\{\begin{bmatrix}1&0&0\\0&1&0\\0&0&1\end{bmatrix}+\delta\varphi\begin{bmatrix}0&-1&0\\1&0&0\\0&0&0\end{bmatrix}+O(\delta\varphi^2)\right\} \quad (5.19)$$

in which $R_z(\delta\varphi)$ denotes the special rotation $R(0\ 0\ \delta\varphi)$ and $O(\delta\varphi^2)$ indicates terms of order $\delta\varphi^2$ and higher. The first matrix on the right describes the identity operator $I = R_z(0)$; the second describes an infinitesimal rotation about the z-axis. We shall write

$$R_z(\delta\varphi) = I + \delta\varphi D_z \quad (5.20)$$

where D_z is the "infinitesimal operator" corresponding to the matrix $\mathbf{D}_z$:

$$\mathsf{D}_z \to \mathbf{D}_z = \begin{bmatrix} 0 & -1 & 0 \\ 1 & 0 & 0 \\ 0 & 0 & 0 \end{bmatrix}.$$

The operator D_z may be described formally as a derivative:

$$\mathsf{D}_z = \underset{\delta\varphi \to 0}{\text{Lt.}} \left(\frac{\mathsf{R}_z(\delta\varphi) - \mathsf{R}_z(0)}{\delta\varphi} \right). \tag{5.21}$$

The existence of this limit may be taken as a definition of "continuity" of the group.

In exactly the same way we can consider rotations about the x- and y-axes, and introduce matrices $\mathbf{D}_x$ and $\mathbf{D}_y$. It is, however, more convenient to employ the Hermitian matrices $\mathbf{J}_x = i\mathbf{D}_x$, $\mathbf{J}_y = i\mathbf{D}_y$, $\mathbf{J}_z = i\mathbf{D}_z$, thus introducing

$$\mathbf{J}_x = i\begin{bmatrix} 0 & 0 & 0 \\ 0 & 0 & -1 \\ 0 & 1 & 0 \end{bmatrix}, \quad \mathbf{J}_y = i\begin{bmatrix} 0 & 0 & 1 \\ 0 & 0 & 0 \\ -1 & 0 & 0 \end{bmatrix}, \quad \mathbf{J}_z = i\begin{bmatrix} 0 & -1 & 0 \\ 1 & 0 & 0 \\ 0 & 0 & 0 \end{bmatrix}. \tag{5.22}$$

Since there is a one-to-one correspondence between the rotations and the matrices, any relations among the matrices imply similar relations among the corresponding Hermitian operators J_x, J_y, J_z (which we still refer to, loosely, as the "infinitesimal operators"). In this way we establish the characteristic properties

$$\begin{aligned} \mathsf{J}_x\mathsf{J}_y - \mathsf{J}_y\mathsf{J}_x &= i\mathsf{J}_z, \\ \mathsf{J}_y\mathsf{J}_z - \mathsf{J}_z\mathsf{J}_y &= i\mathsf{J}_x, \\ \mathsf{J}_z\mathsf{J}_x - \mathsf{J}_x\mathsf{J}_z &= i\mathsf{J}_y. \end{aligned} \tag{5.23}$$

It must be stressed that these commutation properties are essentially those of infinitesimal rotations in ordinary space; in this sense they are more primitive than those of the angular momentum operators which arise in a particular function-space representation of the rotation group.

A finite rotation, φ, can now be expressed as a sequence of rotations through $\delta\varphi = \varphi/n$. Hence, by repetition of (5.20) with $\mathsf{D}_z = -i\mathsf{J}_z$,

$$\mathsf{R}_z(\varphi) = \underset{n \to \infty}{\text{Lt.}} \left[1 - i\left(\frac{\varphi}{n}\right)\mathsf{J}_z \right]^n = \exp(-i\varphi\mathsf{J}_z) \tag{5.24}$$

where the exponential simply means the series:

$$\exp(-i\varphi\mathsf{J}_z) = 1 - i\varphi\mathsf{J}_z - \tfrac{1}{2}\varphi^2\mathsf{J}_z^2 + \dots . \tag{5.25}$$

Since the same reasoning applies to rotations about the other axes, we can write the general rotation, with Euler angles α, β, γ, in terms of the infinitesimal operators:

$$R = \exp\left(-i\alpha J_z\right) \exp\left(-i\beta J_y\right) \exp\left(-i\gamma J_z\right). \tag{5.26}$$

This is the basic correspondence between a three-space rotation and the infinitesimal operators.

To emphasize that the operators just discussed all belong to the three-space, which carries the $(2j+1)$-dimensional representation D_j with $j = 1$, we should, strictly, attach a superscript 1. We now wish to discuss the operators $J_x^{(j)}$, $J_y^{(j)}$, $J_z^{(j)}$, which describe the rotations induced in the representation space of D_j with $j \neq 1$. For two reasons, however, we shall continue to make no notational distinction between rotation operators in ordinary space and those which describe corresponding *induced* rotations in a representation space: first, the meaning is clear from the context; and, second, the operators for each given j have formally identical algebraic properties.

First we consider the case $j = \frac{1}{2}$. From the matrix representation (5.11) we obtain, for the rotations in the space of $D_{1/2}$ induced by infinitesimal rotations about the coordinate axes,

$$R_x(\delta\varphi)(e_1\, e_2) = (e_1\, e_2)\left[\begin{pmatrix} 1 & 0 \\ 0 & 1 \end{pmatrix} + \delta\varphi\begin{pmatrix} 0 & -\frac{1}{2}i \\ -\frac{1}{2}i & 0 \end{pmatrix} + 0(\delta\varphi^2)\right],$$

$$R_y(\delta\varphi)(e_1\, e_2) = (e_1\, e_2)\left[\begin{pmatrix} 1 & 0 \\ 0 & 1 \end{pmatrix} + \delta\varphi\begin{pmatrix} 0 & -\frac{1}{2} \\ \frac{1}{2} & 0 \end{pmatrix} + 0(\delta\varphi^2)\right],$$

$$R_z(\delta\varphi)(e_1\, e_2) = (e_1\, e_2)\left[\begin{pmatrix} 1 & 0 \\ 0 & 1 \end{pmatrix} + \delta\varphi\begin{pmatrix} -\frac{1}{2}i & 0 \\ 0 & \frac{1}{2}i \end{pmatrix} + 0(\delta\varphi^2)\right]. \tag{5.27}$$

The matrices, in the $D_{1/2}$ representation, which correspond to the operators J_x, J_y, J_z, are given by i times the coefficients of $\delta\varphi$:

$$J_x^{(1/2)} = \begin{pmatrix} 0 & \frac{1}{2} \\ \frac{1}{2} & 0 \end{pmatrix} = S_x, \quad J_y^{(1/2)} = \begin{pmatrix} 0 & -\frac{1}{2}i \\ \frac{1}{2}i & 0 \end{pmatrix} = S_y,$$

$$J_z^{(1/2)} = \begin{pmatrix} \frac{1}{2} & 0 \\ 0 & -\frac{1}{2} \end{pmatrix} = S_z. \tag{5.28}$$

The Pauli matrices (5.1) thus appear in a new role; they describe the *infinitesimal rotations* in spin space which correspond to rotations about the axes in real space, and because they must have similar multiplicative properties it might be said that the commutation rules for spin are a manifestation of the non-commutativity of rotations in three dimensions. It is noteworthy that e_1 and e_2 are eigenvectors of J_2.

We now examine the transformations induced in the $(2j+1)$-dimensional space of D_j, showing that the basis vectors are quite generally eigenvectors of J_z and obtaining other important properties of the operators. First we note that for rotations about the z-axis the Cayley–Klein parameters (5.10) take the values $a = e^{-i\varphi/2}, b = 0$. The rotation $R_{z(\varphi)}$ then gives

$$\xi \to \xi' = e^{-i\varphi/2}\xi, \quad \eta \to \eta' = e^{i\varphi/2}\eta, \quad \xi^{j+\mu}\eta^{j-\mu} \to e^{-i\mu\varphi}\xi^{j+\mu}\eta^{j-\mu}.$$

The matrix associated with $R_z(\varphi)$ is thus diagonal and the basis vectors are rotated according to

$$R_z(\varphi)\, e_\mu^{(j)} = \exp\left(-i\mu\varphi\right) e_\mu^{(j)}. \tag{5.29}$$

On expressing $R_z(\varphi)$ in terms of the infinitesimal operator J_z, this becomes

$$[1 - i\varphi J_z + \ldots\,[e_\mu^{(j)} = [1 - i\mu\varphi + \ldots] e_\mu^{(j)},$$

from which it follows that

$$J_z\, e_\mu^{(j)} = \mu\, e_\mu^{(j)}. \tag{5.30}$$

The number μ, which labels the different basis vectors of the representation D_j, is therefore an eigenvalue of the infinitesimal operator J_z: this is true for any j, integral or half-integral.

Next we consider rotations about the x- and y-axes. From (5.28) and (5.12), the operator $R = 1 - i\varphi J_x$ in the $D_{1/2}$ representation space transforms ξ and η according to

$$\xi \to \xi' = \xi - \tfrac{1}{2}i\varphi\eta, \qquad \eta \to \eta' = \eta - \tfrac{1}{2}i\varphi\xi,$$

and the corresponding rotation in the D_j space is described by

$$\frac{\xi^{j+\mu}\eta^{j-\mu}}{[(j+\mu)!(j-\mu)!]^{1/2}} \to \frac{(\xi - \tfrac{1}{2}i\varphi\eta)^{j+\mu}(\eta - \tfrac{1}{2}i\varphi\xi)^{j-\mu}}{[(j+\mu)!(j-\mu)!]^{1/2}}.$$

On expanding, and introducing the monomials (5.13), we obtain to first order

$$f_\mu^{(j)} \to f_\mu^{(j)} - \tfrac{1}{2}i\varphi[(j+\mu)(j-\mu+1)]^{1/2}f_{\mu-1}^{(j)}$$
$$- \tfrac{1}{2}i\varphi[(j-\mu)(j+\mu+1)]^{1/2}f_{\mu+1}^{(j)}.$$

The matrix whose elements appear in the transformation

$$f_\mu^{(j)} \to \sum_\nu D_x^{(j)}(\varphi)_{\mu\nu} f_\nu^{(j)}$$

is thus given by

$$D_x^{(j)}(\varphi)_{\mu\nu} = \delta_{\mu\nu} - \tfrac{1}{2}i\varphi[(j+\mu)(j-\mu+1)]^{1/2}\,\delta_{\mu-1,\nu}$$
$$- \tfrac{1}{2}i\varphi[(j-\mu)(j+\mu+1)]^{1/2}\,\delta_{\mu+1,\nu}. \tag{5.31}$$

F*

The basis vectors of the D_j space are rotated according to (cf. (5.15))

$$\mathbf{e}_\nu{}^{(j)} \to \mathsf{R}_x(\varphi)\,\mathbf{e}_\nu{}^{(j)} = \sum_\mu \mathbf{e}_\mu{}^{(j)}\,D_x{}^{(j)}(\varphi)_{\mu\nu}$$

$$= \mathbf{e}_\nu{}^{(j)} - \tfrac{1}{2}i\varphi[(j+\nu+1)(j-\nu)]^{1/2}\,\mathbf{e}_{\nu+1}{}^{(j)}$$

$$- \tfrac{1}{2}i\varphi[(j+\nu)(j-\nu+1)]^{1/2}\,\mathbf{e}_{\nu-1}{}^{(j)}$$

and, since $\mathsf{R}_x(\varphi) = \mathsf{I} - i\varphi\mathsf{J}_x$, it follows that

$$\mathsf{J}_x\,\mathbf{e}_\nu{}^{(j)} = \tfrac{1}{2}[(j+\nu+1)(j-\nu)]^{1/2}\,\mathbf{e}_{\nu+1}{}^{(j)}$$

$$+ \tfrac{1}{2}[(j-\nu+1)(j+\nu)]^{1/2}\,\mathbf{e}_{\nu-1}{}^{(j)}. \qquad (5.32)$$

An exactly similar treatment of rotations about the y-axis yields

$$\mathsf{J}_y\,\mathbf{e}_\nu{}^{(j)} = -i\tfrac{1}{2}[(j+\nu+1)(j-\nu)]^{1/2}\,\mathbf{e}_{\nu+1}{}^{(j)}$$

$$+ \tfrac{1}{2}i[(j-\nu+1)(j+\nu)]^{1/2}\,\mathbf{e}_{\nu-1}{}^{(j)}. \qquad (5.33)$$

It is then a simple matter to verify explicitly that the operators J_x, J_y, J_z satisfy the commutation relations (5.23), whichever representation space they refer to: the abstract properties of the operators are identical within each space.

It is also possible to define two operators with somewhat simpler properties than J_x and J_y, namely

$$\mathsf{J}^+ = \mathsf{J}_x + i\mathsf{J}_y, \quad \mathsf{J}^- = \mathsf{J}_x - i\mathsf{J}_y. \qquad (5.34)$$

The behaviour of the basis vectors under the infinitesimal operators may then be expressed as follows:

$$\mathsf{J}^+\,\mathbf{e}_\nu{}^{(j)} = [(j+\nu+1)(j-\nu)]^{1/2}\,\mathbf{e}_{\nu+1}{}^{(j)},$$

$$\mathsf{J}^-\,\mathbf{e}_\nu{}^{(j)} = [(j-\nu+1)(j+\nu)]^{1/2}\,\mathbf{e}_{\nu-1}{}^{(j)}, \qquad (5.35)$$

$$\mathsf{J}_z\,\mathbf{e}_\nu{}^{(j)} = \nu\,\mathbf{e}^{(j)}.$$

The "shift" operators J^+ and J^-, which send an eigenvector of J_z into one of its neighbours, with eigenvalue increased or decreased by unity, are of course analogous to the "step-up" and "step-down" operators of Chap. 2. From these properties it is easily established that the J_z-eigenvectors spanning the representation D_j are also eigenvectors of $\mathsf{J}^2 = \mathsf{J}_x{}^2 + \mathsf{J}_y{}^2 + \mathsf{J}_z{}^2$. For, from the commutation relations (5.23)

$$\mathsf{J}^2 = (\mathsf{J}_x + i\mathsf{J}_y)(\mathsf{J}_x - i\mathsf{J}_y) + \mathsf{J}_z{}^2 - \mathsf{J}_z \qquad (5.36)$$

and from (5.35)

$$\mathsf{J}^+[\mathsf{J}^-\,\mathbf{e}_\nu{}^{(j)}] = \mathsf{J}^+[(j-\nu+1)(j+\nu)]^{1/2}\,\mathbf{e}_{\nu-1}{}^{(j)} = [(j-\nu+1)(j+\nu)]\,\mathbf{e}_\nu{}^{(j)}.$$

Hence

$$\mathbf{J}^2\, \mathbf{e}_\nu{}^{(j)} = [(j-\nu+1)(j+\nu)+\nu^2-\nu]\, \mathbf{e}_\nu{}^{(j)}$$

or

$$\mathbf{J}^2\, \mathbf{e}_\nu{}^{(j)} = j(j+1)\, \mathbf{e}_\nu{}^{(j)} \quad (\nu = -j,\ -j+1,\ \dots +j). \tag{5.37}$$

The infinitesimal operators, though completely "geometrical" in origin, therefore exhibit all the properties of angular momentum; we must now discuss the meaning of this result.

5.5. Representations carried by eigenfunctions

The conventions defining "rotated" functions and "rotated" operators have been discussed in Section 4.5. With every spatial rotation of a physical system we may therefore associate a rotation in representation space, operators and functions transforming according to

$$\mathsf{H} \to \mathsf{H}' = \mathsf{R}\mathsf{H}\mathsf{R}^\dagger$$

and

$$\psi \to \psi' = \mathsf{R}\psi.$$

Let us now study a one-electron system whose Hamiltonian is invariant under all three-space rotations. For such a system, we expect from Section 4.6 that the eigenfunctions will fall into degenerate sets carrying irreducible representations; we now verify that this is so and that the basis functions are characterized by *angular momentum eigenvalues*.

We take any one-electron function $\psi(x, y, z)$ and observe that in rotation $\mathsf{R}_z(\varphi)$ about the z-axis

$$\mathsf{R}_z(\varphi)\psi(x, y, z) = \psi(x', y', z') \tag{5.38}$$

where, using (4.35) with the matrix corresponding to a z-axis rotation,

$$x' = x \cos \varphi + y \sin \varphi,$$
$$y' = -x \sin \varphi + y \cos \varphi, \tag{5.39}$$
$$z' = z.$$

For an infinitesimal rotation, substitution of (5.39) in (5.38), followed by Taylor expansion, gives

$$\mathsf{R}_z(\varphi)\psi(x, y, z) = \psi(x, y, z) + \varphi\left(y\frac{\partial \psi}{\partial x} - x\,\frac{\partial \psi}{\partial y} \right) + \mathrm{O}(\varphi^2).$$

Now $\psi(x, y, z)$ is an element of function space, and the infinitesimal rotation of the physical system therefore induces an infinitesimal rotation in function space of the form

$$\mathsf{R}_z(\varphi) = \mathsf{1} - i\varphi\mathsf{L}_z \tag{5.40}$$

where the infinitesimal operator (denoted by J_z in developing the general theory) is

$$\mathsf{L}_z = \frac{1}{i}\left(x\,\frac{\partial}{\partial y} - y\,\frac{\partial}{\partial x}\right). \tag{5.41}$$

The symbol L_z is used for the infinitesimal operator in this particular representation space, carried by functions $\psi(x, y, z)$. Clearly L_z and the operators L_x and L_y which may be defined analogously, using rotations about the other axes, are the angular momentum operators (in dimensionless units) already discussed in Chap. 2. We note that the invariance of H under rotations implies that R and H commute (see p. 135): in particular, H commutes with infinitesimal rotations and therefore with the infinitesimal operators. Commutation of H with the angular momentum operators is therefore essentially a *symmetry* property; the infinitesimal operators must commute with *all* rotationally invariant operators, and hence with L^2 as well as with H. Different infinitesimal operators, however, do not commute, and the commutation rules (2.11) are also evidently dictated by their general properties (5.23).

The standard irreducible representations defined in Section 5.2 are now seen to be carried by sets of wave functions $\psi_m{}^{(l)}$ which, besides being eigenfunctions of H with a common eigenvalue E, are eigenfunctions of the total angular momentum and one of its components. The standard basis has the properties (5.30) and (5.37) which become

$$\mathsf{L}_z\psi_m{}^{(l)} = m\psi_m{}^{(l)}, \tag{5.42}$$

$$\mathsf{L}^2\psi_m{}^{(l)} = l(l+1)\psi_m{}^{(l)} \quad (m = l, l-1, \ldots -l) \tag{5.43}$$

while the properties (2.23) follow from (5.35).

The actual form of the functions $\psi_m{}^{(l)}(x, y, z)$ has already been obtained in Section 2.4. Each is a homogeneous polynomial of degree l in the coordinates, which may be divided by r^l to give a function of the polar angles θ, φ only. Instead of solving the differential equations (2.34) directly, we may obtain the angle-dependent factors by starting from $\psi_m{}^{(l)}$ with $m = l$ and observing that this must transform like $f_l{}^{(l)}$ defined in (5.13): thus, omitting a constant factor, we may say (using $\sim$ for "transforms like")

$$f_l{}^{(l)} \sim (f_1{}^{(1)})^l.$$

But, from (5.18) *et seq.*, $-(x+iy)$ is a function behaving like $f_1^{(1)}$ and consequently

$$\psi_l^{(l)} \sim [-(x+iy)]^l.$$

On dividing by r^l and introducing polar coordinates the factor on the right becomes

$$(-1)^l \sin^l \theta \, e^{il\varphi}.$$

Apart from a normalizing factor, this is the spherical harmonic $Y_{l,l}(\theta, \varphi)$, the factor $(-1)^l$ introducing a definite phase convention. All the other harmonics $Y_{l,m}(\theta, \varphi)$ may be obtained, with phases completely determined, by applying the step-down operator L^-: in polar form

$$\mathsf{L}^- = e^{-i\varphi}\left(-\frac{\partial}{\partial\theta} + i \cot\theta \, \frac{\partial}{\partial\varphi}\right). \tag{5.44}$$

In this way it follows that the degenerate sets of eigenfunctions which carry the standard irreducible representations D_l ($l = 0, 1, 2, \ldots$) contain the angle-dependent factors (after normalization)

$$Y_{l,m}(\theta, \varphi) = (-1)^{(m+|m|)/2}\left(\frac{2l+1}{4\pi}\frac{(l-|m|)!}{(l+|m|)!}\right)^{1/2} P_l^{|m|}(\cos\theta)\, e^{im\varphi}. \tag{5.45}$$

All rotationally invariant systems must have eigenfunctions which exhibit this angle-dependence; only the radial factor depends on the nature of the central potential.

It has been noted that only representations of *integral j* are carried by the spherical harmonics, this being a consequence of the essential single-valuedness of the wave function. On the other hand, the spin space carries $\mathsf{D}_{1/2}$ and, apart from its non-integral value j, exhibits parallel properties: instead of (5.30) and (5.37) we have, using S_z and S^2 for the particular J-operators which belong to spin space,

$$\mathsf{S}_z\theta_m = m\theta_m \qquad (m = \pm\tfrac{1}{2}), \tag{5.46}$$

$$\mathsf{S}^2\theta_m = s(s+1)\theta_m \qquad (s = \tfrac{1}{2}, m = \pm\tfrac{1}{2}) \tag{5.47}$$

where $\theta_{+1/2} = \alpha$, $\theta_{-1/2} = \beta$. The commutation properties of the spin operators (Vol. 1, Section 4.9)—themselves a direct consequence of symmetry considerations—provided the starting point for all the analysis in this chapter. It remains only to show that basis functions carrying representations of non-integral j ($j > \tfrac{1}{2}$) can also be found. Such functions can in fact be constructed by the *coupling* of orbital and spin angular momenta: we therefore consider the group theoretical significance of the procedure used in Section 2.7.

5.6.   The coupling of angular momenta

If the (central-field) Hamiltonian contains spin-dependent terms, as in Section 2.7, its rotational invariance will be preserved only if both space and spin operators are subjected to a common rotation. The invariance property then implies that sets of *space–spin* eigenfunctions will carry representations of the rotation group. Suppose for example that spin is taken into account in discussing the *d*-states of an electron in a central field; in the absence of spin terms in the Hamiltonian the functions

$$d_{+2}\alpha, \quad d_{+2}\beta, \quad d_{+1}\alpha, \quad \dots \quad d_{-2}\beta$$

are all eigenfunctions of H and of the angular momentum operators L^2, L_z, S^2, S_z. These ten space–spin products provide a ten-dimensional representation of the rotation group; but this is clearly the direct product $D = D_2 \times D_{1/2}$ (p. 144), which is not necessarily irreducible. When spin terms are included in the Hamiltonian the individual products are no longer necessarily degenerate eigenfunctions—for only the combinations which, under common space–spin rotations, belong to the same irreducible representation will be essentially degenerate. By reducing the direct product, as in Section 4.6, we can therefore find how spin–orbit coupling resolves the ten-fold degeneracy.

We discuss this problem generally, since it appears very widely, by considering any two sets of quantities

$$\{e_{j_1}^{(j_1)} \dots e_{m_1}^{(j_1)} \dots e_{-j_1}^{(j_1)}\}, \quad \{e_{j_2}^{(j_2)} \dots e_{m_2}^{(j_2)} \dots e_{-j_2}^{(j_2)}\}$$

which carry irreducible representations D_{j_1} and D_{j_2} respectively. The set of all products $e_{m_1}^{(j_1)} e_{m_2}^{(j_2)}$ will carry the direct product representation $D_{j_1} \times D_{j_2}$, which in general may be reducible:

$$D_{j_1} \times D_{j_2} = \sum_j n_j D_j. \tag{5.48}$$

We shall want to make this reduction explicitly, finding the linear combinations of products of each pure symmetry species (j, m), but first ask *which* representations occur and how many times. This follows from (5.48), together with (4.26), once the character systems are known. The characters are easily obtained since they depend only on the rotation angle in R, not on the axial direction (for the axis may be rotated by a unitary transformation—which leaves the trace invariant). From (5.29) the rotation R_φ induced in the space of D_j by a rotation φ in physical space has the character

$$\chi_j(R_\varphi) = \sum_{m=-j}^{j} \exp -im\varphi,$$

which is a geometric series. The sum is

$$\chi_j(R_\varphi) = \frac{\varepsilon^{-j} - \varepsilon^{j+1}}{1 - \varepsilon} \qquad (\varepsilon = e^{i\varphi}). \tag{5.49}$$

The character of R_φ in the direct product representation is then the product

$$\chi(R_\varphi) = \left(\frac{\varepsilon^{-j_1} - \varepsilon^{j_1+1}}{1 - \varepsilon}\right)\left(\frac{\varepsilon^{-j_2} - \varepsilon^{j_2+1}}{1 - \varepsilon}\right)$$

$$= \frac{1}{1 - \varepsilon}\left[\left(\frac{\varepsilon^{j_1+j_2+2} - \varepsilon^{j_1-j_2+1}}{1 - \varepsilon}\right) + \left(\frac{\varepsilon^{-(j_1+j_2)} - \varepsilon^{-j_1+j_2+1}}{1 - \varepsilon}\right)\right].$$

Each term in the square brackets is now seen to be the sum of a geometric progression: the first is $-\varepsilon^{j_1-j_2+1} \times (1 + \varepsilon + \ldots \varepsilon^{2j_2})$ and the second is $\varepsilon^{-(j_1+j_2)} \times (1 + \varepsilon + \ldots \varepsilon^{2j_2})$. Thus, assuming for the moment $j_1 \geq j_2$,

$$(1 - \varepsilon)\chi(R_\varphi) = -[\varepsilon^{(j_1+j_2+1)} + \varepsilon^{(j_1+j_2)} + \ldots \varepsilon^{(j_1-j_2+1)}]$$

$$+ [\varepsilon^{-(j_1+j_2)} + \varepsilon^{-(j_1+j_2)+1} + \ldots \varepsilon^{-(j_1-j_2)}].$$

On putting $j_1 + j_2 = k$ and taking the terms in pairs we obtain

$$\chi(R_\varphi) = (1 - \varepsilon)^{-1}\{(\varepsilon^{-k} - \varepsilon^{k+1})$$

$$+ (\varepsilon^{-(k-1)} - \varepsilon^{(k-1)+1}) \ldots (\varepsilon^{-(k-2j_2)} - \varepsilon^{(k-2j_2)+1})\}$$

or

$$\chi(R_\varphi) = \sum_{j=j_1-j_2}^{j_1+j_2} \left(\frac{\varepsilon^{-j} - \varepsilon^{j+1}}{1 - \varepsilon}\right) = \sum_{j=j_1-j_2}^{j_1+j_2} \chi_j(R_\varphi). \tag{5.50}$$

This is the required result: the character system is a sum of the character systems of the component irreducible representations,

$$\chi(R) = \sum_j n_j \chi_j(R_\varphi),$$

in which each irreducible representation, D_j, with $(j_1 + j_2) \geq j \geq |j_1 - j_2|$ occurs once and once only. For $j_1 = j_2$ the roles of j_1 and j_2 are simply reversed, and hence in general

$$D_{j_1} \times D_{j_2} = \sum_{j=|j_1-j_2|}^{j_1+j_2} D_j. \tag{5.51}$$

This result is of fundamental importance in the theory of atomic spectra where it finds many applications. Before performing the reduction explicitly it is worth clarifying the physical meaning of (5.51). We define operators J_{1x}, J_{1y}, J_{1z} as in (5.29) *et seq.*, in terms of the infinitesimal rotations in the space carrying D_{j_1}; and similarly $J_{2x}, J_{2y},$

J_{2z} working in the space carrying D_{j_2}. A general rotation of a physical system described by the products $\{e_{m_1}{}^{(j_1)} e_{m_2}{}^{(j_2)}\}$ is then effected by the operator, with Euler angles α, β, γ as in (5.7),

$$R_{j_1}R_{j_2} = \exp\left(-i\alpha J_{1x}\right) \exp\left(-i\beta J_{1y}\right) \exp\left(-i\gamma J_{1z}\right)$$

$$\times \exp\left(-i\alpha J_{2x}\right) \exp\left(-i\beta J_{2y}\right) \exp\left(-i\gamma J_{2z}\right).$$

Since, for example, J_{1z} works only on the $e_{m_1}{}^{(j_1)}$ factors and J_{2z} on the $e_{m_2}{}^{(j_2)}$ factors it is evident that the order in which they are applied is immaterial and that the operators with different subscripts (1 and 2) will commute. Hence the rotation of product space is described by the operator

$$R = \exp\left(-i\alpha J_x\right) \exp\left(-i\beta J_y\right) \exp\left(-i\gamma J_z\right) \tag{5.52}$$

where

$$J_x = J_{1x} + J_{2x},$$
$$J_y = J_{1y} + J_{2y}, \tag{5.53}$$
$$J_z = J_{1z} + J_{2z}.$$

Since these are essentially infinitesimal operators for the product space, their commutation properties are given by (5.23) and the argument of earlier sections may be taken over in its entirety. The linear combinations in product space which provide the standard basis D_j are characterized as eigenvectors $E_m{}^{(j)}$ of J_z and J^2 ($=J_x^2 + J_y^2 + J_z^2$) with eigenvalues m and $j(j+1)$ respectively. It follows also, since J_z is a sum of operators, that $m_1 + m_2 = m$ for all the products which appear in $E_m{}^{(j)}$.

When the operators are regarded as quantum-mechanical angular momentum operators the foregoing results may be stated as follows:

> From the eigenvectors $e_{m_1}{}^{(j_1)}$ and $e_{m_2}{}^{(j_2)}$ of two independent commuting pairs of angular momentum operators (J_{1z}, J_1^2) and (J_{2z}, J_2^2), we may construct linear combinations of the products $\{e_{m_1}{}^{(j_1)} e_{m_2}{}^{(j_2)}\}$ which are eigenvectors of the *total* angular momentum operators (J_z, J^2). The allowed eigenvalues of J^2, one for each representation D_j in product space, are $j(j+1)$ with $j = |j_1 - j_2|;\ (|j_1 - j_2| + 1) \ldots (j_1 + j_2)$. The allowed eigenvalues of J_z are $m = -j,\ -j+1,\ \ldots +j$. $\tag{5.54}$

The coupling of angular momenta has already been discussed in Section 2.7 and a "vector model" has been introduced. The group-theoretical significance of the coupling is now clear: it corresponds to the explicit reduction of the direct product representation of the rotation group into its irreducible components. The vector coupling coefficients are therefore determined by symmetry.

Let us now write the coupling constants in the form used in Section 4.7, using

$$\begin{pmatrix} j_1 & j_2 & \Big| & j \\ m_1 & m_2 & \Big| & m \end{pmatrix}$$

to denote the coefficient of the product $e_{m_1}^{(j_1)} e_{m_2}^{(j_2)}$ in the linear combination $E_m^{(j)}$ which transforms like the mth basis vector of irreducible representation D_j. Since, according to (5.51), each

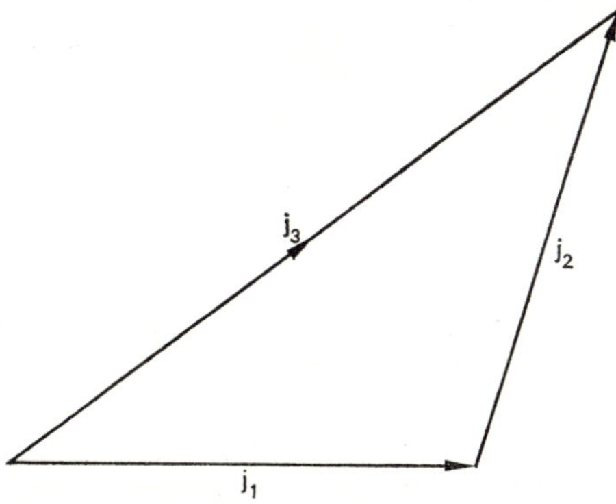

Fig. 5.1. Triangle inequality for angular momentum vectors. Two vectors, j_1 and j_2, can only be coupled to a resultant j, for quantum numbers such that $|j_1+j_2| \geqq j_3 \geqq |j_1-j_2|$.

representation D_j occurs only *once* in $D_{j_1} \times D_{j_2}$, it will be possible to construct only *one* linearly independent vector $E_m^{(j)}$ of given j and m; the notation is therefore adequate, and it is not necessary to add a subscript, as in (4.46), to distinguish different vectors of the same species. We therefore write

$$E_m^{(j)} = \sum_{m_1, m_2} e_{m_1}^{(j_1)} e_{m_2}^{(j_2)} \begin{pmatrix} j_1 & j_2 & \Big| & j \\ m_1 & m_2 & \Big| & m \end{pmatrix} \tag{5.55}$$

and note that this result applies to the vector coupling of *any* quantities which behave under rotations like angular momentum eigenvectors. The fact that compatible values of j, j_1 and j_2 must be related pictorially as in Fig. 5.1 is often referred to as a "triangle rule". The coefficients in (5.55) are the famous "Clebsch–Gordan coefficients" which appear in all discussions of angular momentum coupling. In Section 2.7, for example, the coefficients that appeared in coupling orbital and spin angular momenta, with $l = 1$ and $s = \frac{1}{2}$, to a resultant j ($=\frac{3}{2}$ or $\frac{1}{2}$) were Clebsch–Gordan coefficients of the form

$$\begin{pmatrix} 1 & \frac{1}{2} & \Big| & j \\ m_1 & m_2 & \Big| & m \end{pmatrix}$$

and the algebraic procedure used to generate them can be used quite generally.

5.7. Clebsch–Gordan and Wigner coefficients

There are many methods of actually evaluating the Clebsch–Gordan coefficients; for these the reader is referred to standard works on the theory of angular momentum (see, for example, Edmonds, 1957). It is in fact possible to obtain the closed expression

$$\begin{pmatrix} j_1 & j_2 & \Big| & j \\ m_1 & m_2 & \Big| & m \end{pmatrix} =$$

$$\delta_{m,m_1+m_2} \left[\frac{(2j+1)(j_1+j_2-j)!(j_1-j_2+j)!(-j_1+j_2+j)!}{(j_1+j_2+j+1)!} \right]^{1/2} \sum_r (-1)^r \times$$

$$\frac{[(j_1+m_1)!(j_1-m_1)!(j_2+m_2)!(j_2-m_2)!(j+m)!(j-m)!]^{1/2}}{r!(j_1+j_2-j-r)!(j_1-m_1-r)!(j_2+m_2-r)!(j-j_1-m_2+r)!(j-j_2+m_1+r)!}$$

$$(5.56)$$

in which r takes all values for which the factorials exist (i.e. for which all the arguments are non-negative integers), and the Kronecker delta simply implies that the coupled state (j, m) contains only the products with total z-component $m_1 + m_2 = m$ as noted in the Example in Section 2.7.

The expression (5.56), like all the others that can be derived, is cumbersome to use. The more commonly occurring coefficients have therefore been extensively tabulated (see, for example, Condon and Shortley, 1935, Section 14; Heine, 1960, App. I). It should be noted that possible phase factors have been so chosen that all coefficients are real.

In recent years it has become customary to replace the Clebsch–Gordan coefficients by coefficients which exhibit a higher symmetry: thus the coefficient defined in (5.56) is multiplied by a trivial phase factor $(-1)^{j_1+j_2-j}$ if the subscripts 1 and 2 are interchanged, but similar relationships can be extended to all three angular momenta by introducing new coefficients

$$\begin{pmatrix} j_1 & j_2 & j_3 \\ m_1 & m_2 & m_3 \end{pmatrix} = \frac{(-1)^{j_1-j_2-m_3}}{\sqrt{2j_3+1}} \begin{pmatrix} j_1 & j_2 & \bigg| & j_3 \\ m_1 & m_2 & \bigg| & -m_3 \end{pmatrix}. \qquad (5.57)$$

The quantities defined in this way are called the *Wigner 3j symbols* (Wigner, 1959). They vanish unless $m_1 + m_2 + m_3 = 0$ and have the symmetry properties

(i) an even permutation of columns leaves the numerical value unchanged,

(ii) an odd permutation multiplies the value by $(-1)^{j_1+j_2+j_3}$,

(iii) reversal of the signs of all m's multiplies the value by $(-1)^{j_1+j_2+j_3}$.

Such properties are clearly convenient in the construction of tables and an exhaustive tabulation has been published by Rotenberg *et al.* (1959) for all arguments $j_1, j_2, j_3 < 8$.

Since the coupling coefficients (5.56) define a unitary transformation leading from one orthonormal basis (product functions) to another (vector coupled linear combinations) they may be collected into a unitary matrix whose rows and columns exhibit the usual orthonormality properties. Thus

$$\sum_{m,m_2} \begin{pmatrix} j' & \bigg| & j_1 & j_2 \\ m' & \bigg| & m_1 & m_2 \end{pmatrix} \begin{pmatrix} j_1 & j_2 & \bigg| & j \\ m_1 & m_2 & \bigg| & m \end{pmatrix} = \delta_{j'j}\,\delta_{m'm}, \qquad (5.58a)$$

$$\sum_{j,m} \begin{pmatrix} j_1 & j_2 & \bigg| & j \\ m_1' & m_2' & \bigg| & m \end{pmatrix} \begin{pmatrix} j & \bigg| & j_1 & j_2 \\ m & \bigg| & m_1 & m_2 \end{pmatrix} = \delta_{m_1'm_1}\,\delta_{m_2'm_2}, \qquad (5.58b)$$

where

$$\begin{pmatrix} j & \bigg| & j_1 & j_2 \\ m & \bigg| & m_1 & m_2 \end{pmatrix} = \begin{pmatrix} j_1 & j_2 & \bigg| & j \\ m_1 & m_2 & \bigg| & m \end{pmatrix}^*$$

and the star may be dropped since the coefficients are in this case real. The notation clearly indicates a scalar product $\langle \bar{\phi}_i | \phi_j \rangle$ between a coupled function $(\bar{\phi}_i)$ and a simple product function (ϕ_j), the parts separated by the vertical line corresponding to bra and ket with $\langle \bar{\phi}_i | \phi_j \rangle = \langle \phi_j | \bar{\phi}_i \rangle^*$. The transformation matrix, **U** say, has rows labelled by all pairs of m_1, m_2 values and columns labelled by the j, m values of the coupled states:

$$U_{m_1,m_2;j,m} = \begin{pmatrix} j_1 & j_2 & \bigg| & j \\ m_1 & m_2 & \bigg| & m \end{pmatrix}.$$

Thus (5.58a) expresses the fact that $\mathbf{U}^\dagger\mathbf{U} = \mathbf{1}$ and (5.58b) the fact that

$\mathbf{UU}^{\dagger} = \mathbf{1}$. It should be noted that j in (5.58b) runs only over values compatible with the triangle condition (p. 165).

The $3j$ symbols possess similar orthonormality properties, namely

$$\sum_{j_3, m_3} (2j_3+1)\begin{pmatrix} j_1 & j_2 & j_3 \\ m_1' & m_2' & m_3' \end{pmatrix}\begin{pmatrix} j_1 & j_2 & j_3 \\ m_1 & m_2 & m_3 \end{pmatrix} = \delta_{m_1'm_1}\,\delta_{m_2'm_2}, \quad (5.59a)$$

$$\sum_{m_1, m_2} \begin{pmatrix} j_1 & j_2 & j_3' \\ m_1 & m_2 & m_3' \end{pmatrix}\begin{pmatrix} j_1 & j_2 & j_3 \\ m_1 & m_2 & m_3 \end{pmatrix} = (2j_3+1)^{-1}\,\delta_{j_3'j_3}\,\delta_{m_3'm_3} \quad (5.59b)$$

where again the j values must conform to the triangle inequalities. The m values in (5.59) are of course also restricted since

$$m_1 + m_2 + m_3 = m_1' + m_2' + m_3' = 0.$$

It is now a simple matter to construct explicitly sets of space–spin angular momentum eigenstates carrying the irreducible representations

TABLE 5.2

The coefficients $\begin{pmatrix} l & \frac{1}{2} & \big| & j \\ m_1 & m_2 & \big| & m \end{pmatrix}$

	$m_2 = \frac{1}{2}$	$m_2 = -\frac{1}{2}$
$j = l+\frac{1}{2}$	$\left(\dfrac{l+m+\frac{1}{2}}{2l+1}\right)^{1/2}$	$\left(\dfrac{l-m+\frac{1}{2}}{2l+1}\right)^{1/2}$
$j = l-\frac{1}{2}$	$-\left(\dfrac{l-m+\frac{1}{2}}{2l+1}\right)^{1/2}$	$\left(\dfrac{l+m+\frac{1}{2}}{2l+1}\right)^{1/2}$

D_j with j half an odd integer. We consider angular momenta l (integral) and $s\ (=\frac{1}{2})$ and note from (5.51) that we can construct two sets of states, with $j = l+\frac{1}{2}$ and $j = l-\frac{1}{2}$ respectively. The Clebsch–Gordan coefficients in this case are given in Table 5.2. Consequently, from orbital states $\{\phi_{l,m}\}$ and spin states $\{\alpha, \beta\}$ we can obtain eigenstates of *total* angular momentum

$$\psi_{j,m} = \left(\frac{l+m+\frac{1}{2}}{2l+1}\right)^{1/2}\phi_{l,m-\frac{1}{2}}\,\alpha + \left(\frac{l-m+\frac{1}{2}}{2l+1}\right)^{1/2}\phi_{l,m+\frac{1}{2}}\,\beta$$

$$(j = l+\tfrac{1}{2}: m = j, j-1, \ldots, -j) \quad (5.60)$$

and a similar set for $j = l-\frac{1}{2}$. These are the states constructed from first principles, for $l = 1$, in Section 2.7: we now appreciate, at last, the full group theoretical significance of the coupling procedure—which allows us to pass from any representation with integral quantum numbers to a neighbouring representation with half-integral quantum numbers. These results are fundamental to the theory of atomic spectra.

5.8. Spherical tensors and the Wigner–Eckart theorem

The general matrix element theorem (4.48) takes the simple form (4.49) for the rotation group, where each $\mathbf{D}_j$ $(j_1+j_2 \geqslant j \geqslant |j_1-j_2|)$ occurs only once. The symmetry species are indicated by the angular momentum eigenvalues, and on replacing the species labels (α, i) by (j, m) we obtain in general (adding n_1, n_2 to symbolize any further labels defining the two states)

$$\langle n_1 j_1 m_1 | \mathsf{T}_m{}^{(j)} | n_2 j_2 m_2 \rangle = \begin{pmatrix} j & j_2 & j_1 \\ m & m_2 & m_1 \end{pmatrix} A \qquad (5.61)$$

where $\mathsf{T}_m{}^{(j)}$ is an irreducible tensor operator (p. 143) under the full rotation group and the bra and ket correspond to arbitrary states of species (j_1, m_1) and (j_2, m_2) respectively. The "quantum numbers" n_1 and n_2 are irrelevant except in determining the constant A, which needs to be evaluated only once in order to obtain the matrix elements for *all* allowed values of m_1, m_2, m_3. This result is so widely used that it is worth mentioning some of the more important types of operator that occur.

Invariants $(j = 0)$

The Hamiltonian for any spherically symmetric system is an invariant operator with respect to the full rotation group. When spin is admitted, as in Section 2.7, the rotation which yields $\mathsf{H} \to \mathsf{H}'$ must, of course, be applied to spin operators as well as spatial operators. It is easily verified that all terms, and in particular the scalar product $\mathbf{L} \cdot \mathbf{S}$, are invariant when space and spin operators change simultaneously according to (4.36). Thus $\mathsf{H}' = \mathsf{H}$, a tensor operator of rank zero. The theorem (5.61) then states that

$$\langle n_1 j_1 m_1 | \mathsf{H} | n_2 j_2 m_2 \rangle = \delta_{j_1, j_2}\, \delta_{m_1, m_2} A \qquad (5.62)$$

since the coupling coefficients are trivial when the operator corresponds to zero angular momentum "coupling" yielding only a state $\mathsf{H}|j_2 m_2\rangle$ with identical eigenvalues. The result is, however, non-trivial: it states that the expectation value of the energy $E_{jm} = \langle jm|\mathsf{H}|jm\rangle = A$ is the same for all allowed values of m. The energy levels with spin–orbit coupling are therefore $(2j+1)$fold degenerate: this degeneracy can be resolved only by breaking the spherical symmetry of the system, e.g. by applying a magnetic field and hence imposing *axial* symmetry.

Vector operators $(j = 1)$

The irreducible components of a *vector* operator carry representation D_1 (p. 151) and are related to Cartesian components by

$$V_{+1}{}^{(1)} = -(V_x + iV_y)/\sqrt{2}, \quad V_0{}^{(1)} = V_z, \quad V_{-1}{}^{(1)} = (V_x - iV_y)/\sqrt{2} \quad (5.63)$$

the superscript indicating behaviour corresponding to $j = 1$.

An example is provided by the application of an electric field along the z-axis of a spherically symmetrical one-electron system: the perturbation operator $V_0{}^{(1)} = ez$ (cf. the Example on p. 7) is then one component of a vector operator and (5.61) becomes, disregarding any spin terms,

$$\langle n_1 l_1 m_1{}'|V_0{}^{(1)}|n_2 l_2 m_2\rangle = \begin{pmatrix} 1 & l_2 & \bigm| & l_1 \\ 0 & m_2 & \bigm| & m_1 \end{pmatrix} A.$$

The triangle condition (cf. Fig. 5.1) shows immediately that the matrix elements will vanish unless $(l_2 + 1) \geq l_1 \geq |l_2 - 1|$ and hence that a state will be "connected", through the perturbation, only with states corresponding to $\Delta l = 0, \pm 1$. Thus, as in the Example referred to, an s state will be perturbed by admixture of p-functions but not (to first order) of d-functions ($\Delta l = 2$). The same considerations apply in discussing the absorption of radiation, in which case the dipole perturbation is time-dependent and the probability of transition between two states depends on similar matrix elements. The Wigner–Eckart theorem thus provides the fundamental basis for all spectroscopic *selection rules*.

The rules for vector operators are so frequently used that we list them in full in Table 5.3 for states with quantum numbers n, l, m and n', l', m', inserting the known values of the Clebsch–Gordan coefficients. The "constants" (A) have here been replaced by "double bar matrix elements" $\langle n'j'||V||nj\rangle$ which indicate their dependence on the two particular families of states considered (n, j, and n', j'), and common factors from the Clebsch–Gordan coefficients have been absorbed into the definition of the constants.[‡] Thus to obtain all matrix elements connecting the $(n', j+1)$ states with the (n, j) states we need evaluate only *one*, e.g. choosing $m = j$ in the first line of Table 5.3: this fixes the

---

[‡] It should be noted that the "double bar matrix element" is not strictly a matrix element at all, but rather a convenient notation for the quantity that determines a whole set of matrix elements.

value of $\langle n', j+1||\mathsf{V}||n, j\rangle$ and the first two lines of the table then give all other matrix elements for all allowed values of m and for all three vector components.

The "double-bar matrix elements" are often used in conjunction with the $3j$ symbols, with a more precise normalization: they are defined so that

$$\langle n'j'm'|\mathsf{T}_q^{(k)}|njm\rangle = (-1)^{j'-m'}\begin{pmatrix} j' & k & j \\ -m' & q & m \end{pmatrix}\langle n'j'||\mathsf{T}^{(k)}||nj\rangle \quad (5.64)$$

but the procedure for using the result is unchanged.

TABLE 5.3

Matrix elements of vector operators

$$
\begin{aligned}
\langle n', j+1, m\pm 1|\mathsf{V}_{\pm 1}|n, j, m\rangle &= \mp[(j\pm m+1)(j\pm m+2)]^{1/2} \Big\} \\
\langle n', j+1, m|\mathsf{V}_0|n, j, m\rangle &= [(j+m+1)(j-m+1)]^{1/2} \Big\} \times\langle n', j+1||\mathsf{V}||n, j\rangle \\[2mm]
\langle n', j, m\pm 1|\mathsf{V}_{\pm 1}|n, j, m\rangle &= [(j\mp m)(j\pm m+1)]^{1/2} \Big\} \\
\langle n'j, m|\mathsf{V}_0|n, j, m\rangle &= m \Big\} \times\langle n', j||\mathsf{V}||n, j\rangle \\[2mm]
\langle n', j-1, m\pm 1|\mathsf{V}_{\pm 1}|n, j, m\rangle &= [(j\pm m)(j\mp m-1)]^{1/2} \Big\} \\
\langle n', j-1, m|\mathsf{V}_0|n, j, m\rangle &= [(j+m)(j-m)]^{1/2} \Big\} \times\langle n', j-1||\mathsf{V}||n, j\rangle
\end{aligned}
$$

Tensors of higher rank[‡]

From two vectors A and B, we may form nine products $\mathsf{A}_\alpha\mathsf{B}_\beta$ $(\alpha, \beta = x, y, z)$ of the Cartesian components; these products have a well-defined transformation law under rotation of axes and are the components of a second-rank Cartesian tensor. Examples are the products xx, xy, ..., whose expectation values determine the quadrupole moments of a system; and the components of the electric polarizability, which relate the three components of an induced electric dipole to the three components defining an applied electric field, and which have exactly similar transformation properties.

From such products we may form *irreducible* or "spherical" tensors, namely linear combinations providing bases for irreducible representations of the rotation group. To do this we first introduce the spherical

---

[‡] Discussions of tensors are available in other volumes of this series (Hall, 1963; McWeeny, 1963).

components (5.63) and form the products $A_\mu^{(1)}B_{\mu'}^{(1)}$. From (5.55) the required combinations are

$$T_m^{(j)} = \sum_{\mu,\mu'} A_\mu^{(1)}B_{\mu'}^{(1)} \begin{pmatrix} 1 & 1 & \bigg| & j \\ \mu & \mu' & \bigg| & m \end{pmatrix} \tag{5.65}$$

and, by the triangle rule, these fall into sets with $j = 0, 1, 2$. There will thus be one invariant ($j = 0$); one set of three quantities transforming like *vector* components; and one set of five quantities ($j = 2$). If we put in the appropriate coefficients and introduce the Cartesian tensor components $T_{\alpha\beta} = A_\alpha B_\beta$ ($\alpha\beta = xx, xy, \ldots$) it follows easily that

$$T_{\pm2}^{(2)} = \tfrac{1}{2}(T_{xx} \pm iT_{xy} \pm iT_{yx} - T_{yy})$$

$$T_{\pm1}^{(2)} = \mp\tfrac{1}{2}(T_{zx} \pm iT_{zy} \pm iT_{yz} + T_{xz})$$

$$T_0^{(2)} = \frac{1}{\sqrt{6}}(2T_{zz} - T_{xx} - T_{yy})$$

$$\rule{10cm}{0.4pt}$$

$$T_{\pm1}^{(1)} = \tfrac{1}{2}(T_{zx} \pm iT_{zy} \mp iT_{yz} - T_{xz})$$

$$T_0^{(1)} = \frac{i}{\sqrt{2}}(T_{xy} - T_{yx})$$

$$\rule{10cm}{0.4pt}$$

$$T_0^{(0)} = -\frac{1}{\sqrt{3}}(T_{xx} + T_{yy} + T_{zz})$$

$$\tag{5.66}$$

It is clear that $T_0^{(0)}$ is, apart from normalization and phase, the scalar product of the two vectors **A** and **B**; this is the only invariant bilinear form in the components. On the other hand, $T_{+1}^{(1)}$, $T_0^{(1)}$ and $T_{-1}^{(1)}$ are spherical components of a vector; this is the "vector product" $\mathbf{A} \times \mathbf{B}$ of elementary vector algebra. The five remaining components form an "irreducible tensorial set of rank 2"—in other words, they transform under rotations exactly like a set of five spherical harmonics $Y_{l,m}$ with $l = 2$.

It is clear that the reduction of the representation carried by the Cartesian components $\mathbf{T}_{\text{cart}} = (T_{xx}, T_{xy}, \ldots T_{zz})$ is achieved by means of a unitary transformation defined by (5.66). If the irreducible components are collected into a row matrix $\mathbf{T}_{\text{irred}}$, then

$$\mathbf{T}_{\text{irred}} = \mathbf{T}_{\text{cart}}\, \mathbf{U} \tag{5.67}$$

where (with appropriate labelling of rows and columns)

$$
\mathbf{U} =
\begin{array}{ccccc|ccc|c}
+2 & +1 & 0 & -1 & -2 & +1 & 0 & -1 & 0 \\
\tfrac{1}{2} & 0 & \dfrac{-1}{\sqrt{6}} & 0 & \tfrac{1}{2} & 0 & 0 & 0 & \dfrac{-1}{\sqrt{3}} & xx \\[2mm]
\tfrac{1}{2}i & 0 & 0 & 0 & -\tfrac{1}{2}i & 0 & \dfrac{-i}{\sqrt{2}} & 0 & 0 & xy \\[2mm]
0 & -\tfrac{1}{2} & 0 & \tfrac{1}{2} & 0 & -\tfrac{1}{2} & 0 & -\tfrac{1}{2} & 0 & xz \\[2mm]
\tfrac{1}{2}i & 0 & 0 & 0 & -\tfrac{1}{2}i & 0 & \dfrac{-i}{\sqrt{2}} & 0 & 0 & yx \\[2mm]
-\tfrac{1}{2} & 0 & \dfrac{-1}{\sqrt{6}} & 0 & -\tfrac{1}{2} & 0 & 0 & 0 & \dfrac{-1}{\sqrt{3}} & yy \\[2mm]
\hline
0 & -\tfrac{1}{2}i & 0 & -\tfrac{1}{2}i & 0 & -\tfrac{1}{2}i & 0 & \tfrac{1}{2}i & 0 & yz \\[2mm]
0 & -\tfrac{1}{2} & 0 & \tfrac{1}{2} & 0 & \tfrac{1}{2} & 0 & \tfrac{1}{2} & 0 & zx \\[2mm]
0 & -\tfrac{1}{2}i & 0 & -\tfrac{1}{2}i & 0 & \tfrac{1}{2}i & 0 & -\tfrac{1}{2}i & 0 & zy \\[2mm]
\hline
0 & 0 & \dfrac{2}{\sqrt{6}} & 0 & 0 & 0 & 0 & 0 & \dfrac{-1}{\sqrt{3}} & zz
\end{array}
\tag{5.68}
$$

Since $\mathbf{U}$ is unitary it is easy to reverse the transformation and express the Cartesian tensors in terms of the irreducible components:

$$\mathbf{T}_{\mathrm{Cart}} = \mathbf{T}_{\mathrm{irred}}\,\mathbf{U}^{\dagger}. \tag{5.69}$$

This is an explicit example of the general transformation matrix referred to (p. 167) in connection with the orthonormality properties of the Clebsch–Gordan coefficients.

The Wigner–Eckart theorem is applied in obtaining matrix elements of tensor operators in exactly the same way as for vector operators and needs no further discussion. Tensors of higher rank ($j > 2$) occur less frequently, but may be constructed and handled by similar methods.

The scalar product of two tensors

Tensor operators are frequently encountered in expressions which are *invariant* against rotations. The scalar product of two vectors is a special case, but a scalar product of two tensorial sets may be defined analogously. Let us consider the effect of rotation R upon the quantity

$$\sum_{m} \mathbf{e}_{m}^{(j)*}\,\mathbf{e}_{m}^{(j)} \tag{5.70}$$

where $e_m^{(j)*}$ is a basis vector conjugate to $e_m^{(j)}$ and, by definition,

$$R\, e_m^{(j)} = \sum_{m'} e_{m'}^{(j)}\, D_j(R)_{m'm}.$$

It follows (remembering $D_j R$) is a unitary matrix) that

$$\sum_m e_m^{(j)*}\, e_m^{(j)} \to \sum_m \sum_{m'} \sum_{m''} e_{m''}^{(j)*}\, e_{m'}^{(j)}\, D_j(R)_{m''m}{}^*\, D_j(R)_{m'm}$$

$$= \sum_m \sum_{m'} \sum_{m''} e_{m''}^{(j)*}\, e_{m'}^{(j)}\, D_j(R)_{m'm}\, D_j(R^{-1})_{mm''} = \sum_{m'} \sum_{m''} e_{m''}^{(j)*}\, e_{m'}^{(j)}\, \delta_{m'm''}$$

$$= \sum_{m'} e_{m'}^{(j)*}\, e_{m'}^{(j)}$$

and hence that (5.70) is an invariant. This important invariant may be written in another form by referring to the relations (5.30), (5.32) and (5.33), and proving that

$$e_m^{(j)*} \sim (-1)^{j-m}\, e_{-m}^{(j)} \tag{5.71}$$

($\sim$ meaning "transforms like"). It then follows that

$$\sum_m e_m^{(j)*}\, e_m^{(j)} = \sum_m (-1)^{j-m}\, e_{-m}^{(j)}\, e_m^{(j)} = \text{invariant} \tag{5.72}$$

and that, being a *sum* of $2j+1$ basis bectors of a product space, the invariant may be normalized in the usual sense by division by $\sqrt{(2j+1)}$.

In general, from any two tensorial sets $\{A_{m_1}^{(j_1)}\}$ and $\{B_{m_2}^{(j_2)}\}$ we may form a third $\{C_m^{(j)}\}$ which is a "direct product". We indicate this formally by writing

$$C_m^{(j)} = [\mathbf{A}^{(j_1)} \times \mathbf{B}^{(j_2)}]_m^{(j)}. \tag{5.73}$$

The construction of an invariant is a special case, with $j = m = 0$, and from (5.72) we infer

$$[\mathbf{A}^{(j)} \times \mathbf{B}^{(j)}]_0^{(0)} = \sum_m (2j+1)^{-1/2}(-1)^{j-m} A_{-m}^{(j)} B_m^{(j)} \tag{5.74}$$

where this process, yielding a tensor of lower rank, is referred to as "contraction". Reference to (5.65) shows that we have in effect evaluated the Clebsch–Gordan coefficients which appear in the coupling of two states of the same angular momentum j to a third with zero angular momentum:

$$\begin{pmatrix} j & j \,\Big|\, 0 \\ -m & m \,\Big|\, 0 \end{pmatrix} = \frac{(-1)^{j-m}}{\sqrt{(2j+1)}}. \tag{5.75}$$

For $j = 1$ the invariant (5.72) takes the form of $T_0^{(0)}$ in (5.66). When j

is integral the factor independent of m is often dropped, the resultant invariant being called the scalar product of the two tensors.

$$\mathbf{A}^{(j)} \cdot \mathbf{B}^{(j)} = \sum_m (-1)^m \mathsf{A}_{-m}{}^{(j)} \mathsf{B}_m{}^{(j)} \tag{5.76}$$

and for $j = 1$ this becomes the scalar product of two vectors in the usual sense.

EXAMPLE. *Dipole–dipole interactions.* As an illustration of the tensor operator expression of an invariant we consider the dipole–dipole interaction between an electron (spin $\mathbf{S}$) at point $\mathbf{r}$ and a nucleus (spin $\mathbf{I}$, say) at the origin. This term is usually written in elementary vector form as

$$g\beta g_n \beta_n \{ 3r^{-5}(\mathbf{S}\cdot\mathbf{r})(\mathbf{I}\cdot\mathbf{r}) - r^{-3}\mathbf{S}\cdot\mathbf{I} \}$$

and is clearly invariant, containing only scalar products. The first term in the brackets may be expanded in terms of four-index Cartesian quantities, $\mathsf{S}_\alpha r_\beta \mathsf{I}_\gamma r_\delta$, and must therefore be expressible in terms of invariants formed by contraction from two sets of two-index quantities of the type defined in (5.66). If we write

$$(\mathbf{S}\cdot\mathbf{r})(\mathbf{I}\cdot\mathbf{r}) = \sum_{\alpha\beta} r_\alpha r_\beta \mathsf{S}_\alpha \mathsf{I}_\beta = \sum_{jm} (-1)^m A_{-m}{}^{(j)} [\mathbf{S}\times\mathbf{I}]_m{}^{(j)},$$

the $[\mathbf{S}\times\mathbf{I}]_m{}^{(j)}$ being determined as in (5.66), the $A_m{}^{(j)}$ may be identified by comparing coefficients. Thus, from (5.69) the xx column of $\mathbf{U}^\dagger$ (xx *row* of $\mathbf{U}$ in (5.68)) contains the expansion coefficients of the xx component,

$$\mathsf{S}_x \mathsf{I}_x = \tfrac{1}{2}[\mathbf{S}\times\mathbf{I}]_2{}^{(2)} - \frac{1}{\sqrt{6}}[\mathbf{S}\times\mathbf{I}]_0{}^{(2)} + \tfrac{1}{2}[\mathbf{S}\times\mathbf{I}]_{-2}{}^{(2)} - \frac{1}{\sqrt{3}}[\mathbf{S}\times\mathbf{I}]_0{}^{(0)}$$

and the other $\mathsf{S}_\alpha \mathsf{I}_\beta$ may be expressed similarly. If these expressions are inserted in the left-hand side of the above expression for $(\mathbf{S}\cdot\mathbf{r})(\mathbf{I}\cdot\mathbf{r})$, and coefficients of each irreducible tensor are compared, we obtain

$$(\mathbf{S}\cdot\mathbf{r})(\mathbf{I}\cdot\mathbf{r}) = \sum_m (-1)^m A_m{}^{(2)}[\mathbf{S}\times\mathbf{I}]_m{}^{(2)} + A_0{}^{(0)}[\mathbf{S}\times\mathbf{I}]_0{}^{(0)}$$

with $A_m{}^{(l)}$ defined as in (5.66).[‡] The last term on the right is simply $\tfrac{1}{3}r^2(\mathbf{S}\cdot\mathbf{I})$ and hence the spin–spin dipole interaction becomes

$$3r^{-5}(\mathbf{S}\cdot\mathbf{r})(\mathbf{I}\cdot\mathbf{r}) - r^{-3}(\mathbf{S}\cdot\mathbf{I}) = r^{-3} \sum_m (-1)^m A_{-m}{}^{(2)}[\mathbf{S}\times\mathbf{I}]_m{}^{(2)},$$

which is simply the scalar product of two second-rank spherical tensors.

Irreducible tensor methods are of great value in applications which we consider later: they also form the starting point for the derivation of powerful matrix element decomposition theorems, for which the reader is referred to the specialized monographs (see, for example, Edmonds, 1957; Wigner, 1959; Judd, 1963; Talmi and de Shalit, 1963.

[‡] Thus, since $r_\alpha = x, y, z$, $A_2{}^{(2)} = \tfrac{1}{2}(x^2 + 2ixy - y^2)$, etc.

REFERENCES

BETHE, H. A. (1929) *Ann. Phys.* **3**, 133.

CONDON, E. U. and SHORTLEY, E. (1935) *The Theory of Atomic Spectra*, Cambridge University Press, London.

EDMONDS, A. R. (1957) *Angular Momentum in Quantum Mechanics*, Princeton University Press.

HALL, G. G. (1963) *Vectors, Matrices and Tensors*, Pergamon Press.

HEINE, V. (1960) *Group Theory in Quantum Mechanics*, Pergamon Press, London.

JUDD, B. R. (1963) *Operator Techniques in Atomic Spectroscopy*, McGraw-Hill, New York.

McWEENY, R. (1963) *Symmetry—An Introduction to Group Theory*, Pergamon Press.

OPECHOWSKI, W. (1940) *Physica* **7**, 552.

ROTENBERG, M., BIVINS, R., METROPOLIS, N. and WOOTEN, J. K. Jr. (1959) *The 3-j and 6-j Symbols*, Technology Press, M.I.T., Cambridge, Mass.

TALMI, I. and DE SHALIT, A. (1963) *Nuclear Shell Theory*, Academic Press, London and New York.

WIGNER, E. P. (1959) *Group Theory and Its Application to the Quantum Mechanics of Atomic Spectra*, Academic Press, London and New York.

CHAPTER 6

EFFECTS OF ELECTRIC AND MAGNETIC FIELDS

6.1. The classical approach

So far, we have been concerned mainly with one electron moving in
a static electric field due only to a set of fixed nuclei and have assumed
a Hamiltonian of the form (1.2a). It is frequently necessary, however,
to consider the interaction of the electron with a general electromagnetic field, including in particular the small effects arising from
electron spin and its associated magnetic moment. To treat such effects
fully it is necessary to employ field theory and relativistic quantum
mechanics, topics which lie outside the scope of this work.

Fortunately, it is possible to describe quite simply the results of the
more complete treatment, and to incorporate them in the Schrödinger–
Pauli approximation without much trouble. First, however, it is
instructive to consider in a purely classical way the introduction of
electric and magnetic fields. The *electric field strength* $\mathbf{F}$[‡] and the
magnetic flux density $\mathbf{B}$ may be derived from a scalar potential, φ, and
a three-component vector potential, $\mathbf{A}$, according to

$$\mathbf{F} = -\operatorname{grad} \varphi - \frac{\partial \mathbf{A}}{\partial t}, \qquad \mathbf{B} = \operatorname{curl} \mathbf{A}. \tag{6.1a}$$

Instead of using the vector operators "grad" and "curl", it is convenient to use the component forms, writing the equations in the form

$$F_\alpha = -\frac{\partial \varphi}{\partial r_\alpha} - \frac{\partial A_\alpha}{\partial t}, \quad B_\alpha = \frac{\partial}{\partial r_\beta}(A_\gamma) - \frac{\partial}{\partial r_\gamma}(A_\beta) \quad (\alpha\beta\gamma \text{ cyclic}) \tag{6.1b}$$

where α, β, γ label the Cartesian components ($r_\alpha = x, y, z$) and the
cyclic order[§] in the expression for B_α is reminiscent of the rule for

---

[‡] It is convenient to use $\mathbf{F}$, instead of the conventional $\mathbf{E}$, to avoid confusion with
energies.

[§] In tensor theory it is usual to write

$$B_\alpha = \varepsilon_{\alpha\beta\gamma}(\partial A_\gamma/\partial r_\beta)$$

when summation over the repeated indices (β, γ) is understood and $\varepsilon_{\alpha\beta\gamma} = \pm 1$ according
as $\alpha\beta\gamma$ is an even or odd permutation of $xyz(= 0$ otherwise). It is convenient in (6.1b) to
use the more explicit form.

177

angular momentum components. Thus, for a uniform static field $\mathbf{B}$ we require

$$B_x = \frac{\partial A_z}{\partial y} - \frac{\partial A_y}{\partial z} = \text{constant}$$

which is satisfied when $A_z = \frac{1}{2}yB_x +$ (term independent of y), and $A_y = -\frac{1}{2}zB_x +$ (term independent of z). By examining the other components we find a solution

$$A_\alpha = \tfrac{1}{2}(B_\beta r_\gamma - B_\gamma r_\beta) \quad (\alpha, \beta, \gamma \text{ cyclic}) \tag{6.2}$$

corresponding to the vector potential $\mathbf{A} = \frac{1}{2}\mathbf{B} \times \mathbf{r} \;(= -\frac{1}{2}\mathbf{r} \times \mathbf{B})$. As in the case of the electrostatic potential, this solution is not unique: for with a change of origin the r_α for a given point in space would assume different values. It can in fact be verified that any change of description in which

$$A_\alpha \to A_\alpha{}' = A_\alpha + \partial U/\partial r_\alpha, \quad \varphi \to \varphi' = \varphi - (\partial U/\partial t) \tag{6.3}$$

leads to exactly the same fields, whatever the form of the potentials. A change of type (6.3) is referred to as a "change of gauge" for the electromagnetic potentials. Clearly the new potential energy terms must be incorporated in the Hamiltonian in such a way that all *physical* quantities remain independent of choice of gauge. At the same time the wave function may be permitted to change, provided $\psi^*\psi$ remains invariant, and we must now find the nature of this change. We study the case of a one-electron system, indicating generalizations where appropriate.

The required invariance under change of gauge is somewhat more general than that dealt with in Vol. 1, App. 1, where only a change of coordinate system was considered. Let us return to the Cartesian formulation, using mv_α for components of linear momentum; these are observables, and must be invariant against change of gauge, but in the presence of fields they will not necessarily coincide with the generalized momenta conjugate to the r_α in the classical Lagrangian form of the equations of motion. The generalized momenta are defined by

$$p_\alpha = \frac{\partial L}{\partial \dot{r}_\alpha} \tag{6.4}$$

and the problem is to find a Lagrangian function $L = L(\mathbf{r}, \mathbf{p})$ with which the equations of motion may be written in the form

$$\dot{p}_\alpha = \frac{\partial L}{\partial r_\alpha}. \tag{6.5}$$

Once we have found L, and consequently the classical Hamiltonian function

$$H(\mathbf{r}, \mathbf{p}) = \sum_\alpha p_\alpha \dot{r}_\alpha - L(\mathbf{r}, \mathbf{p}), \tag{6.6}$$

we shall apply the usual prescription (cf. App. 1 of Vol. 1)

$$p_\alpha \rightarrow \mathsf{p}_\alpha = \frac{\hbar}{i} \frac{\partial}{\partial r_\alpha} \tag{6.7}$$

in order to derive the Hamiltonian operator.

The derivation of H is purely classical (see, for example, Slater, 1960, Vol. I, App. 4) and leads to the following results:

$$L = \frac{1}{2m}(m\mathbf{v})^2 + e\varphi - e(\mathbf{v} \cdot \mathbf{A}), \tag{6.8}$$

$$p_\alpha = \frac{\partial L}{\partial v_\alpha} = mv_\alpha - eA_\alpha, \tag{6.9}$$

$$H = \frac{1}{2m}(\mathbf{p} + e\mathbf{A})^2 - e\varphi \tag{6.10}$$

—the last expression being simply $(m\mathbf{v})^2/2m + V$ as in the absence of the field. Here the p_α are the generalized momenta which will be replaced by operators according to (6.7), but it is the quantity $mv_\alpha = p_\alpha + eA_\alpha$ which is the Cartesian α-component of the actual linear momentum. The Hamiltonian operator, including electromagnetic potentials, is now

$$\mathsf{H} = \frac{1}{2m}(\mathbf{p} + e\mathbf{A})^2 - e\varphi \tag{6.11}$$

where the quantity in brackets is sometimes called the "gauge invariant" momentum operator and denoted by $\boldsymbol{\pi}$:

$$\boldsymbol{\pi} = (\mathbf{p} + e\mathbf{A}). \tag{6.12}$$

If we wish to discuss current flow (depending on the velocity operator $\mathbf{v}$) or expectation values of the angular momentum, we must define these quantities using $\boldsymbol{\pi}$ instead of the $\mathbf{p}$ alone—which serves only in the case $\mathbf{A} = 0$.

Finally, we verify that expectation values of $\boldsymbol{\pi}$ are indeed invariant under change of gauge provided the wave function ψ is multiplied by an appropriate phase factor. We take new potentials as in (6.3) and suppose the corresponding new function is

$$\psi' = e^{i\lambda}\psi.$$

Then

$$\pi_\alpha' \psi' = \left(\pi_\alpha + e\,\frac{\partial U}{\partial r_\alpha}\right) e^{i\lambda}\psi = e^{i\lambda}\left(\pi_\alpha + \hbar\,\frac{\partial \lambda}{\partial r_\alpha} + e\,\frac{\partial U}{\partial r_\alpha}\right)\psi$$

and π_α' can be "carried through" the $e^{i\lambda}$ and replaced by π_α, giving

$$\langle \psi'|\pi_\alpha'|\psi'\rangle = \langle \psi|\pi_\alpha|\psi\rangle$$

provided we choose

$$\lambda = -\frac{e}{\hbar}\,U. \tag{6.13}$$

The gauge-invariant wave function, consistent with gauge change (6.2), is thus obtained by the replacement

$$\psi \to \psi' = \psi\,\exp\,[-i(e/\hbar)U]. \tag{6.14}$$

The expectation values of other operators, such as the angular momentum components and the Hamiltonian itself, are shown to be invariant in a similar way.

The time development of the state ψ' requires further discussion. The analogues of $\mathsf{H}\psi$ and $(\hbar/i)\partial\psi/\partial t$ are now, respectively, from (6.3)

$$\mathsf{H}'\psi' = \left(\frac{1}{2m}\,\pi'^2 - e\varphi + e\,\frac{\partial U}{\partial t}\right)\exp\left(-i\,\frac{e}{\hbar}\,U\right)\psi$$

$$= \exp\left(-i\,\frac{e}{\hbar}\,U\right)\left(\frac{1}{2m}\,\pi^2 - e\varphi + e\,\frac{\partial U}{\partial t}\right)\psi$$

where the π' operator is carried through the exponential (as above), and

$$\frac{\hbar}{i}\,\frac{\partial \psi'}{\partial t} = \frac{\hbar}{i}\,\exp\left(-i\,\frac{e}{\hbar}\,U\right)\frac{\partial \psi}{\partial t} - \left(\frac{\hbar}{i}\,i\,\frac{e}{\hbar}\,\frac{\partial U}{\partial t}\right)\exp\left(-i\,\frac{e}{\hbar}\,U\right)\psi.$$

Thus,

$$\mathsf{H}'\psi' + \frac{\hbar}{i}\,\frac{\partial \psi'}{\partial t} = \exp\left(-i\,\frac{e}{\hbar}\,U\right)\left(\mathsf{H}\psi + \frac{\hbar}{i}\,\frac{\partial \psi}{\partial t}\right). \tag{6.15}$$

If ψ' is an exact solution of the time-dependent equation the right-hand side will vanish and it follows that the time development of ψ' is determined by an equation of exactly the same form as the one satisfied by ψ, but with the new choice of gauge. Thus, in general, the gauge invariant solutions are determined by the usual equation

$$\mathsf{H}\psi = i\hbar(\partial\psi/\partial t) \tag{6.16}$$

where H is the Hamiltonian (6.11) which includes the vector potential. Whereas, however, the invariance of expectation values does not require that ψ be an *exact* wave function at the instant t, the expression for the time development of ψ does refer to an exact function, and properties involving time derivatives must be treated with special care. Clearly if ψ is a truncated expansion $H\psi$ may lie outside the domain of ψ and the right-hand side of (6.15) would not then vanish on invoking the time-dependent Schrödinger equation.

This means, in particular, that any direct derivation (cf. Vol. 1, Section 1.3) of an expression for the current density, and hence an induced magnetic moment, would only apply to an *exact* solution—which would not be available for a system acted upon by general electromagnetic fields. To avoid this difficulty, it seems to be more satisfactory to infer the form of the current density, and the fields to which it gives rise, by a less direct method which we now develop.

6.2. Current density

Since, from classical physics, application of a magnetic field is known to set up currents, it might appear necessary to start from the time-dependent Schrödinger equation and investigate the time development of induced currents. Fortunately, for static fields, this is not so; we may pass directly to the steady state situation by looking for stationary state eigenfunctions of the final, time-dependent Hamiltonian. Although exact solutions may not be available, variation theory may then be applied; and in this way we are able to infer the existence of a current density indirectly, obtaining expressions which apply to either exact or approximate wave functions. In order to do this, we insert an infinitesimal magnetic dipole at an arbitrary point, calculate its energy of interaction with the electron distribution, and show that the result is always consistent with a certain current density expression, just as if the electron distribution were a smeared-out charge containing induced currents.

First, a remark is necessary about the expansion of the π term in the Hamiltonian operator (6.11). Retaining the order of $\mathbf{A}$ and $\mathbf{p}$, the latter being an operator, we obtain

$$\pi^2 = (\mathbf{p}+e\mathbf{A})^2 = \{\mathbf{p}^2+e\mathbf{p}\cdot\mathbf{A}+e\mathbf{A}\cdot\mathbf{p}+e^2\mathbf{A}^2\}.$$

Thus, in the second term, $\mathbf{p}$ will act not only on the wave function but also on the $\mathbf{A}$ factor which precedes it: $\mathsf{p}_\alpha(A_\alpha\psi) = \psi(\mathsf{p}_\alpha A_\alpha)+A_\alpha(\mathsf{p}_\alpha\psi)$. In the interests of simplicity, it is usually convenient to choose the arbitrary gauge so that $\sum \mathsf{p}_\alpha A_\alpha$ vanishes (i.e. $\psi(\mathbf{p}\cdot\mathbf{A}) = 0$), the gauge then being referred to as "Coulombic". With this choice, which will henceforth be assumed,

$$\pi^2 = \mathbf{p}^2+2e\mathbf{A}\cdot\mathbf{p}+e^2\mathbf{A}^2. \tag{6.17}$$

G

It is well known (see, for example, Griffith, 1961, p. 42) that the gauge can always be chosen in this way, and that even then considerable freedom still remains.

We now consider a vector potential

$$\mathbf{A} = \mathbf{A}^{(0)} + \mathbf{A}^{(\mu)} \tag{6.18}$$

where

$$\mathbf{A}^{(0)} = \tfrac{1}{2}\mathbf{B} \times \mathbf{r}, \tag{6.18a}$$

$$\mathbf{A}^{(\mu)} = \mu_0(\boldsymbol{\mu} \times \mathbf{r})/4\pi r^3. \tag{6.18b}$$

Here $\mathbf{A}^{(0)}$ represents an arbitrary applied field and $\mathbf{A}^{(\mu)}$ is the well-known vector potential[‡] for a "test dipole" $\boldsymbol{\mu}$, whose position is chosen as the origin ($\mathbf{A} = 0$ for $\mathbf{r} = 0$). Let us suppose that ψ is a "best approximate" wave function for the case $\mathbf{A} = \mathbf{A}^{(0)}$, corresponding to absence of the dipole. The expectation energy, that of the system in the presence of a uniform static field $\mathbf{B}$, is then

$$E^{(0)} = \langle\psi|\,\frac{1}{2m}\,\mathbf{p}^2 + \frac{e}{m}\,\mathbf{A}^{(0)}\cdot\mathbf{p} + \frac{e^2}{2m}\,\mathbf{A}^{(0)2} + V|\psi\rangle \tag{6.19}$$

and this expression may be assumed stationary against any variation $\psi \to \psi + \delta\psi$, preserving normalization. If we now add to H the first-order contribution due to the infinitesimal dipole, we obtain

$$\mathsf{H}' = \frac{e}{m}\,\mathbf{A}^{(\mu)}\cdot\mathbf{p} + \frac{e^2}{2m}\,2\mathbf{A}^{(0)}\cdot\mathbf{A}^{(\mu)} = \frac{e}{m}\,\mathbf{A}^{(\mu)}\cdot\boldsymbol{\pi}^{(0)} \tag{6.20}$$

where $\boldsymbol{\pi}^{(0)}$ is the gauge invariant momentum operator for the unperturbed system. It follows easily[§] that the corresponding first-order energy change will be

$$E_{\mathrm{dip}} = \langle\psi|\mathsf{H}'|\psi\rangle$$

or, using (6.20) with the dipole vector potential given in (6.18b),

$$E_{\mathrm{dip}} = \frac{e}{m}\,\frac{\mu_0}{4\pi}\,\langle\psi|\frac{\boldsymbol{\mu} \times \mathbf{r}\cdot\boldsymbol{\pi}^{(0)}}{r^3}|\psi\rangle. \tag{6.21}$$

[‡] Note that in SI units the permeability of free space ($\mu_0/4\pi$) occurs in the expression for the vector potential (and hence flux density) due to a dipole.

[§] If $\delta\psi$ is the first-order change of ψ due to H', then

$$E = \langle\psi + \delta\psi|\mathsf{H}_0 + \mathsf{H}'|\psi + \delta\psi\rangle = E_0 + \langle\psi|\mathsf{H}'|\psi\rangle + \ldots$$

up to terms linear in μ, owing to the stationary property of the variation function ψ.

This may now be written, integrating over spin and noting that the operator is spinless,

$$E_{\text{dip}} = -\frac{\mu_0}{4\pi} \int \left(\frac{\boldsymbol{\mu} \times \mathbf{r}}{r^3} \right) \cdot (-e\hat{\mathbf{J}}) \, d\mathbf{r} \tag{6.22}$$

where (dropping the superscript zero, used only to label the unperturbed system) the scalar product is a sum of three terms and the λ-component of $\hat{\mathbf{J}}$ is defined by

$$\hat{J}_\lambda = \int \psi^*(\pi_\lambda/m)\psi \, \gamma s \quad (\lambda = x, y, z). \tag{6.23}$$

It now appears that $-e\hat{\mathbf{J}}$ plays the part of an electric current density, and that $\hat{\mathbf{J}}$ therefore corresponds to a probability current density (Vol. 1, Section 1.3). Thus, on interchanging dot and multiple in the triple scalar product we obtain[‡]

$$E_{\text{dip}} = -\boldsymbol{\mu} \cdot \mathbf{B}^{(\text{ind})} \tag{6.24}$$

where the "induced field" has components

$$B_\lambda^{\text{ind}} = \frac{\mu_0}{4\pi} \int \frac{[\mathbf{r} \times (-e)\hat{\mathbf{J}}]_\lambda}{r^3} \, d\mathbf{r}. \tag{6.25}$$

The last two equations are well known in classical physics: they describe the energy of a dipole in the field $\mathbf{B}^{\text{ind}}$ due to induced currents flowing in a continuous medium, the current density being $-e\hat{\mathbf{J}}$.

The result just established is valid for any position of the test dipole, and the interpretation of $\hat{J}_\lambda$ in (6.23) as a component of probability current density in the presence of a magnetic field is therefore justified. Since the expression does not depend on the availability of an exact wave function it is of very general utility, giving the best results available with any given variational function. To see that (6.23) provides a generalization of the current density expression in the absence of a magnetic field, we need only insert $\pi^{(0)} = \mathbf{P} + e\mathbf{A}^{(0)}$ and note that $\hat{J}_\lambda$ in the integrand of any expression such as (6.22) may be replaced by its real part. If we introduce

$$J_\lambda = \text{Re}\hat{J}_\lambda = \tfrac{1}{2}(\hat{J}_\lambda + \hat{J}_\lambda^*) \tag{6.26}$$

we obtain a *real* density. For a system in a definite spin state $\psi = \phi\eta$ ($\eta = \alpha$ or β) this yields

$$J_\lambda = \frac{\hbar}{2mi} \left(\phi^* \frac{\partial \phi}{\partial r_\lambda} - \phi \frac{\partial \phi}{\partial r_\lambda} \right) + \frac{e}{m} A_\lambda \phi^* \phi, \tag{6.27}$$

[‡] Use of the term "induced" does not preclude the existence of "permanent" currents (still determining by (6.23)) in cases where there is a non-zero angular momentum (see, for example, p. 194).

which reduces to the usual field-free result (Vol. 1, Section 1.3) when $\mathbf{A} = 0$. This important result shows that in the presence of a magnetic field the current density contains a term depending explicitly on the vector potential; this term is essential in preserving the gauge invariance of all computed properties.

6.3. Addition of relativistic terms

We now turn to a very brief description of the results derived from the Dirac equation for an electron moving in a field with scalar and vector potentials φ and $\mathbf{A}$. The Dirac equation still introduces the *field* in a semi-classical way (and for that reason still needs very small quantum electrodynamic corrections) but has the great merit of recognizing the requirements of special relativity theory, thereby taking account of the electronic mass variation at high velocity and leading to other remarkable results—notably the spin of the electron and the prediction of the positron. Since many of these by-products are of special importance in "non-relativistic" quantum mechanics, particularly in quantum chemistry and solid state physics, it is necessary to list and discuss them in some detail.

Dirac's equation employs a *four*-component wave function, whereas the simple description of spin effects has so far required the use only of a two-component "spin-orbital"

$$\psi = \phi_\alpha \alpha + \phi_\beta \beta. \tag{6.28}$$

Fortunately, however, only two of the four components are large and the small components may be eliminated, their effects being absorbed into a suitable "effective" Hamiltonian (by a procedure reminiscent of that developed in Section 1.5). The resultant "Pauli approximation" describes the effects of applied electromagnetic fields and introduces the relativistic correction terms with an accuracy which is well defined and sufficient for most purposes: in addition the Pauli Hamiltonian works on a *two*-component wave function and may therefore be handled without fundamental extension of the methods developed so far. Rather complete discussions are available elsewhere (Slater, 1960; Bethe and Salpeter, 1957; McWeeny and Sutcliffe, 1969): here we simply record the results and indicate the meaning of the new terms in the Hamiltonian.

For one electron, in a field characterized by the scalar and vector potentials φ and $\mathbf{A}$, the Pauli approximation to the Dirac equation becomes

$$\mathsf{H}\psi = \frac{\hbar}{i} \frac{\partial \psi}{\partial t}. \tag{6.29}$$

As in the non-relativistic case, stationary states occur as the eigen-functions of H,

$$H\psi = E\psi, \tag{6.30}$$

where E is the energy excluding the mass-energy term $E_0 = mc^2$. The wave function is now the usual two-component function (6.28) but the effective Hamiltonian, replacing (6.11), is given by

$$H = \left[\frac{\pi^2}{2m} + V\right] + \left[-\frac{\pi^4}{8m^3c^2} + \frac{e\hbar}{m}\,\mathbf{B}\cdot\mathbf{S} + \frac{e\hbar}{4m^2c^2}\,(\mathbf{S}\cdot\mathbf{F}\times\boldsymbol{\pi} - \mathbf{S}\cdot\boldsymbol{\pi}\times\mathbf{F})\right.$$

$$\left. + \frac{e\hbar^2}{8m^2c^2}\,\operatorname{div}\mathbf{F}\right]. \tag{6.31}$$

In terms of the electric potential φ, $V = -e\varphi$ (the usual potential energy term), while all other symbols have their usual meanings. The operators S_α, denoted collectively by $\mathbf{S}$ in the usual vector notation, have properties identical with those of the Pauli spin operators (introduced in Section 4.9 of Vol. 1) which we have used extensively in Chaps. 2 and 5: we continue to refer to them simply as "the spin operators". Again, they are expressed in dimensionless form, so that the components $\hbar S_\alpha$ are in angular momentum units. It should be noted that the π operators work on everything standing to the right of them but that div $\mathbf{F}$ is simply a multiplicative factor.

The Pauli–Dirac Hamiltonian (6.31), used with the two-component function $\psi = \phi_\alpha\alpha + \phi_\beta\beta$, leads to results mathematically identical with those from the Dirac equation up to terms in $1/c^2$ and appears to be correct to this order (for a discussion see McWeeny and Sutcliffe, 1969, App. 4), though many inaccurate versions appear in the literature.

The first bracket in (6.31) contains the non-relativistic part of the interaction with the fields included exactly as in Section 6.1. The second bracket contains the terms which are essentially relativistic in origin. The first term arises from the variation of mass with velocity, while the last has no direct interpretation, arising simply from Lorentz invariance requirements: both these terms are usually discarded because, although they give minute shifts in the energy levels, they do not resolve degeneracies and consequently are of comparatively small experimental interest. The second and third terms, on the other hand, are of great importance because they introduce in a fairly rigorous way the small spin couplings which are known to split multiplet levels and to be responsible for many easily observed effects; they represent, respectively, the *spin–orbit coupling* and the *Zeeman coupling* between the spin and the applied magnetic field—both of which have been introduced (Section 2.6), on phenomenological and semi-classical grounds, as a means of accounting qualitatively for the observed multiplet structure of atomic spectra.

On discarding the terms in π^4 and in div $\mathbf{F}$ and introducing the Bohr magneton

$$\beta = \frac{e\hbar}{2m} \tag{6.32}$$

the Hamiltonian (6.31) may be written

$$\mathsf{H} = \frac{1}{2m}\,\pi^2 + V + g\beta\mathbf{B}\cdot\mathbf{S} + \frac{g\beta}{4mc^2}\,(\mathbf{S}\cdot\mathbf{F}\times\pi - \mathbf{S}\cdot\pi\times\mathbf{F}) \tag{6.33}$$

where g, predicted here to have the value 2, is called the "electronic g factor". In fact, the Dirac theory needs a very slight correction to allow for quantization of the field; this quantum electrodynamic correction may be included simply by giving g the value 2.0023, which appears to be in exact agreement with the best experimental value. The fact that g is predicted to be almost exactly 2, as compared with the value unity which arises from a classical model of the electron as a spinning charge distribution (cf. p. 61), was one of the great successes of the Dirac theory. It is also worth noting that for practical purposes it is convenient to use atomic units, in which case (6.33) becomes

$$\mathsf{H} = \tfrac{1}{2}\pi^2 + V + \tfrac{1}{2}g(\mathbf{B}\cdot\mathbf{S}) + \tfrac{1}{8}g\alpha^2(\mathbf{S}\cdot\mathbf{F}\times\pi - \mathbf{S}\cdot\pi\times\mathbf{F}) \tag{6.33a}$$

where α is the fine structure constant ($\alpha = e^2/4\pi\varepsilon_0\hbar c$, using SI units) which has a value of almost exactly 1/137. The use of atomic units, when electric and magnetic quantities also occur, is discussed in App. 1.

The small terms in (6.33) are usually allowed for by perturbation methods, the unperturbed system being that which results when applied fields are absent and spin effects are neglected. Before considering applications, however, it is useful to ask how the expression for the field at any point in space—already interpreted in terms of currents induced in the electron distribution—must be modified to take account of electron spin.

6.4. Magnetic effects due to the spin distribution

Let us again place a "test dipole" at an arbitrary point in space. This produces an infinitesimal extra term in the vector potential, of the form given in (6.18b). The derivation of a dipole interaction energy (p. 182) is then unaffected, except that (6.19) now contains the Hamiltonian (6.33), with its spin terms. The dipole perturbation then contains an *extra* term coming from $g\beta(\mathbf{B}\cdot\mathbf{S})$ when $\mathbf{B}$ is augmented by the test-dipole field. This latter term (working to first order in the dipole field) gives an extra contribution to (6.21) of the form

$$E_{\text{dip}}^{(\text{spin})} = g\beta \langle | \sum_\lambda B_\lambda^{(\mu)} \mathsf{S}_\lambda | \psi \rangle \tag{6.34}$$

where $B_\lambda^{(\mu)}$ is the field due to the test dipole. This expression is independent of the applied field and does not arise from flow of charge. It may, however, be described in terms of a distribution of magnetic dipoles. To show this, we complete the spin integration in (6.34), remembering that $\int d\mathbf{x}$ indicates integration over both space and spin variables ($d\mathbf{x} = d\mathbf{r}\, ds$), and introduce the three quantities

$$M_\lambda = -g\beta \int (\phi_\alpha \alpha + \phi_\beta \beta)^* \mathsf{S}_\lambda (\phi_\alpha \alpha + \phi_\beta \beta)\, ds \quad (\lambda = x, y, z). \tag{6.35}$$

From the properties of the spin operators, in the form $\mathsf{S}_x \alpha = \tfrac{1}{2}\beta$, $\mathsf{S}_x \beta = \tfrac{1}{2}\alpha$, etc., corresponding to the matrix representation (5.1)[‡] it follows at once that

$$M_x = -\tfrac{1}{2}g\beta(\phi_\alpha^* \phi_\beta + \phi_\beta^* \phi_\alpha),$$

$$M_y = -\tfrac{1}{2}ig\beta(-\phi_\alpha^* \phi_\beta + \phi_\beta^* \phi_\alpha), \tag{6.36}$$

$$M_z = -\tfrac{1}{2}g\beta(\phi_\alpha^* \phi_\alpha - \phi_\beta^* \phi_\beta).$$

On writing (6.34) in terms of these quantities and inserting the field components derived from the dipole vector potential (6.18b), we finally obtain for the coupling energy between the electron spin and the test dipole

$$E_{\text{dip}}^{(\text{spin})} = \frac{\mu_0}{4\pi} \int \frac{r^2(\boldsymbol{\mu}\cdot\mathbf{M}) - 3(\boldsymbol{\mu}\cdot\mathbf{r})(\mathbf{r}\cdot\mathbf{M})}{r^5}\, d\mathbf{r} \tag{6.37}$$

—an integral over ordinary three-dimensional space. This expression is again familiar from classical physics: it is the energy of a dipole in the magnetic field produced by a continuous distribution of magnetization, in which $M_\lambda = M_\lambda(\mathbf{r})$ is the λ-component of the magnetic moment per unit volume at point $\mathbf{r}$.

The expressions in (6.36) for the components of the density of magnetization are in complete agreement with those which may be derived directly from the Dirac equation in its original four-component form (see, for example, Slater, 1960, Vol. II, Sections 23–26), thus confirming the internal consistency of the Pauli approximation.

---

‡ The definition of the representation is such that, for example, $\mathsf{S}_x(\alpha\,\beta) = (\alpha\,\beta)\mathsf{S}_x$ where S_x is the 2×2 matrix in (5.1).

6.5. Charge, current and spin density

It is already clear that the properties of a one-electron system, embodied in a probability distribution function $\psi\psi^*$, may to some extent be described in a semi-classical way; the forces exerted on nuclei may be calculated (p. 91) as if the electron were equivalent to a smeared-out distribution of negative charge; the response to an applied magnetic field may be discussed in terms of induced currents set up in this distribution; and even the magnetic effects of electron spin may be calculated by assuming the distribution to possess an appropriate density of magnetization. Semi-classical concepts of this kind, particularly when their validity can be rigorously established, are extremely valuable as an aid to the pictorial interpretation of complicated situations; at this point, therefore, it is useful to review the statistical interpretation of the wave function and the definitions of charge and current densities, and to show how a "spin density" may be introduced along similar lines.

For an arbitrary one-electron state, represented by a two-component spin-orbital (6.28), the expectation value of any function $V(\mathbf{r})$ of electronic position—let us say the potential energy—is given by

$$\langle V \rangle = \int \psi^*(\mathbf{x}) V(\mathbf{r}) \psi(\mathbf{x})\, d\mathbf{x} = \int V(\mathbf{r})\left(|\phi_\alpha(\mathbf{r})|^2 + |\phi_\beta(\mathbf{r})|^2\right) d\mathbf{r}.$$

This is consistent with the interpretation[‡] of

$$P(\mathbf{r}) = |\phi_\alpha(\mathbf{r})|^2 + |\phi_\beta(\mathbf{r})|^2 \tag{6.38}$$

either as a probability per unit volume of finding an electron (arbitrary spin) at $\mathbf{r}$, or as a density of charge (expressed in electrons/unit volume). With the latter interpretation the integrand clearly represents the potential energy of $P(\mathbf{r})\, d\mathbf{r}$ electrons in volume element $d\mathbf{r}$ at $\mathbf{r}$ ($V(\mathbf{r})$ being that of *one* electron at the same point). The charge density is in general made up of two parts, the "up-spin" density plus the "down-spin". As noted in Section 3.3, the spatial density of charge $P(\mathbf{r})$ may be derived from the density function $\rho(\mathbf{x}) = \psi(\mathbf{x})\psi^*(\mathbf{x})$, which includes the spin, by integration over the spin variable as in (3.22).

The expectation value of a general space-spin operator, $\mathbf{A}$ say, cannot be expressed in terms of $P(\mathbf{r})$ or even $\rho(\mathbf{x})$, because the operator always comes *between* ψ^* and ψ and works only on the latter. We can, however, write

$$\langle A \rangle = \int \psi^*(\mathbf{x}) A \psi(\mathbf{x})\, d\mathbf{x} = \int [A\psi(\mathbf{x})\psi^*(\mathbf{x}')]_{\mathbf{x}' = \mathbf{x}}\, d\mathbf{x} \tag{6.39}$$

---

‡ It may be useful to consult Vol. 1 (pp. 101–3) at this point.

where the complex conjugate function has been taken over to the right, but its variables have been changed from $\mathbf{x}$ to $\mathbf{x}'$ to make it immune from the action of A (which by convention only acts on the given variables $\mathbf{x}$). The subscript means that $\mathbf{x}'$ is replaced by $\mathbf{x}$ *after* applying the operator. The quantity $\psi(\mathbf{x})\psi^*(\mathbf{x}')$ is the *density matrix*[‡] for a one-particle system in state $\psi(\mathbf{x})$. As a generalization of $\rho(\mathbf{x})$ we therefore introduce

$$\rho(\mathbf{x}; \mathbf{x}') = \psi(\mathbf{x})\psi^*(\mathbf{x}').$$

It may then be asserted that *all* expectation values are determined by the single function $\rho(\mathbf{x}; \mathbf{x}')$, whose "diagonal element"

$$\rho(\mathbf{x}) = [\rho(\mathbf{x}; \mathbf{x}')]_{\mathbf{x}'=\mathbf{x}}$$

has the usual significance as the probability density function $|\psi(\mathbf{x})|^2$. Thus, (6.39) becomes

$$\langle A \rangle = \langle \psi|\mathsf{A}|\psi \rangle = \int [\mathsf{A}\rho(\mathbf{x}; \mathbf{x}')]_{\mathbf{x}'=\mathbf{x}}\, d\mathbf{x}. \tag{6.40}$$

In the special case of a spinless operator, the spin integration $\int [\ ]_{s'=s}\, ds$ may be completed at once to give

$$\langle A \rangle = \langle \psi|\mathsf{A}|\psi \rangle = \int [\mathsf{A}P(\mathbf{r}; \mathbf{r}')]_{\mathbf{r}'=\mathbf{r}}\, d\mathbf{r} \tag{6.41}$$

where $P(\mathbf{r}; \mathbf{r}')$ is a corresponding generalization of the charge density $P(\mathbf{r})$. Thus

$$P(\mathbf{r}; \mathbf{r}') = \int [\rho(\mathbf{x}; \mathbf{x}')]_{s'=s}\, ds \tag{6.42}$$

and the diagonal element $(\mathbf{r}' = \mathbf{r})$ yields $P(\mathbf{r})$ as defined in (6.38).

The functions $\rho(\mathbf{x}; \mathbf{x}')$ and $P(\mathbf{r}; \mathbf{r}')$ are the prototypes of corresponding density matrices which may be defined for a *many*-electron system; $P(\mathbf{r}; \mathbf{r}')$ has already been found useful for a system of independent particles (Section 3.3), and in general such functions provide a very convenient connection between results established for one-electron systems and corresponding results for N-electron systems, as will be evident later (Vol. 3). The immediate aim, however, is to show how the results of the preceding sections may be expressed, through $\rho(\mathbf{x}; \mathbf{x}')$, in terms of the current density and a spin density.

Let us first take the current density defined by (6.23) and (6.26). This becomes, since π_λ is a spinless operator,

$$J_\lambda = \mathrm{Re} \int \psi^*(\mathbf{x})(\pi_\lambda/m)\psi(\mathbf{x})\, ds = \mathrm{Re} \int [(\pi_\lambda/m)\rho(\mathbf{x}; \mathbf{x}')]_{s'=s}\, ds.$$

[‡] The term "matrix" is used because $\mathbf{x}$ and $\mathbf{x}'$ are in some ways analogous to row and column indices labelling the elements of a matrix. Some of the formal properties of density matrices have been indicated in Vol. 1, Section 5.7.

But from (6.42) this gives at once[‡]

$$J_\lambda = \mathrm{Re}[(\pi_\lambda/m)P(\mathbf{r};\mathbf{r}')]_{r'=r}. \tag{6.43}$$

With a general two-component wave function, the density function is

$$P(\mathbf{r};\mathbf{r}') = \int [\phi_\alpha(\mathbf{r})\alpha(s)+\phi_\beta(\mathbf{r})\beta(s)]^*[\phi_\alpha(\mathbf{r}')\alpha(s)+\phi_\beta(\mathbf{r}')\beta(s)]\,ds$$

$$= \phi_\alpha(\mathbf{r})\phi_\alpha{}^*(\mathbf{r}')+\phi_\beta(\mathbf{r})\phi_\beta{}^*(\mathbf{r}').$$

Consequently, J_λ is a sum of "up-spin" and "down-spin" currents

$$J_\lambda = J_\lambda{}^{(\alpha)}+J_\lambda{}^{(\beta)} \tag{6.44}$$

where the two components are

$$J_\lambda{}^{(\alpha)} = \mathrm{Re}[(\pi_\lambda/m)\phi_\alpha(\mathbf{r})\phi_\alpha{}^*(\mathbf{r}')]_{\mathbf{r}'=\mathbf{r}}, \quad J_\lambda{}^{(\beta)} = \mathrm{Re}[(\pi_\lambda/m)\phi_\beta(\mathbf{r})\phi_\beta{}^*(\mathbf{r}')]_{\mathbf{r}'=\mathbf{r}}. \tag{6.45}$$

Each component may be written in an explicit form, corresponding to (6.26), if desired: thus, from the first expression,

$$J_\lambda{}^{(\alpha)} = \frac{\hbar}{2mi}\left[\phi_\alpha{}^*\frac{\partial}{\partial r_\lambda}\phi_\alpha-\phi_\alpha\frac{\partial}{\partial r_\lambda}\phi_\alpha{}^*\right]+\frac{e}{m}A_\lambda\phi_\alpha{}^*\phi_\alpha \tag{6.46}$$

with a similar expression for $J_\lambda{}^{(\beta)}$. The importance of (6.43) lies in the fact that (with the appropriate definition of the density matrix) it applies to a many-electron system with any kind of wave function, exact or approximate, whereas an expression such as (6.46) refers specifically to a one-electron system.

Finally, the components of the magnetization density (6.36) may be given a more transparent interpretation by introducing the concept of *spin density*. To do this we define the functions

$$Q_\lambda(\mathbf{r};\mathbf{r}') = \int [S_\lambda\rho(\mathbf{x};\mathbf{x}')]_{s'=s}\,ds \quad (\lambda = x, y, z) \tag{6.47}$$

(primed and unprimed spin variables identified *after* operation with S_λ) with diagonal elements

$$Q_\lambda(\mathbf{r}) = [Q_\lambda(\mathbf{r};\mathbf{r}')]_{r'=r}. \tag{6.48}$$

From (6.35) it follows that the components of the magnetization density may be expressed very simply as

$$M_\lambda(\mathbf{r}) = -g\beta Q_\lambda(\mathbf{r}) \tag{6.49}$$

and that $Q_\lambda(\mathbf{r})$ may thus be regarded as the λ-component of a *density of spin angular momentum*. This interpretation is obviously consistent

---

[‡] Again (cf. p. 183) it is the *real* parts that give the conventional probability currents: the generalized current is real only across surfaces orthogonal to those of constant charge density.

with the fact that the expectation value of the λ-component of spin angular momentum is, using (6.40) together with the definition (6.47),

$$\langle S_\lambda \rangle = \langle \psi | \mathsf{S}_\lambda | \psi \rangle = \int Q_\lambda(\mathbf{r})\, d\mathbf{r} \tag{6.50}$$

just as if the angular momentum contributions were "smeared out" with a density $Q_\lambda(\mathbf{r})$.

The results we have established in previous sections may now be collected in the following statement:

When an electron is described by a Pauli-type (two-component) wave function the magnetic field it produces at any point may be computed "classically" as that due to a distribution of *charge*, the current density at any point having components

$$-eJ_\lambda(\mathbf{r}) = -\frac{e}{m}\,\mathrm{Re}\,[(\mathsf{p}_\lambda + eA_\lambda)P(\mathbf{r};\mathbf{r}')]_{\mathbf{r}'=\mathbf{r}}$$

together with a distribution of magnetization, the moment/unit volume having components

$$M_\lambda(\mathbf{r}) = -g\beta Q_\lambda(\mathbf{r})$$

where the (generalized) charge and spin densities are defined in (6.42) and (6.47).

On taking the point at which the field is required as origin of co-ordinates, this means

$$B_\lambda = \frac{\mu_0}{4\pi} \int \frac{[\mathbf{r} \times (-e\mathbf{J})]_\lambda}{r^3}\, d\mathbf{r} - \int \frac{[r^2 M_\lambda - 3(\mathbf{M}\cdot\mathbf{r})r_\lambda]}{r^5}\, d\mathbf{r}. \tag{6.51}$$

These results are in complete accord with those derived directly from Dirac theory, in terms of a relativistic four-vector whose elements are the charge density and the three components of current density. The Pauli approximation therefore provides a satisfactory and very convenient method of introducing relativistic corrections.

Examples

We conclude this section with one or two simple examples to illustrate the concepts of charge and spin density.

(i) *Electron in an s-orbital. S_z define*

For an electron in spin-orbital $\psi = \phi\alpha$, where ϕ is a spherically symmetric central field eigenfunction, there is only one non-zero component in (6.28) and $\rho(\mathbf{x}; \mathbf{x}') = \phi(\mathbf{r})\phi^*(\mathbf{r}')\alpha(s)\alpha^*(s')$. Integration gives

$$P(\mathbf{r}; \mathbf{r}') = \phi(\mathbf{r})\phi^*(\mathbf{r}'),$$

$$Q_\lambda(\mathbf{r}; \mathbf{r}') = \tfrac{1}{2}\phi(\mathbf{r})\phi^*(\mathbf{r}'),$$

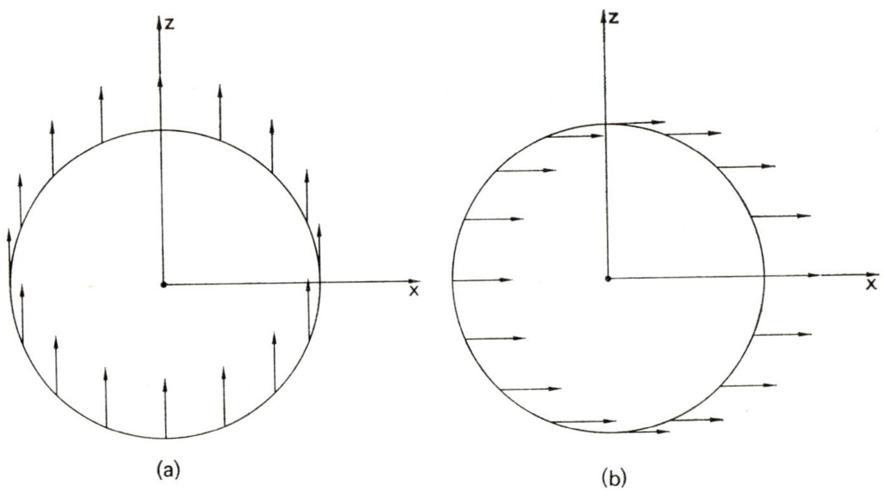

Fig. 6.1. Spin density for electron in an *s*-orbital. The spin angular momentum density is indicated by arrows, for points on a contour of constant electron density (a) for definite *z*-component of spin, (b) for definite *x*-component. In each case the spin is constant over any spherical surface, and the distribution has axial symmetry about the axis of quantization. The magnitude of the spin angular momentum density has the same radial dependence as the electron density.

with Q_x and Q_y vanishing identically. The probability density $P(\mathbf{r})$ integrates to unity and $Q_z(\mathbf{r})$ integrates to give $\tfrac{1}{2}$, which by (6.50) is the expectation value of S_z. The expectation values of S_x and S_y are zero and the vanishing of Q_x and Q_y reflects the fact that contributions to $\langle S_x \rangle$ and $\langle S_y \rangle$ vanish at all points in space. The spin vector is at all points directed along the *z*-axis (Fig. 6.1a). When ϕ is spherically symmetric, and there is no external field ($\mathbf{A} = 0$), (6.43) immediately yields zero for all components of the current density.

(ii) *Electron in an s-orbital. S_x definite*

If we take $\psi = (\phi\alpha + \phi\beta)/\sqrt{2}$ which is of the form (6.28) with $\phi_\alpha = \phi_\beta$, we obtain $P(\mathbf{r};\mathbf{r}') = \phi(\mathbf{r})\phi^*(\mathbf{r}')$ but now find $Q_z(\mathbf{r};\mathbf{r}') = 0$. Instead, however,

$$Q_x(\mathbf{r};\mathbf{r}') = \tfrac{1}{2}\int \phi(\mathbf{r})\phi^*(\mathbf{r}')\{S_x[\alpha(s)+\beta(s)][\alpha(s')+\beta(s')]^*\}_{s'=s}\,ds$$

$$= \tfrac{1}{2}\phi(\mathbf{r})\phi^*(\mathbf{r}')\int \tfrac{1}{2}[\beta(s)+\alpha(s)][\alpha(s)+\beta(s)]^*\,ds$$

and hence

$$Q_x(\mathbf{r};\mathbf{r}') = \tfrac{1}{2}\phi(\mathbf{r})\phi^*(\mathbf{r}').$$

In this case Q_x is the only non-zero component of spin density, which means that although the electron density is the same as in case (i), the spin distribution is "polarized" entirely along the x-direction (Fig. 6.1b). This is a consequence of the fact (p. 148) that the spin function $(\alpha+\beta)/\sqrt{2}$ represents a state in which S_x is definite instead of S_z.

(iii) *Electron in p-orbital. J^2, J_z definite*

Let us consider a central field eigenfunction (2.67) with orbital and spin angular momenta ($l = 1$, $s = \tfrac{1}{2}$) coupled to give a resultant with $j = m = \tfrac{1}{2}$. Then, denoting the three degenerate p-orbitals by $p_{0'}\,p_{\pm1'}$

$$\rho(\mathbf{x};\mathbf{x}') = \tfrac{1}{3}[p_0(\mathbf{r})\alpha(s) - \sqrt{2}p_{+1}(\mathbf{r})\beta(s)][p_0(\mathbf{r}')\alpha(s') - \sqrt{2}p_{+1}(\mathbf{r}')\beta(s')]^*$$

and we obtain easily

$$P(\mathbf{r};\mathbf{r}') = \tfrac{1}{3}[p_0(\mathbf{r})p_0^*(\mathbf{r}) + 2p_{+1}(\mathbf{r})p_{+1}^*(\mathbf{r})],$$

$$Q_x(\mathbf{r};\mathbf{r}') = -\frac{\sqrt{2}}{6}\,[p_0(\mathbf{r})p_{+1}^*(\mathbf{r}') + p_{+1}(\mathbf{r})p_0^*(\mathbf{r}')],$$

$$Q_y(\mathbf{r};\mathbf{r}') = \frac{-i\sqrt{2}}{6}\,[p_0(\mathbf{r})p_{+1}^*(\mathbf{r}') - p_{+1}(\mathbf{r})p_0^*(\mathbf{r})],$$

$$Q_z(\mathbf{r};\mathbf{r}') = \tfrac{1}{6}[p_0(\mathbf{r})p_0^*(\mathbf{r}') - 2p_{+1}(\mathbf{r})p_{+1}^*(\mathbf{r}')].$$

It is left as an exercise for the reader to show that the spin distribution has the general form shown in Fig. 6.2, the arrows again indicating the average magnitude and orientation of the spin vector at different points in space. It is also a straightforward matter to obtain the components of the current density, starting from $P(\mathbf{r};\mathbf{r}')$ and using equations (6.43) and (6.26): the current flows in circles around the z-axis as indicated in Fig. 6.3.

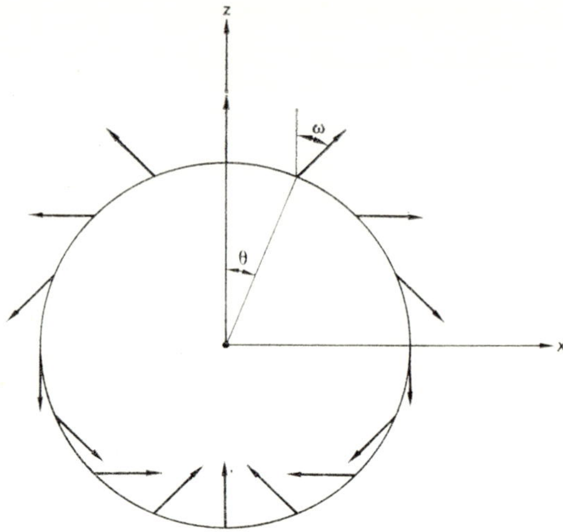

FIG. 6.2. Spin density for an electron in a p-orbital, with ls coupling. Spin angular momentum density indicated as in Fig. 6.1 for state with $j = m = \frac{1}{2}$. The electron density is spherically symmetrical; the magnitude of the spin angular momentum is constant over any spherical surface, but its inclination to the axis of quantization is $\omega = 2\theta$. The magnitude of the spin angular momentum density shows the same radial dependence as the electron density.

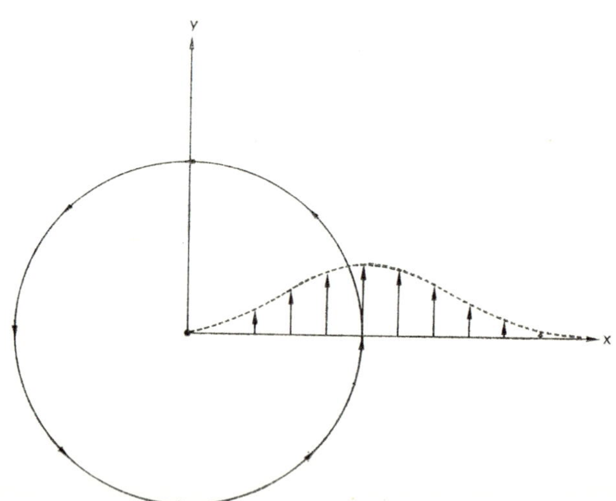

FIG. 6.3. Current density for an electron in a p-orbital, with ls coupling. The current density vector **J** is indicated by arrows, for points on a contour of constant electron density, the section shown passing through the nucleus, normal to the axis of quantization ($j = m = \frac{1}{2}$). The profile shows the magnitude of **J** as a function of axial distance.

REFERENCES

BETHE, H. A. and SALPETER, E. E. (1957) *Quantum Mechanics of One- and Two-Electron Atoms*, Springer-Verlag, Berlin.

GRIFFITH, J. S. (1961) *The Theory of Transition-Metal Ions*, Cambridge University Press, London.

McWEENY, R. and SUTCLIFFE, B. T. (1969) *Methods of Molecular Quantum Mechanics*, Academic Press, London and New York.

SLATER, J. C. (1960) *Quantum Theory of Atomic Structure*, Vols. 1 and 2, McGraw-Hill, New York.

CHAPTER 7

SPINS AND FIELDS. ILLUSTRATIVE EXAMPLES

7.1. Significance of small terms in the Hamiltonian

From Chap. 6 we now have a rather complete one-electron Hamiltonian including electric and magnetic fields and the electron spin. With this Hamiltonian it is possible to discuss in detail a large number of experimentally significant situations. In the sections that follow we consider a variety of problems, not aiming at completeness but rather at gaining further insight into quantum-mechanical concepts and methods.

First we take the Hamiltonian (6.33) expanding the π^2 operator with the vector potential $\mathbf{A} = \frac{1}{2}\mathbf{B} \times \mathbf{r}$ for a *uniform field*. This choice means that $\mathbf{A} = 0$ at the origin ($\mathbf{r} = 0$) and, as we know from (6.3), is not unique; for example, $\mathbf{A} = \frac{1}{2}\mathbf{B} \times (\mathbf{r} - \mathbf{R})$ would be equally acceptable, defining a choice of "gauge" such that $\mathbf{A} = 0$ or $\mathbf{r} = \mathbf{R}$. For a central field system the former choice is most convenient, though in a molecule further considerations may be necessary. The result of the expansion may be written in the form

$$\mathsf{H} = \mathsf{H}_0 + \mathsf{H}_{\text{ext}} + \mathsf{H}_P + \mathsf{H}_D + \mathsf{H}_Z + \mathsf{H}_{SL} \tag{7.1}$$

where

$$\mathsf{H}_0 = \frac{\mathsf{p}^2}{2m} + V_0 \qquad \text{(no field or spin terms),}$$

$$\mathsf{H}_{\text{ext}} = V_{\text{ext}} \qquad \text{(potential energy due to applied, i.e. "external" electric, fields),}$$

$$\mathsf{H}_P = \beta \mathbf{B} \cdot \mathbf{L} \qquad \text{(term linear in the field, leading to paramagnetism),}$$

$$\mathsf{H}_D = \frac{e^2}{8m}(\mathbf{B} \times \mathbf{r})^2 \qquad \text{(term quadratic in the field, leading to diamagnetism),}$$

$$\mathsf{H}_Z = g\beta \mathbf{B} \cdot \mathbf{S} \qquad \text{(the Zeeman term),} \tag{7.2}$$

$$\mathsf{H}_{SL} = \frac{g\beta}{4mc^2}(\mathbf{S} \cdot \mathbf{F} \times \boldsymbol{\pi} - \mathbf{S} \cdot \boldsymbol{\pi} \times \mathbf{F}) \qquad \text{(the spin–orbit coupling).}$$

Further terms may be added, to represent the possible presence of a nucleus possessing both a magnetic moment and an electric quadrupole moment, and in this way it is possible to account for the *hyperfine* structure of energy levels. Here, however, attention will be confined to one electron in an arbitrary field, the necessary generalizations being taken up in Vol. 3 (see also McWeeny and Sutcliffe, 1969, Chap. 8; and other references on p. 195.

It is usual to deal with the small terms in (7.1) using perturbation theory (or a combination of variation and perturbation methods) with H_0 defining the unperturbed system. The division is of course to some extent arbitrary and if a good wave function were available for a system in the presence of external fields (first four terms in (7.2)) it would be appropriate to regard the spin terms alone as the perturbation: alternatively it would be possible to apply double perturbation theory (Section 1.7). We illustrate various possibilities in the following sections.

7.2. Spin–orbit coupling (no external fields)

We are now in a position to give a more rigorous discussion of the spin–orbit interactions, arising from the last term in (7.2) and dealt with in a qualitative way in Section 2.6 for the case of an electron in a central field. In the absence of applied electric and magnetic fields, the Hamiltonian takes the form

$$H = \frac{1}{2m}\,\mathbf{p}^2 + V + \frac{g\beta}{4mc^2}\,(\mathbf{S}\cdot\mathbf{F}\times\mathbf{p} - \mathbf{S}\cdot\mathbf{p}\times\mathbf{F}) = H_0 + H_{SL} \qquad (7.3)$$

where H_0 refers to the usual spinless Hamiltonian and H_{SL} to the remaining term. For illustration, we consider a central field system in which the electric potential is $\varphi = \varphi(r)$. The potential energy and electric field strength are then, respectively, $V = -e\varphi$ and

$$\mathbf{F} = -\nabla\phi = \frac{1}{e}\frac{1}{r}\frac{\partial V}{\partial r}\,\mathbf{r},$$

and it is easily shown[‡] that the term $-\mathbf{S}\cdot\mathbf{p}\times\mathbf{F}$ may be identified with $\mathbf{S}\cdot\mathbf{F}\times\mathbf{p}$, so long as $\partial\mathbf{B}/\partial t = 0$: thus since ($\beta = e\hbar/2m$)

$$H_{SL} = \frac{g\beta}{2emc^2}\frac{1}{r}\frac{\partial V}{\partial r}\,(\mathbf{S}\cdot\mathbf{r}\times\mathbf{p}) = \frac{g\beta^2}{e^2c^2}\left(\frac{1}{r}\frac{\partial V}{\partial r}\right)\mathbf{S}\cdot\mathbf{L} = f(r)\mathbf{S}\cdot\mathbf{L} \qquad (7.4)$$

---

‡ Write out the component forms of both operators and remember that $\mathrm{p}_\beta F_\gamma\varphi = F_\gamma(\mathrm{p}_\beta\varphi) + \varphi(\mathrm{p}_\beta F_\gamma)$.—From (6.1a) curl $\mathbf{F} = 0$ when $\delta\mathbf{B}/\delta t = 0$.

since $\mathbf{r} \times \mathbf{p} = \hbar\mathbf{L}$, the angular momentum about the origin. For motion in the Coulomb field of a nucleus of atomic number Z we easily find

$$f(r) = \frac{g\beta^2}{\kappa_0 c^2}\frac{Z}{r^3}\,, \tag{7.5}$$

so the importance of spin–orbit coupling will increase with atomic number—not only directly, through the Z factor, but also because the average value of $1/r^3$ varies in proportion to Z^3 (as follows from the orbital forms obtained in Section 2.5).

Let us again consider the spin–orbit splitting of a multiplet, comprising the set of states associated with given values of the quantum numbers L and S, and try to justify the procedure used in Section 2.6 where H_{SL} was represented by a term of the form $\lambda\mathbf{S}\cdot\mathbf{L}$, λ being an empirical constant. We could of course start from first principles, considering the degenerate set of space-spin products, evaluating all matrix elements individually, and applying the perturbation theory of Section 1.3. This would, however, obscure the generality of the result, which in fact carries over essentially unchanged to many-electron systems, so long as spherical symmetry of the potential is preserved. We therefore use the group theoretical results available to us from Chaps. 4 and 5, rewriting (7.4) in the form

$$\mathsf{H}_{SL} = f(r) \sum (-1)^m \mathsf{S}_{-m}\mathsf{L}_m \tag{7.6}$$

where $m\ (= +1, 0, -1)$ runs over the spherical components of the spin and orbital angular momentum operators, as defined in (5.63). This will allow us to use the Wigner–Eckart theorem (5.61) to reduce the number of matrix elements to be computed.

We take as a basis of states the product functions $\phi_{L,M_L}\theta_{S,M_S}$ or, with a ket notation $|LM_L, SM_S\rangle$, assuming these to be eigenstates of the Hamiltonian H_0 with common eigenvalue E_0. To determine the splitting of the levels we must solve the secular equations (1.22) in which the general matrix element is a sum of three terms ($m = 0, \pm1$)

$$\langle LM_L', SM_S'|f(r)\mathsf{S}_{-m}\mathsf{L}_m|LM_L, SM_S\rangle.$$

But we know from (5.61) that matrix elements in which the three parts (function, operator, function) transform according to irreducible representations of the rotation group must be proportional to certain "geometrically" determined coupling constants which in this case are the Clebsch–Gordan coefficients. In the present case there is a spatial operator $f(r)\mathsf{L}_m$ transforming according to the representation D_1 (the L_m being *vector* operators) while the left- and right-hand functions behave, under spatial rotations, like basis vectors ($\mathbf{e}_{M_L'}$ and $\mathbf{e}_{M_L}$, say) of the representation D_L. Application of rotation operations to the spatial factors alone therefore indicates that the matrix element is proportional to

$$\begin{pmatrix} L & 1 & L \\ M_L & m & M_L' \end{pmatrix},$$

as M_L', m and M_L run over their allowed values, the remaining factor being independent of these three quantum numbers. Similar considerations apply to the spin-dependent factors and give a further proportionality factor

$$\begin{pmatrix} S & 1 & S \\ M_S & -m & M_S' \end{pmatrix},$$

and we may therefore write

$$\langle LM_L', SM_S'|f(r)S_{-m}L_m|LM_L, SM_S\rangle = \begin{pmatrix} L & 1 & L \\ M_L & m & M_L' \end{pmatrix}\begin{pmatrix} S & 1 & S \\ M_S & -m & M_S' \end{pmatrix}$$
$$\times \begin{pmatrix} \text{factor independent of} \\ M_L,\ M_L',\ M_S,\ M_S',\ m \end{pmatrix}.$$

$$(7.7)$$

Thus, if we can identify the "constant" on the right-hand side, then we can obtain all matrix elements simply by multiplying this constant by the appropriate tabulated Clebsch–Gordan coefficients. The simplest way of getting the constant is usually to evaluate just *one* matrix element, e.g. the one with $M_L = M_L' = L$, $M_S = M_S' = S$ and $m = 0$ (the matrix element vanishing unless $M_L + m = M_L'$, $M_S - m = M_S'$): it then follows immediately that

$$\langle LM_L', SM_S'|f(r)S_{-m}L_m|LM_L, SM_S\rangle = \begin{bmatrix} L & 1 & L \\ M_L & m & M_L' \end{bmatrix}\begin{bmatrix} S & 1 & S \\ M_S & -m & M_S' \end{bmatrix}$$
$$\times \langle LL, SS|f(r)S_0L_0|LL, SS\rangle$$

$$(7.8)$$

where each square-bracket coefficient is simply a ratio of coupling coefficients for the two cases considered. Thus, for any three representations D_{j_1}, D_{j_2}, D_{j_3}, the square-bracket coefficients are defined by

$$\begin{bmatrix} j_1 & j_2 & j_3 \\ m_1 & m_2 & m_3 \end{bmatrix} = \begin{pmatrix} j_1 & j_2 & j_3 \\ m_1 & m_2 & m_3 \end{pmatrix}\begin{pmatrix} j_1 & j_2 & j_3 \\ j_1 & (j_3 - j_1) & j_3 \end{pmatrix}^{-1}. \qquad (7.9)$$

It is a simple matter to evaluate these coefficients from the tables of Clebsch–Gordan coefficients (or we may proceed in an entirely parallel way using the Wigner $3j$ symbols) and thus to evaluate all matrix elements in the secular problem in terms of the *single* matrix element appearing on the right-hand side of (7.8). This procedure depends only on the rotational transformation properties of the operators and wave functions concerned, not on their specific forms, and is therefore exceedingly general. On the other hand, it does not seem to give a transparent explanation of the success of the simple procedure in Section 2.6 where it was assumed that H_{SL} could be represented by a term $\lambda \mathbf{L} \cdot \mathbf{S}$, λ being merely a numerical parameter, and where a simple algebraic argument then led to a general formula for the energies of the perturbed eigenstates.

Let us therefore examine the validity of the simpler approach: to do this we must show that the matrix elements of $f(r)\mathsf{S}_{-m}\mathsf{L}_m$ are identical with those of $\lambda \mathsf{S}_{-m}\mathsf{L}_m$ provided the numerical parameter is given the right value, and hence that $f(r)\mathbf{S} \cdot \mathbf{L}$ is *equivalent* to $\lambda \mathbf{S} \cdot \mathbf{L}$. That this is so follows easily on applying the Wigner–Eckart theorem (5.61) to the angular momentum operators alone: thus, for L_m without the factor $f(r)$, we obtain for any set of angular momentum eigenstates

$$\langle LM_L'|\mathsf{L}_m|LM_L\rangle = \begin{bmatrix} L & 1 & L \\ M_L & m & M_L' \end{bmatrix} \langle LL|\mathsf{L}_0|LL\rangle \qquad (7.10)$$

in which the coefficient is identical with that introduced in (7.8) and defined in (7.9): but $|LL\rangle$ is an eigenstate of L_0 ($=\mathsf{L}_z$) with eigenvalue $M_L = L$ and hence, from (7.10), the first coefficient in (7.8) can be written as a matrix element of L_m divided by the quantum number L. Similarly, the second coefficient can be expressed as a matrix element of S_{-m}, between states with spin quantum numbers M_S' and M_S', divided by S. We now introduce product kets $|LM_L, SM_S\rangle = |LM_L\rangle|SM_S\rangle$, which are simultaneous eigenfunctions of both orbital and spin angular momentum: in terms of these we can write the product of the coefficients in (7.8) as

$$\begin{bmatrix} L & 1 & L \\ M_L & m & M_L' \end{bmatrix}\begin{bmatrix} S & 1 & S \\ M_S & -m & M_S' \end{bmatrix}(=LS)^{-1}\langle LM_L', SM_S'|\mathsf{S}_{-m}\mathsf{L}_m|LM_L, SM_S\rangle.$$

$$(7.11)$$

If we insert this result in (7.8) and sum over m, after adding the coefficient $(-1)^m$, we obtain finally

$$\langle LM_L', SM_S'|f(r)\mathbf{S} \cdot \mathbf{L}|LM_L, SM_S\rangle = \langle LM_L', SM_S'|\lambda \mathbf{S} \cdot \mathbf{L}|LM_L, SM_S\rangle$$

$$(7.12)$$

where λ is a numerical parameter with the value

$$\lambda = (LS)^{-1}\langle LL, SS|f(r)\mathbf{S} \cdot \mathbf{L}|LL, SS\rangle \qquad (7.13)$$

which is determined by evaluating just the one matrix element. In other words, we have justified the representation of the spin–orbit coupling by an operator $\lambda \mathbf{S} \cdot \mathbf{L}$ and have shown how the parameter may be determined, via (7.13) and (7.4), from a knowledge of the field in which the electron moves.

It is easily verified that the coupling term (7.4) still commutes with $\mathbf{L}^2$ and $\mathbf{S}^2$, and the eigenstates may consequently still be classified by quantum numbers J and M, and constructed by angular momentum coupling exactly as in Section 2.6. The derivation may then be completed in the same way, yielding the Landé interval rule. But it is now clear that the parameter λ is a measure of the expectation value of the function $f(r)$ in (7.4); for only the $\mathsf{S}_0\mathsf{L}_0$ part of $\mathbf{S} \cdot \mathbf{L}$ in (7.13) gives a non-zero term (multiplying the ket by $M_L M_S \; (= LS)$) and after completing the spin integration we are left with

$$\lambda = \frac{g\beta^2}{e^2 c^2} \int \phi_{LL}{}^*(\mathbf{r}) \, \frac{1}{r} \frac{\partial V}{\partial r} \, \phi_{LL}(\mathbf{r}) \, d\mathbf{r}. \qquad (7.14)$$

This result could obviously have been obtained by using the spin-orbitals $\phi_{LM_L}\theta_{SM_S}$ explicitly from the start, separating each matrix element into a product of orbital and spin factors; but this would not have shown the generality of the derivation, which depends only on the spherical symmetry of the system and may readily be extended to many-electron systems—for which the basis functions can no longer be written as simple products of space and spin factors and the elementary procedure is no longer applicable.

The procedure by which we have justified the use of $\lambda \mathbf{L} \cdot \mathbf{S}$ instead of the more complete operator $f(r)\mathbf{L} \cdot \mathbf{S}$ is sometimes called the "method of operator equivalents". The operators are not of course identical; equivalence means that *within a certain set of states* the operators $\mathbf{L} \cdot \mathbf{S}$ and $f(r)\mathbf{L} \cdot \mathbf{S}$ have matrix elements that differ only by a constant factor λ. The group theoretical basis of this result is the observation that two sets of operators, $\{\mathsf{A}_m{}^{(j)}\}$ and $\{\mathsf{B}_m{}^{(j)}\}$ say, that provide bases for the same irreducible representation D_j of some symmetry group must have matrix elements (between functions belonging to irreducible representations of the same group) such that, for all values of m, M_1, M_2,

$$\langle \Phi_{M_1}{}^{(J_1)} | \mathsf{A}_m{}^{(j)} | \Phi_{M_2}{}^{(J_2)} \rangle = \text{constant} \times \langle \Phi_{M_1}{}^{(J_1)} | \mathsf{B}_m{}^{(j)} | \Phi_{M_2}{}^{(J_2)} \rangle \qquad (7.15)$$

as follows immediately from (5.61). In the present application we have simply replaced $\mathsf{A}_m{}^{(1)} = f(r)\mathsf{L}_m{}^{(1)}$ by $\mathsf{B}_m{}^{(1)} = \mathsf{L}_m{}^{(1)}$ since the latter operator is easier to handle: the corresponding matrix elements then differ only by a numerical factor (λ) whose value has ultimately been fixed by calculating just one complete matrix element. The result (7.15) is also known as the "replacement theorem". Its use requires great care; the operators should, strictly, always be expressed in irreducible tensor form (otherwise ambiguities can arise) and, of course, the equivalence may break down if some matrix elements vanish (e.g. if $\mathsf{B}_m{}^{(j)} = \mathsf{J}_m$ and J_1 and J_2 in (7.15) differ, then the matrix elements on the right all vanish and the equivalence cannot in fact be set up). In cases of doubt it is always safest to return to the basic theorem (5.61).

In conclusion, it should be noted that for an accurately Coulombic central field, as for a hydrogen-like system, there is an additional

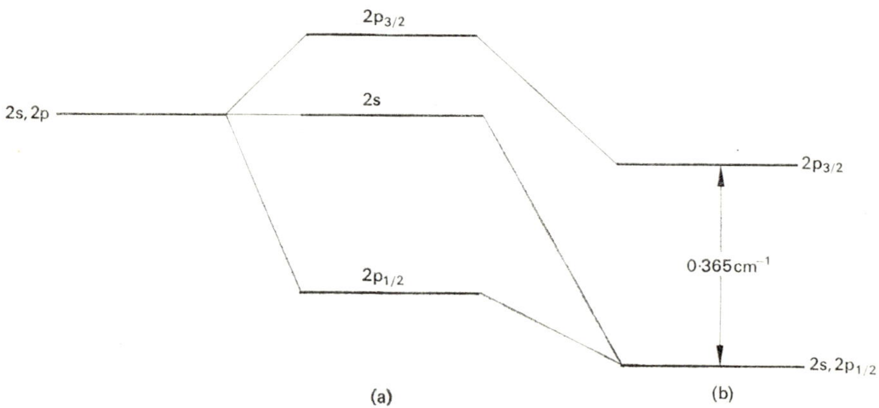

FIG. 7.1. Splitting of hydrogen $2s$ and $2p$ levels (schematic): (a) effect of spin–orbit coupling, (b) further effect of relativistic mass variation.

degeneracy since the energy is independent of the quantum number L. Nevertheless, the operator H_{SL} does not mix functions with different L values, and different sets may still be treated independently as above. Figure 7.1(a) shows the effect of spin–orbit coupling for the hydrogen atom states with $n = 2$, the s-state (as would be expected) being unaffected. In this case, however, another remarkable degeneracy arises on admitting the relativistic variation of electronic mass with velocity. When the term $-\mathbf{p}^4/8m^3c^2$ in the Dirac–Pauli Hamiltonian (6.31) is admitted, using first-order perturbation theory, there is a further shift of the levels (Fig. 7.1b), the energies of the ^2S and ^2P states now coalescing. This result is *almost* in agreement with experiment: there is, however, an extremely small discrepancy in that the two states are separated by about $0{\cdot}035$ cm⁻¹. This failure, like the fact that the electronic g value is not exactly equal to 2, goes back to the Dirac equation itself; the S levels suffer a "Lamb shift" (Lamb, 1941) associated with quantization of the field. Such minute shifts lie far beyond the limits of accuracy of molecular calculations and—although of great fundamental significance—are of little practical importance in the quantum theory of atoms and molecules.

7.3. The Zeeman effect (uniform magnetic field)

The states of a central field system remain degenerate, even in the presence of spin–orbit coupling; for each value of J ($J = L+S$, $L+S-1$, ... $|L-S|$) there are $2J+1$ distinct states. Application of the term H_Z in (7.2) is sufficient to resolve this degeneracy; if the applied field is weak we may regard H_Z as a further small perturbation mixing the states of given J, and can account for the observed *Zeeman effect*; if, on the other hand, the field is strong compared with H_{SL} the coupling of L and S to a resultant J may be broken, leading to the *Paschen–Back effect*. Other cases (e.g. "intermediate" coupling) may be distinguished according to the relative magnitudes of the various perturbing terms in (7.2) and must be considered individually.

(i) *Normal Zeeman effect*

With non-zero *orbital* angular momentum the term H_P (orbital paramagnetism) in (7.2) also makes an important contribution to the interaction with the field, and we therefore introduce a *total* (orbital plus spin) Zeeman term

$$\mathsf{H}_Z{}^{\mathrm{tot}} = \mathsf{H}_P + \mathsf{H}_Z = \beta B(\mathsf{L}_z + g\mathsf{S}_z) \tag{7.16}$$

where the field direction has been used as z-axis. It is at once clear that the states of given J, M, *referred to this axis*, are correct zero-order eigenfunctions (p. 9) of $\mathsf{H}_0 + \mathsf{H}_{SL}$ in the sense that the additional small perturbation $\mathsf{H}_Z{}^{\mathrm{tot}}$ will have zero matrix elements between states of different M_J and will therefore cause no mixing. It therefore remains only to calculate the energy shift for each state separately, namely

$$\Delta E_M = \langle JM|\mathsf{H}_Z{}^{\mathrm{tot}}|JM\rangle = B\beta\langle JM|(\mathsf{L}_z + g\mathsf{S}_z)|JM\rangle \tag{7.17}$$

according to first-order perturbation theory for a degenerate level (Section 1.3).

To obtain the matrix elements of L_z and S_z we note that L_m and S_m are spherical components of vector operators and, in a rotation affecting both space and spin operators, each would transform by exactly the same rule as $\mathsf{J}_m = \mathsf{L}_m + \mathsf{S}_m$. In matrix elements between $|J, M\rangle$ states we may therefore write

$$\langle JM'|\mathsf{S}_m|JM\rangle = a\langle JM'|\mathsf{J}_m|JM\rangle,$$
$$\langle JM'|\mathsf{L}_m|JM\rangle = b\langle JM'|\mathsf{J}_m|JM\rangle. \tag{7.18}$$

The operators are "equivalent" in the sense of equation (7.15), and a and b are simply proportionality constants. If now we observe that $\mathsf{J}_m = \mathsf{L}_m + \mathsf{S}_m$ and add the last two equations we obtain

$$\langle JM'|\mathsf{J}_m|JM\rangle = (a+b)\langle JM'|\mathsf{J}_m|JM\rangle$$

and hence $a+b = 1$. Let us work in terms of a: to identify this quantity we use the first equation of (7.18), together with the identity

$$\mathbf{L}^2 = \mathbf{J}^2 + \mathbf{S}^2 - 2\mathbf{J}\cdot\mathbf{S}$$

and the fact that the kets $|JM\rangle$ are also eigenstates of $\mathbf{L}^2$ and $\mathbf{S}^2$.

Thus, on taking the expectation value of $\mathbf{L}^2$ in state $|JM\rangle$, we obtain

$$L(L+1) = J(J+1) + S(S+1) - 2\langle JM|\mathbf{J}\cdot\mathbf{S}|JM\rangle. \tag{7.19}$$

The $\mathbf{J}\cdot\mathbf{S}$ matrix element may now be expressed, through (7.18), in terms of known matrix elements, enabling us to eliminate the unknown a. To see this we note that the set of $(2J+1)$ kets $\{|JM\rangle\}$ is closed under the J and S operators, so a matrix element of a product operator may be calculated by the usual matrix product rule

$$\langle M|\mathsf{AB}|M\rangle = \sum_{M'} \langle M|\mathsf{A}|M'\rangle\langle M'|\mathsf{B}|M\rangle :$$

hence

$$\langle JM| \sum_m (-1)^m \mathsf{J}_{-m}\mathsf{S}_m|JM\rangle = \sum_m (-1)^m \sum_{M'}\langle JM|\mathsf{J}_{-m}|JM'\rangle\langle JM'|\mathsf{S}_m|JM\rangle$$

with all matrix elements known except that of S_m. On using (7.18) we then introduce the proportionality factor a, obtaining

$$\langle JM|\mathbf{J}\cdot\mathbf{S}|JM\rangle = a \sum_m (-1)^m \sum_{M'} \langle JM|\mathsf{J}_{-m}|JM'\rangle\langle JM'|\mathsf{J}_m|JM\rangle$$
$$= a\langle JM|\mathbf{J}^2|JM\rangle. \tag{7.20}$$

The matrix elements of $\mathbf{J}^2$ and $\mathbf{J}\cdot\mathbf{S}$ are known, however, and hence a is identified through (7.19) and (7.20). On noting that $\langle JM|\mathsf{J}_0|JM\rangle = M$, and remembering that $b = 1-a$, we obtain from (7.18) the required diagonal elements

$$\langle JM|\mathsf{S}_0|JM\rangle = M\left[\frac{J(J+1)+S(S+1)-L(L+1)}{2J(J+1)}\right],$$
$$\langle JM|\mathsf{L}_0|JM\rangle = M\left[\frac{J(J+1)-S(S+1)+L(L+1)}{2J(J+1)}\right]. \tag{7.21}$$

It follows at once from (7.17), on inserting the value $g = 2$, that

$$\Delta E_M = \beta g_J M B = \langle JM|\beta g_J \mathbf{J} \cdot \mathbf{B}|JM\rangle \tag{7.22}$$

where the *Landé g-factor* is defined by

$$g_J = 1 + \frac{J(J+1) - L(L+1) + S(S+1)}{2J(J+1)}. \tag{7.23}$$

In other words, the combined angular momentum $\mathbf{J} = \mathbf{L} + \mathbf{S}$ is magnetically equivalent to a dipole $g_J \beta \mathbf{J}$ in which g_J lies between the values 1 and 2 associated with orbital and spin angular momenta separately. This result is not unexpected and is often derived non-rigorously (see, for example, Slater, 1960, Vol. I, Section 10.4), using a vector diagram (cf. Fig. 2.7) to represent the coupling and adding the magnetic moment vectors in a similar way.

(ii) *Paschen–Back effect*

If the applied field is very strong, so that $\mathsf{H}_Z^{\text{tot}}$ is the dominant perturbation, we must proceed in the reverse order, first finding the correct zero-order eigenfunctions corresponding to $\mathsf{H}_{SL} \to 0$. Clearly the unperturbed states in the absence of coupling are the kets $|LM_L, SM_S\rangle$, which are simultaneous eigenvectors of L_z and S_z and hence of $(\mathsf{H}_0 + \mathsf{H}_Z^{\text{tot}})$. The corresponding energy, relative to that in the absence of the field, is

$$\beta B(M_L + g M_S).$$

Such states may still be degenerate (e.g. since $g \simeq 2$ the states $|M_L+1, -\tfrac{1}{2}\rangle$ and $|M_L-1, +\tfrac{1}{2}\rangle$ will have the same energy), but the remaining perturbation term H_{SL} has zero matrix elements between them and they are therefore correct zero-order functions of the full Hamiltonian. When H_{SL} is admitted, using the form $\lambda \mathbf{S} \cdot \mathbf{L}$ to which we know it is equivalent, we obtain the first-order energies, relative to E_{LS} of the degenerate multiplet, as the diagonal matrix element

$$\Delta E_{M_L M_S} = \langle LM_L SM_S|\mathsf{H}_Z^{\text{tot}} + \mathsf{H}_{SL}|LM_L, SM_S\rangle$$
$$= \beta B(M_L + g M_S) + \lambda M_L M_S. \tag{7.24}$$

This formula correctly describes the high-field splitting known as the Paschen–Back effect.

(iii) *Intermediate case*

Starting from either limit (Zeeman or Paschen–Back) it is possible to derive formulae valid to the *second* order in the smaller of the two perturbations and in that way to obtain some account of the behaviour of the splitting in the "intermediate range" where the two perturbations are of comparable magnitude. Very often, however, as in the case $M_S = \frac{1}{2}$ with which we are concerned here, the secular equations are small and may be solved exactly.

It is easily verified that for $M_S = \frac{1}{2}$ the full perturbation $\mathsf{H}_Z{}^{\text{tot}} + \mathsf{H}_{SL}$ has off-diagonal matrix elements only between functions $|M_L, \frac{1}{2}\rangle$ and $|M_L+1, -\frac{1}{2}\rangle$, i.e. with the same value of $M = M_L+M_S = M_L+\frac{1}{2}$. Taking these kets as a basis the matrix of the perturbation is easily found to be (using $g = 2$), for a pair of given M,

$$B\beta\begin{bmatrix} M+\tfrac{1}{2} & 0 \\ 0 & M-\tfrac{1}{2} \end{bmatrix} + \frac{\lambda}{2}\begin{bmatrix} (M-\tfrac{1}{2}) & \sqrt{\{(L+\tfrac{1}{2})^2 - M^2\}} \\ \sqrt{\{(L+\tfrac{1}{2})^2 - M^2\}} & -(M+\tfrac{1}{2}) \end{bmatrix}.$$

The resolution of the degeneracy (which arises for $B = 0,\ \lambda = 0$) may then be followed over the whole range by solving, for each pair of basis functions, the 2×2 secular equation

$$\begin{vmatrix} (a-E) & c \\ c & (b-E) \end{vmatrix} = 0 \tag{7.25}$$

in which

$$a = \beta B(M+\tfrac{1}{2}) + \tfrac{1}{2}(M-\tfrac{1}{2})\lambda,$$

$$b = \beta B(M-\tfrac{1}{2}) - \tfrac{1}{2}(M+\tfrac{1}{2})\lambda,$$

$$c = \tfrac{1}{2}\lambda\sqrt{\{(L+\tfrac{1}{2})^2 - M^2\}}. \tag{7.26}$$

The results of solving the quadratic equation $(a-E)(b-E)-c^2 = 0$ for all states of a P multiplet are indicated in Fig. 7.2, the high- and low-field limits appearing at either side of the diagram.

7.4. Diamagnetism and paramagnetism

If the applied magnetic field is exceedingly strong the "diamagnetic term" in (7.1) which we have neglected so far may become appreciable. The full perturbation may then be written

$$\mathsf{H}' = \mathsf{H}_{SL} + \mathsf{H}_Z + \mathsf{H}_P + \mathsf{H}_D$$

where the first three terms give the Paschen–Back energy formulae (7.24). The additional term simply causes a small "diamagnetic shift" of each level, given as the expectation value

$$\Delta E_{\text{dia}} = \langle M_L M_S| \frac{e^2(\mathbf{B} \times \mathbf{r})^2}{8m} |M_L M_S\rangle. \tag{7.27}$$

With the field in the z direction $(\mathbf{B} \times \mathbf{r})$ is a vector in the xy-plane and the operator reduces to $B^2(e^2/8m)r^2 \sin^2 \theta$ where θ is measured from, say, the x-axis: we then obtain

$$\Delta E_{\text{dia}} = \frac{e^2 B^2}{8m} \langle r^2 \sin^2 \theta \rangle,$$

which may easily be evaluated with central field wave functions. This result allows us to define a "diamagnetic polarizability" α_m, analogous

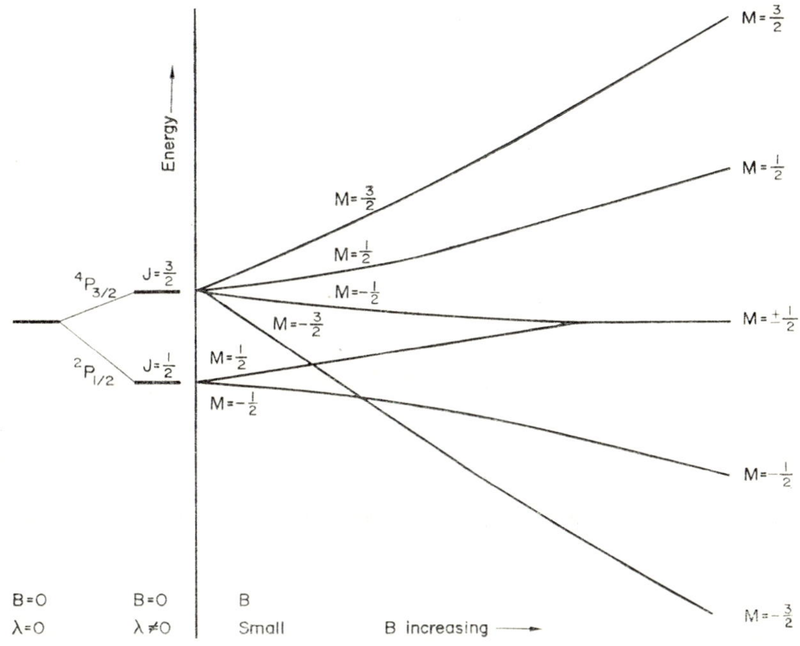

Fig. 7.2. Zeeman splitting of $^4P_{3/2}$ and $^2P_{1/2}$ states; showing transition from weak to strong field limit. (The splitting has been exaggerated in the weak field region.)

to the electric polarizability α introduced in a previous Example (p. 20), by writing ΔE_{dia} in the form

$$\Delta E_{\text{dia}} = -\tfrac{1}{2}\alpha_m B^2. \tag{7.28}$$

The implication of (7.28) is that the system possesses a field-dependent magnetic moment $\alpha_m \mathbf{B}$, as will be verified presently.

We shall shortly interpret the magnetic moment in terms of "currents" in the electron distribution: first, however, we recall the classical definitions relating to permanent and induced moments and collect the results obtained so far. If the energy of a system (which at present we assume isotropic) is written

$$E(\mathbf{B}) \;=\; E_0 - \boldsymbol{\mu}_m \cdot \mathbf{B} - (\tfrac{1}{2}!)\alpha_m \mathbf{B} \cdot \mathbf{B} + \dots \qquad (7.29)$$

we call $\boldsymbol{\mu}_m$ the *permanent magnetic moment* and α_m the *diamagnetic polarizability*.

The permanent moment $\boldsymbol{\mu}_m$ has already been evaluated, for a free atom, during the discussion of the Zeeman effect (Section 7.3). In the case of strong spin–orbit coupling, the linear term (7.22) indicates that $\boldsymbol{\mu}_m$ has components

$$\mu_{m\kappa} \;=\; -g_J\beta\langle J_\kappa\rangle \quad (\kappa = x, y, z) \qquad (7.30\text{a})$$

while in the Paschen–Back limit (coupling "broken")

$$\mu_{m\kappa} \;=\; -\beta\langle L_\kappa\rangle - 2\beta\langle S_\kappa\rangle. \qquad (7.30\text{b})$$

With quantization along the field direction (z) the relevant component ($\kappa = z$) is, of course, determined by quantum number M (or M_L, M_S).

On the other hand, α_m measures the moment induced by an applied field of unit flux density. This interpretation follows because $E(\mathbf{B})$ may be interpreted classically as the energy of a dipole of moment

$$\mathbf{d}_m \;=\; \boldsymbol{\mu}_m + \alpha_m\mathbf{B} \qquad (7.31)$$

in the field $\mathbf{B}$: for the interaction energy of a dipole of moment $\mathbf{d}_m$ is $-\mathbf{d}_m \cdot \mathbf{B}$ and if $\mathbf{d}_m$ is field-dependent according to (7.31) then integration from flux density zero up to its final value $\mathbf{B}$ immediately yields (7.29). From equation (7.27) *et seq.* it follows that the diamagnetic polarizability is

$$\alpha_m \;=\; -\frac{e^2}{4m}\,\langle r^2\sin^2\theta\rangle. \qquad (7.32)$$

Since the induced moment $\alpha_m B$ does not become comparable with μ_m until B is of the order of 1 atomic unit ($\sim 10^9$ gauss) it is clear that diamagnetism is normally completely masked by the presence of any permanent magnetic moment. Diamagnetism is an important effect only for substances not showing paramagnetism.

We note in passing that for a large number of systems (N) at any finite temperature (T) the mean moment of the sample may be written[‡] $\mathbf{M} = (\chi_m/\mu_0)\mathbf{B}$ and is obtained by averaging over all degenerate quantum states. The *paramagnetic susceptibility* χ_m is a sum of two

---

[‡] The permeability of free space μ_0 appears only because χ_m is traditionally defined by $\mathbf{M} = \chi_m\mathbf{H}$ where $\mathbf{H} = \mathbf{B}/\mu_0$ is the magnetic field strength. It is now generally recognized that the flux density $\mathbf{B}$ is the more primitive quantity (see App. 1).

parts; one is associated with the permanent moment μ_m, arising from the tendency of the field to align the dipoles along it (in competition with thermal motion which would give a random distribution with no resultant moment), and gives

$$\chi_m^{\,p} = N\mu_0\mu_m^{\,2}/3kT$$

where k is Boltzmann's constant. In cases where paramagnetism is absent, on the other hand, there is a total induced moment in the sample, which is again proportional to the field with a proportionality constant

$$\chi_m^{\,d} = N\mu_0\bar{\alpha}_m$$

where $\bar{\alpha}_m$ is the induced moment per system, again suitably averaged if necessary (e.g. in the case of molecules, by "tumbling"), and the *diamagnetic susceptibility* $\chi_m^{\,d}$ is temperature independent. It is therefore a simple matter to obtain information about μ_m and α_m from susceptibility measurements on bulk samples.

Current density interpretation

Let us now try to interpret diamagnetism and paramagnetism in terms of the probability currents discussed in Section 6.2. The energy term associated with H_P in the Paschen–Back limit where orbital and spin contributions are clearly separated is

$$\Delta E_{\text{para}} = \beta M_L B$$

corresponding to a magnetic dipole of moment $-\beta M_L$ along the field direction. This dipole arises from the permanently circulating current corresponding to M_L units of angular momentum about the field axis.

To see how the current is distributed we may consider a p_{+1} state ($L = 1, M_L = 1$) in which ϕ has the form

$$\phi = (x + iy)F(r).$$

Let us use the field as the z-axis and evaluate the current density components. Thus

$$\pi_x\phi = (\mathsf{p}_x + eA_x)\phi$$

and on inserting $A_x = \frac{1}{2}[\mathbf{B} \times \mathbf{r}] = -By$ we may evaluate $\hat{J}_x = \phi^*\pi_x\phi/m$. On introducing the axial distance $d = (x^2 + y^2)^{1/2}$, we obtain (using the real form of the current density given in (6.26))

$$J_x = \tfrac{1}{2}(\hat{J}_x + \hat{J}_x{}^*) = -m^{-1}(\hbar + \tfrac{1}{2}eBd^2)|F(r)|^2 y.$$

In the same way we find $J_z = 0$ but

$$J_y = \tfrac{1}{2}(\hat{J}_y + \hat{J}_y{}^*) = m^{-1}(\hbar + \tfrac{1}{2}eBd^2)|F(r)|^2 x.$$

In the absence of an applied field the electric current ($\mathbf{j} = -e\mathbf{J}$) thus flows in circles about the axis of quantization, its magnitude at distance r from the axis being

$$j_0 = (e\hbar/m)d|F(r)|^2 = (e\hbar/m)r \sin \theta |F(r)|^2. \tag{7.33}$$

This distribution of currents is essentially as indicated in Fig. 6.3. The effect of the field is to add a field proportional term, giving

$$j = (e/m)(\hbar + \tfrac{1}{2}eBr^2 \sin^2 \theta)r \sin \theta |F(r)|^2.$$

It is easily verified that for an orbital with -1 unit of angular momentum about the axis the effect would be to subtract a similar term. Again, remembering that one atomic unit of $\mathbf{B}$ corresponds to about 2×10^9 gauss, it is clear that the current j_0, which gives the system its "permanent" magnetic moment, is scarcely affected by normal laboratory fields—which speed up or slow down the circulation about the axis by only a minute amount. It is the magnetic moment associated with j_0 that is responsible for the orbital paramagnetism of atoms in states of non-zero angular momentum.

We now turn to the field-proportional part of $\mathbf{j}$, which arises from the $\mathbf{A}$ term in (6.27) and has components[‡]

$$j_\lambda^{\text{ind}} = -(e^2/m)\phi^* A_\lambda \phi.$$

It should be possible to show that this induced current accounts for the energy term ΔE_{dia} defined in (7.27). We know that the interaction energy with external magnetic fields may be computed classically from the current distribution, and therefore require the energy of interaction between the induced currents and the applied field $\mathbf{B}$. This may be obtained by calculating the work done on the current distribution by building up the field to a final flux density $\mathbf{B}$, starting from zero. During this process the changing magnetic field produces an electric field $\mathbf{F} = -\partial \mathbf{A}/\partial t$ and the work done on the current $\mathbf{j}$ (i.e. the moving charge density) per unit volume and unit time is $\mathbf{F} \cdot \mathbf{j}$. This becomes, when the components of $\mathbf{j}$ are defined by the last equation,

$$\phi^*(e^2/m)\mathbf{A} \cdot (\partial \mathbf{A}/\partial t)\phi$$

and integration with respect to t gives (since $\int_0^t x(\partial x/\partial t)\,dt = \int_0^x x\,dx = \tfrac{1}{2}x^2$)

$$\phi^* \tfrac{1}{2}(e^2/m)\mathbf{A}^2 \phi.$$

With a uniform field $\mathbf{A} = \tfrac{1}{2}\mathbf{B} \times \mathbf{r}$ and the work done in building up the

---

[‡] This is true with our present choice of gauge (p. 196): otherwise the central field functions acquire a field-dependent phase factor and the first part of (6.27) must be included to obtain a gauge invariant result.

current distribution, obtained by integrating over all space, is thus (cf. (7.27))

$$\int \phi^* \frac{e^2}{8m} (\mathbf{B} \times \mathbf{r})^2 \phi \, d\mathbf{r},$$

which is just the diamagnetic energy ΔE_{dia}. We have thus obtained a detailed and consistent picture of the origin of the diamagnetism described by the $\mathbf{B}^2$ term in (7.29): it represents the interaction between the applied field $\mathbf{B}$ and the induced currents built up as it is turned on.

Generalizations

The above derivation of the effect of the perturbation $\mathsf{H}_P + \mathsf{H}_D$ in (7.1) is valid only for a central-field system, and depends on the fact that the gauge (i.e. the "origin" of the vector potential, where $\mathbf{A} = 0$) may be chosen so that the term H_P has zero matrix elements between the state of interest and all other unperturbed states. This is because H_D is *quadratic* in the field $\mathbf{B}$ and in working to order $\mathbf{B}^2$ we should allow for the fact that the linear term H_P may also produce $\mathbf{B}^2$ terms in the *second* order of perturbation theory. Thus, treating for simplicity only the perturbation $\mathsf{H}' = \mathsf{H}_P + \mathsf{H}_D$, the energy of the state $|n\rangle$ (a correct zero-order function) is given up to terms in $\mathbf{B}^2$ by

$$\Delta E_n = \langle n|\mathsf{H}_P + \mathsf{H}_D|n\rangle + \sum_{(m \neq n)} \frac{\langle n|\mathsf{H}_P|m\rangle \langle m|\mathsf{H}_P|n\rangle}{(E_n - E_m)} . \qquad (7.34)$$

In the central-field case, we took $\mathsf{H}_P = (e/m)\mathbf{B} \cdot \mathbf{L} = (e/m)BL_z$ and the sum therefore vanished, all kets being eigenstates of H_P. More generally, however, the sum cannot be made to vanish, and its evaluation raises all the usual difficulties of second-order perturbation theory. Moreover, the division of ΔE_n into the two parts is arbitrary owing to the non-uniqueness of the choice of gauge: even in the central-field case, the choice $\mathbf{A} = \frac{1}{2}\mathbf{B} \times (\mathbf{r} - \mathbf{R})$, which makes no *physical* difference, changes the balance of the two terms in ΔE_n, only their *sum* being unaffected.

Except for systems of high symmetry, a further generalization is also necessary. This arises when the induced moment differs in direction from that of the field $\mathbf{B}$; the response of the system is then *anisotropic* and instead of (7.31) we obtain

$$d_{m\kappa} = \mu_{m\kappa} + \sum_\lambda \alpha_{m\kappa\lambda} B_\lambda \qquad (7.35)$$

where the set of coefficients $\alpha_{m\kappa\lambda}$ comprise a *polarizability tensor*. The energy up to terms quadratic in the field is then given by

$$\bar{E}(\mathbf{B}) = \bar{E}_0 - \sum_\kappa \mu_{m\kappa} B_\kappa - \frac{1}{2!} \sum_{\kappa,\lambda} \alpha_{m\kappa\lambda} B_\kappa B_\lambda + \cdots, \qquad (7.36)$$

which is a corresponding generalization of (7.29). Expressions for the coefficients may be derived by picking out terms in (7.34), according to the field components they contain, but for such developments the reader is referred elsewhere (see, for example, Davies, 1967).

7.5. Crystal field effects (applied electric field)

In Sections 1.2 and 1.3 we illustrated, by means of examples, some of the effects of applying a uniform electric field to a central-field system. The consequent resolution of degenerate electronic levels is the *Stark effect*. Basically, the effect of the field is to produce a polarization of the electron distribution along the field direction and to first approximation linear in the field strength, and the interaction between the field and this induced electric dipole then leads to an energy term quadratic in the field. If the electric dipole moment is (cf. (7.31))

$$\mathbf{d} = \boldsymbol{\mu} + \alpha\mathbf{F} + \ldots \tag{7.37}$$

the energy takes the form

$$E(\mathbf{R}) = E_0 - \boldsymbol{\mu}\cdot\mathbf{F} - \tfrac{1}{2}\alpha\mathbf{F}\cdot\mathbf{F} + \ldots \tag{7.38}$$

corresponding to the work done in building up the dipole $\mathbf{d}$ as the field is increased from zero to its final value $\mathbf{F}$ (cf. (7.29) *et seq.*). Thus, $\boldsymbol{\mu}$ represents a *permanent* electric moment of the system and $\alpha\mathbf{F}$ an *induced* moment corresponding to *electric polarizability* α. For a system lacking spherical symmetry, the response to an applied field may be anisotropic; in this case the field may be defined with respect to system-fixed axes (e.g. axes of symmetry of a molecule) and the vectors $\boldsymbol{\mu}$ and $\mathbf{F}$ need not even lie in the same direction. The more general formulae are then

$$d_\kappa = \mu_\kappa + \sum_\lambda \alpha_{\kappa\lambda} F_\lambda, \tag{7.39}$$

$$E(\mathbf{F}) = E_0 - \sum_\kappa \mu_\kappa F_\kappa + \tfrac{1}{2} \sum_{\kappa,\lambda} \alpha_{\kappa\lambda} F_\kappa F_\lambda + \ldots \tag{7.40}$$

where the coefficients $\alpha_{\kappa\lambda}$ define a polarizability tensor. These definitions and formulae are entirely analogous to those describing the response to a magnetic field (Section 7.8), but here there are none of the underlying complications connected with velocity-dependent perturbations, induced currents and gauge invariance.

There is no difficulty in extending the perturbation theory of Chap. 1 to the anisotropic situation and to the calculation not only of the Stark effect but also of related bulk properties of molecular assemblies—such as dielectric constant and refractive index. We therefore turn at once to the effects of *non-uniform* applied fields.

One of the most interesting examples of a non-uniform field perturbation has been referred to already in Section 4.1: in many types of crystal each ion is surrounded by neighbouring ions which produce an electric field of rather high symmetry (e.g. tetrahedral or octahedral). An electron in such an ion therefore experiences a dominant central field and a perturbing *crystal field*. To illustrate the effect of such a field we shall consider a perturbation of orthorhombic symmetry and indicate how the discussion opened in Chap. 4 may be extended to obtain information on the actual positions of the resolved energy levels.

Let us consider a central field system (at the origin of coordinates) in the presence of equal point charges located at $x = \pm X$, $y = \pm Y$, $z = \pm Z$. The field, or more conveniently the potential, may then be expanded about the origin and expressed in terms of spherical harmonics. The leading term will be spherically symmetrical and may be included as part of the central field: the next non-zero term has the form

$$\Delta V = f(r)[ax^2 + by^2 + cz^2] \tag{7.41}$$

where $a + b + c = 0$ and this may be written in terms of real harmonics d_{z^2} and $d_{x^2-y^2}$ (p. 58) or equivalently in the form

$$\Delta V = F(r)[A Y_{2,0} - (iB/\sqrt{6})(Y_{2,2} - Y_{2,-2})] \tag{7.42}$$

where $A = -(a+b)$ and $B = (a-b)$. The values of the constants are determined by the charges and distances. To investigate the splitting of the central-field states[‡] $|LM\rangle$ ($M = L, L-1, \ldots, -L$), due to the perturbing potential ΔV, we must evaluate the matrix elements $\langle LM'|\Delta V|LM\rangle$ and then solve a secular problem in the usual way.

As, however, both the wave functions and the perturbation terms are classified according to symmetry species under the full rotation group, the discussion may be simplified by constructing an equivalent operator with which to represent ΔV. This may be formed from angular momentum operators, which have a known effect on the kets $|LM\rangle$, but must be *quadratic* in L_x, L_y, L_z because the terms in ΔV are *rank* 2 spherical tensors. In Section 5.8 we saw how tensor operators of higher rank could be formed by coupling, using appropriate Clebsch–Gordan coefficients, and constructed rank 2 tensors $[\mathbf{S} \times \mathbf{S}]_m^{(2)}$ from the spin operators. The corresponding results for orbital angular momentum (expressed in terms of spherical rather than Cartesian components) give, as may be verified using (5.66),

$$[\mathbf{L} \times \mathbf{L}]_0^{(2)} = [L_z^2 - L^2]/\sqrt{6},$$

$$[\mathbf{L} \times \mathbf{L}]_{\pm 2}^{(2)} = (L_{\pm 1})^2 = \tfrac{1}{2}(L^{\pm})^2,$$

---

‡ Initially we may consider a one-electron system without spin. In Vol. 3 we shall find that exactly similar considerations apply to general many-electron systems.

where $\mathsf{L}^{\pm}$ are the step-up and step-down operators of Section 2.2. The replacement theorem (p. 202) then states that matrix elements of these operators within a manifold $|LM\rangle$ (L fixed) differ only by a numerical factor from those of $Y_{2,0}$ and $Y_{2,\pm 2}$ respectively.

On removing a factor $\sqrt{6}$ a suitable equivalent operator is apparently

$$\mathsf{V}_{\mathrm{eq}} = A[3\mathsf{L}_z{}^2 - \mathsf{L}^2] - \tfrac{1}{2}iB[(\mathsf{L}^+)^2 - (\mathsf{L}^-)^2] \tag{7.43}$$

whose matrix elements are easily found using (2.23). Thus, for three p-states, $L = 1$ and we obtain for the perturbation matrix $\mathbf{H}^{(1)}$ of Section 1.3.

$$\mathbf{H}^{(1)} = k\mathbf{V}_{\mathrm{eq}} = k\begin{bmatrix} A & 0 & -iB \\ 0 & -2A & 0 \\ iB & 0 & A \end{bmatrix}. \tag{7.44}$$

Solution of the corresponding secular equations (1.30) shows that for a system with axial symmetry ($B = a - b = 0$) the $p_{\pm 1}$ states remain degenerate but are raised in energy by an amount kA, while p_0 is depressed by $2kA$. Here k is the numerical factor in the operator equivalence and may be fixed (cf. p. 200) by evaluating just one matrix element. When $B \neq 0$ the $p_{\pm 1}$ states will be mixed and their degeneracy resolved: the splitting follows from the quadratic secular equation

$$(A - \Delta E)^2 - B^2 = 0,$$

which gives

$$\Delta E^{\pm} = A \pm B. \tag{7.45}$$

The resultant energy levels are indicated in Fig. 7.3. The corresponding states, in the presence of the orthorhombic field, are

$$|0\rangle = |1, 0\rangle, \quad |+\rangle = \frac{|1, +1\rangle + i|1, -1\rangle}{\sqrt{2}}, \quad |-\rangle = \frac{|1, +1\rangle - i|1, -1\rangle}{\sqrt{2}}, \tag{7.46}$$

the $|\pm\rangle$ states corresponding to the *real* p-functions pointing along the x and y directions: this could have been inferred from group theory (cf. the Example on p. 136) but the *energies* of the states, and their dependence on the parameters characterizing the applied field, could not. Figure 7.3 also indicates the splitting of states with $L = 2$, results which should be checked by the reader.

Addition of a magnetic field

Finally, we consider the effect of the crystal field on the angular momentum of the central-field states (i.e. on the currents which give a permanent magnetic moment) and on any additional currents that may be induced by an applied *magnetic* field. It will be sufficient to use the same Example, starting from the states (7.46) and computing the current density using (6.26). It is convenient to denote the states by p_x, p_y, p_z.

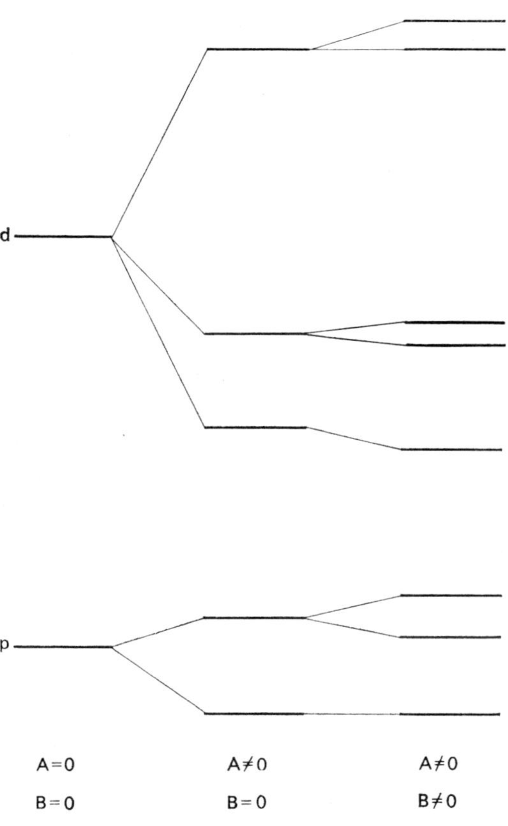

FIG. 7.3. Splitting of p- and d-levels by a crystal field. A and B are the parameters in equation (7.43).

In the absence of an external field **B** we obtain in the p_x state a current with components $J_\lambda = \operatorname{Re} \hat{J}_\lambda$. Thus, with $p_x = xf$ and $f' = df/dr$,

$$\hat{J}_x = m^{-1} p_x{}^* \pi_x p_x = \frac{\hbar}{im}\left(xf^*f - \frac{x_3}{r}f^*f'\right)$$

and $J_x = 0$ at every point in space. It is easy to see that with no applied field the current density must vanish identically whenever the wave function is real: for then

$$\hat{J}_x = (\hbar/im)\phi^* \frac{\partial}{\partial x} \phi$$

and is purely imaginary. Another implication of the suppression of the current density is the reduction to zero of the angular momentum; for the expectation value of any component L_λ is a measure of the *moment* of the current about the λ-axis, and in states for which the current vanishes everywhere we likewise find

$$\langle L_x \rangle = \langle L_y \rangle = \langle L_z \rangle = 0.$$

In such a case we usually say that the crystal field, which forces us to adopt *real* linear combinations of the angular momentum eigenfunctions as the zero-order functions (in the limit of zero crystal field), has "quenched" the angular momentum.

What is the situation when an external magnetic field is applied? As the flux builds up, one direction of electron circulation will be favoured (cf. changing flux through a circuit setting up an e.m.f.) and we expect this will to some extent "unquench" the angular momentum. This is exactly what happens. If we assume that p_x, say, is an acceptable starting point for a perturbation calculation of the effect of B (it may indeed be quite a good non-degenerate wave function for a system in a weak crystal field with $B = 0$) we may calculate the current exactly as above but with $\pi_\lambda = \mathsf{p}_\lambda + eA_\lambda$. We then obtain a current

$$\hat{J}_\lambda{}^0 = m^{-1}p_x{}^*\tfrac{1}{2}e(\mathbf{B} \times \mathbf{r})_\lambda p_x = \frac{e}{2m}\,[xf(r)]^2(B_\mu r_\nu - B_\nu r_\mu)$$

where $\lambda\mu\nu$ is a cyclic permutation of x, y, z. A superscript 0 has been added to indicate use of the unperturbed field-independent wave function. This expression is already in real form, giving at once

$$J_x{}^0 = \frac{e}{2m}\,[xf(r)]^2(B_y z - B_z y), \quad J_y{}^0 = \frac{e}{2m}\,[xf(r)]^2(B_z x - B_x z), \quad (7.47)$$

and we know from Section 7.4 that this part of the current leads, in a free atom, to diamagnetism. On taking the field along the z-axis, to compare with earlier sections, we obtain (with $B_z = B$)

$$j_x{}^0 = \frac{e^2 B}{2m}\,y[xf(r)]^2, \quad j_y{}^0 = -\frac{e^2 B}{2m}\,x[xf(r)]^2, \quad (7.48)$$

which resembles that depicted in Fig. 6.3 except that the density follows the pattern $|p_x|^2$ instead of $|p_{+1}|^2$. In the present case, however, this is only *part* of the induced current.[‡] We know that the perturbation H_P will give a perturbed wave function proportional to **B** and therefore an additional energy term in $\mathbf{B}^2$; to get the complete diamagnetic current we must therefore obtain the field-dependent modification of p_x, p_y and p_z and then compute the current again, taking care to get all the terms linear in **B**.

We shall assume that only the neighbouring states (degenerate in the absence of the crystal field) need be included in the perturbation calculation. The first-order change in p_x follows from simple perturbation theory and the amounts of p_y and p_z mixed into p_x depend on the matrix elements $\langle p_y|H_P|p_x\rangle$, $\langle p_z|H_P|p_x\rangle$: these are easily obtained on writing

$$H_P = \beta\mathbf{B\cdot L} = \beta[B_z L_z + \tfrac{1}{2}(B^+L^- + B^-L^+)]$$

where $B^\pm = B_x \pm iB_y$, and remembering that $p_x = (p_{+1}+p_{-1})/\sqrt{2}$, $p_y = -i(p_{+1}-p_{-1})/\sqrt{2}$. Also, we continue to assume the field is in the z direction; $B_x = B_y = 0$, $B_z = B$. Then, for example,

$$\langle p_y|H_P|p_x\rangle = \tfrac{1}{2}i\langle(p_{+1}-p_{-1})|H_P|(p_{+1}+p_{-1})\rangle$$

and on expansion, since $\langle p_{\pm 1}|H_P|p_{\pm 1}\rangle = \pm\beta B$ for eigenfunctions with $M_L = \pm 1$, we obtain

$$\langle p_y|H_P|p_x\rangle = i\beta B.$$

The other matrix element $\langle p_z|H_P|p_x\rangle$ vanishes and the only effect of the field along the z-axis is therefore to mix the p_x and p_y functions. According to perturbation theory for a non-degenerate state (Section 1.2), the function p_x becomes

$$p_x{}' = p_x + \frac{\langle p_y|H_P|p_x\rangle}{E_x - E_y}\, p_y.$$

Thus

$$p_x{}' = p_x - i\varepsilon p_y, \quad \varepsilon = \beta B/\Delta \tag{7.49}$$

where Δ is the energy separation (Fig. 7.3) of the p_x and p_y states.

The current components $J_x{}^0$ and $J_y{}^0$ in (7.47) are therefore augmented by new terms which arise from the p_x and p_y operators in π_x and π_y; the vector potential term $e\mathbf{A}$ may evidently be disregarded since it is already linear in **B** and would therefore lead to contributions of order $\mathbf{B}^2$. The new terms are thus, remembering $p_x = xf(r)$ and $p_y = yf(r)$,

$$\Delta J_x = \mathrm{Re}\,(p_x - i\varepsilon p_y)^* \frac{\hbar}{mi}\frac{\partial}{\partial x}(p_x - i\varepsilon p_y) = \varepsilon\hbar y|f(r)|^2/m,$$

$$\Delta J_y = \mathrm{Re}\,(p_x - i\varepsilon p_y)^* \frac{\hbar}{mi}\frac{\partial}{\partial y}(p_x - i\varepsilon p_y) = -\varepsilon\hbar x|f(r)|^2/m,$$

---

‡ Indeed, this cannot be the whole current since it would indicate a continual build-up of probability density in some regions of space and a continual outflow from others, contrary to the fact that we are considering a stationary state.

after noting the cancellation of terms containing derivatives of the radial function $f(r)$. The components of $\mathbf{j} = -e\mathbf{J}$, given in (7.48), are thus changed by correction terms, due to change in the wave function,

$$\Delta j_x = -(e\hbar\beta B/m\Delta)y|f(r)|^2, \quad \Delta j_y = (e\hbar\beta B/m\Delta)x|f(r)|^2. \tag{7.50}$$

The extra currents are in this case small: in atomic units $e\hbar\beta = \frac{1}{2}$ and even with near degeneracy, $\Delta \sim 10^{-3}$ H say, the numerical coefficient would be appreciable only for, say, $B \sim 10^{-4}$ a.u. or $\sim 2 \times 10^5$ gauss. Nevertheless, they are significant in the calculation of magnetic properties, which inevitably involve very small energy differences.

We note that the *paramagnetic* current, corresponding to a p_{+1} wave function, must follow if we replace ε ($= \beta B/\Delta$) in (7.50) by -1 and divide by 2 (to obtain the particular mixture of p_x and p_y giving the function p_{+1}). This limit is also appropriate in a crystal field of square planar symmetry ($a = b$ in (7.41)) for which the true degeneracy of p_x and p_y functions reappears ($E_x = E_y$) and the perturbation theory breaks down. The resultant circulating current is given by (7.33).[‡] Comparison with (7.50), which applies in the quenched case ($E_x \neq E_y$), yields the interesting result

$$\begin{pmatrix}\text{Induced current due to} \\ \text{partial ``unquenching''}\end{pmatrix} = \frac{\beta B}{\Delta}\begin{pmatrix}\text{permanent current for} \\ \text{unquenched angular momentum}\end{pmatrix},$$

which shows that quenching is effective provided $\Delta \gg \beta B$.

The part of the induced current associated with the field-dependent perturbation of the wave function, as distinct from the "diamagnetic" part which arises from the vector potential term in π working on the *unperturbed* function, is often called the "paramagnetic" part of the current density; but this is unfortunate terminology because the two parts really have no independent existence. *Both* parts are required in computing any gauge invariant quantity, such as the whole current density or the energy of the system, and the relative contributions from both sources are strongly dependent on the choice of gauge; for example, if the origin of the vector potential is chosen some distance away from the origin of the central field, by using $\mathbf{A} = \frac{1}{2}\mathbf{B} \times (\mathbf{r} - \mathbf{R})$, the "paramagnetic" and "diamagnetic" terms both become large and the invariant result arises as a small difference. When there is no single "natural" origin, as in a polynuclear potential field, the calculation of magnetic properties becomes a somewhat precarious exercise.

---

‡ Note that in (7.33) F is the radial factor in a normalized p_{+1} orbital; it is related to f in (7.50) by $F = f/\sqrt{2}$.

With *approximate* unperturbed functions there is an additional difficulty in that the gauge invariance is no longer automatically satisfied and the results are therefore non-unique, depending on the arbitrary choice of origin. Even in the simple example treated in this section, serious inconsistencies arise from the use of unperturbed functions which are not exact eigenfunctions for the system in the absence of the magnetic field (they are exact only when the crystal field is also removed). As already noted (p. 217), the current does not possess the property div $\mathbf{j} = 0$ and this would indicate a net inflow of current into some regions and a net outflow from others: in a stationary state this is nonsense. The predicted currents are in fact only approximate and to get an accurate result it would be necessary to admit mixing (under the influence of the crystal field, and again under the magnetic field) of all the functions of a complete set—not just the three p-functions corresponding to an initially degenerate level. In general an adequate discussion of the properties considered in the last few sections, even for a single electron, provides a stringent test of wave functions and perturbation techniques.

7.6. Emission and absorption of radiation

As a final example of the effect of the vector potential terms in the Hamiltonian, let us consider the interaction between an electronic system and an oscillating electromagnetic field. In this preliminary discussion we focus attention, as usual, on a *one*-electron system, but in due course (Vol. 3) we find the same results carry over almost without change to a general many-electron system. We also confine ourselves to the *semi-classical* treatment, in which the electromagnetic field is regarded merely as a perturbation (dictating the form of the scalar and vector potential in the Schrödinger equation of the electronic system) rather than as a dynamical system composed of photons—which, to do it full justice, requires a *quantum field theory*. We wish to find how the presence of radiation can induce transitions between different states of the system and to find expressions for *transition probabilities*.

The electromagnetic field in free space is determined by the two vectors $\mathbf{F}$ and $\mathbf{B}$, which in general vary from point to point in space (describing a *vector field*) and are time dependent. They satisfy Maxwell's equations in the form

$$\text{div } \mathbf{F} = 0, \qquad \text{div } \mathbf{B} = 0,$$
$$\text{curl } \mathbf{F} = -\partial \mathbf{B}/\partial t, \qquad \text{curl } \mathbf{B} = (\mu_0 \varepsilon_0)\, \partial \mathbf{F}/\partial t, \tag{7.51}$$

where ε_0 and μ_0 are the permittivity and permeability, respectively, of free space (App. 1).

Straightforward manipulation of these equations shows that $\mathbf{F}$ and $\mathbf{B}$ must each satisfy a wave equation:

$$\nabla^2\mathbf{F} = (\varepsilon_0\mu_0)\,\partial^2\mathbf{F}/\partial t^2, \qquad \nabla^2\mathbf{B} = (\varepsilon_0\mu_0)\,\partial^2\mathbf{B}/\partial t^2. \qquad (7.52)$$

Since the wave equation $\nabla^2\phi = (1/c^2)\,\partial^2\phi/\partial t^2$ describes propagation of the disturbance represented by ϕ with velocity c, it follows that the electromagnetic field is propagated with the velocity c given by

$$c^2 = 1/\varepsilon_0\mu_0. \qquad (7.53)$$

Moreover, $\mathbf{F}$ and $\mathbf{B}$ arise from the potentials $\mathbf{A}$ and φ (p. 177) and at a sufficient distance from the oscillating charges that produce a radiation field $\varphi = 0$: the vector potential $\mathbf{A}$ then determines both $\mathbf{F}$ and $\mathbf{B}$, through (6.1), and itself is found to satisfy

$$\nabla^2\mathbf{A} = (\varepsilon_0\mu_0)\,\partial^2\mathbf{A}/\partial t^2. \qquad (7.54)$$

Different kinds of radiation (e.g. differing in frequency, plane of polarization, etc.) are all determined by particular special forms of $\mathbf{A}$ as a function of position and time.

The energy of the electromagnetic field may be regarded as distributed throughout space. The amount of energy per unit volume is given by

$$\rho = \tfrac{1}{2}\varepsilon_0\mathbf{F}^2 + \tfrac{1}{2}\bar{\mu}_0\mathbf{B}^2, \qquad (7.55)$$

where $\bar{\mu}_0 = \mu_0^{-1}$ has been introduced to emphasize the parallel between $\mathbf{F}$ and $\mathbf{B}$, and by considering the conservation of energy as the field changes it is possible to define a rate of flow of energy by the *Poynting vector*

$$\mathbf{S} = \bar{\mu}_0\mathbf{F} \times \mathbf{B}. \qquad (7.56)$$

The field energy flows in the direction of the vector $\mathbf{S}$ and the rate of flow, per unit time and across unit area normal to $\mathbf{S}$, is just the magnitude of $\mathbf{S}$.

As a simple example we consider the radiation field corresponding to a plane wave. It is easily verified that (7.54) is satisfied by the vector potential

$$\mathbf{A} = \mathrm{Re}\,\mathbf{A}_0 \exp i(\mathbf{k}\cdot\mathbf{r} - \omega t) \quad (k = \omega/c), \qquad (7.57)$$

which exhibits a simple harmonic variation both in time, with frequency $\nu = \omega/2\pi$, and in space (along the direction of the propagation vector $\mathbf{k}$, which is perpendicular to A) with wave number $\lambda^{-1} = \omega/c = k/2\pi$. In general $\mathbf{A}_0$ is a constant vector, possibly with complex components, whose form determines the amplitude of the corresponding wave train and the nature of the polarization. It follows at once, from (6.1), that the electric and magnetic vectors are given by

$$\mathbf{F} = \mathrm{Re}\,[ick\,\mathbf{A}_0 \exp i(\mathbf{k}\cdot\mathbf{r} - \omega t)],$$
$$\mathbf{B} = \mathrm{Re}\,[i(\mathbf{k}\times\mathbf{A}_0)\exp i(\mathbf{k}\cdot\mathbf{r} - \omega t)]. \tag{7.58}$$

Thus if $\mathbf{k}$ and $\mathbf{A}_0$ are along the z- and x-axes, respectively, $\mathbf{k} = k\mathbf{e}_3$ and $\mathbf{A}_0 = A_0\mathbf{e}_1$, the fields become

$$\mathbf{F} = -ckA_0 \sin(kz - \omega t)\,\mathbf{e}_1, \quad \mathbf{B} = -kA_0 \sin(kz - \omega t)\,\mathbf{e}_2. \tag{7.59}$$

The electric and magnetic vectors are thus perpendicular to each other and to the direction of propagation, lying always in the x and y directions, respectively. This solution defines *plane* polarized radiation. It is not difficult to show that the alternative choice $\mathbf{A}_0 = A_x\,\mathbf{e}_1 {\pm} 2 A_z\,\mathbf{e}_2$ would describe *circularly polarized* radiation in which the planes containing $\mathbf{F}$ and $\mathbf{B}$ rotate (in a positive or negative sense according to sign) as one moves along the direction of propagation.

We shall consider interaction between the electronic system and a plane wave with vector potential (7.57). We shall also need expressions for the energy density and its rate of flow, time-averaged over one period, so that we can relate the transition probabilities to the intensity of the radiation. On making use of the fact that the two terms in (7.55) give equal average values, we obtain easily (noting that only the constant terms give non-zero average values)

$$\bar{\rho} = \tfrac{1}{2}\varepsilon_0\omega^2 A_0{}^2 \tag{7.60}$$

where

$$A_0{}^2 = \mathbf{A}_0{}^*\cdot\mathbf{A}_0 \tag{7.61}$$

and is evidently a real quantity. A similar treatment of (7.56) gives an average energy flow in the direction of propagation and of magnitude

$$\overline{S} = \tfrac{1}{2}\bar{\mu}_0(\omega^2/c)A_0{}^2 = \tfrac{1}{2}\varepsilon_0 c\omega^2 A_0{}^2. \tag{7.62}$$

This quantity is the *intensity* (I) of the radiation, for a monochromatic wave train with frequency factor ω and vector potential of complex amplitude $A_0 = A_0(\omega)$. If there are many incoherent wave trains, let us say $n(\omega)\,d\omega$ in frequency range $d\omega$, $A_0{}^2$ will be replaced by a sum of contributions $A_0{}^2(\omega)n(\omega)\,d\omega$ and the intensity contribution per unit frequency range will become

$$I(\omega) = \tfrac{1}{2}\varepsilon_0 c\omega^2 A_0{}^2(\omega)n(\omega). \tag{7.63}$$

We now examine separately the cases of monochromatic and polychromatic radiation.

H*

Monochromatic radiation

Strictly monochromatic radiation cannot induce transitions between discrete energy levels unless there is a virtually exact frequency coincidence, $\omega = |\omega_f - \omega_i|$. The time-dependent perturbation theory (Section 1.8) applies directly, the perturbation operator arising from the vector potential terms in (6.11). Thus, disregarding spin effects,

$$\mathsf{H}^{(1)} = \frac{1}{2m}\,[2e\mathbf{A}\cdot\mathbf{p} + e^2\mathbf{A}^2] \tag{7.64}$$

where $\mathbf{A}$ is given by (7.57). The term in $\mathbf{A}^2$ is normally negligible and will be discarded. Starting from an initial state‡

$$\Psi_i = \Phi_i \exp\left(-iE_i t/\hbar\right)$$

we obtain for the rate of growth of the coefficient $c_f{}^{(1)}$ of the final state, according to (1.73),

$$\frac{dc_f{}^{(1)}}{dt} = -\frac{i}{\hbar}\int_0^t H_{fi}{}^{(1)}\, e^{i\omega_{fi}t}\, dt \tag{7.65}$$

where $H_{fi}{}^{(1)} = \langle\Phi_f|\mathsf{H}^{(1)}|\Phi_i\rangle$ and $\omega_{fi} = (E_f - E_i)/\hbar$. On substituting (7.57) in (7.64), we obtain

$$H_{fi}{}^{(1)}(t) = H_{fi}{}'\, e^{-i\omega t} + H_{fi}{}''\, e^{i\omega t} \tag{7.66}$$

where the time-independent factors in the matrix element are

$$H_{fi}{}' = \frac{e}{2m}\int \Phi_f{}^* \, e^{i\overline{\mathbf{k}\cdot\mathbf{r}}}(\mathbf{A}_0\cdot\mathbf{p})\Phi_i \, d\mathbf{x},$$

$$H_{fi}{}'' = \frac{e}{2m}\int \Phi_f{}^* \, e^{-i\mathbf{k}\cdot\mathbf{r}}(\mathbf{A}_0\cdot\mathbf{p})\Phi_i \, d\mathbf{x}; \tag{7.67}$$

when (7.66) is used in (7.65) and the integration is completed, it follows that

$$c_f(t) = -\frac{1}{\hbar}H_{fi}{}'\left[\frac{e^{i(\omega_{fi}-\omega)t}-1}{\omega_{fi}-\omega}\right] - \frac{1}{\hbar}H_{fi}{}''\left[\frac{e^{i(\omega_{fi}+\omega)t}-1}{\omega_{fi}+\omega}\right]. \tag{7.68}$$

The first term predominates for $\omega \simeq \omega_{fi}$, the second being negligible; the situation is reversed for $\omega \simeq -\omega_{fi}$ (corresponding to ω_{fi} negative, and the final state of lower energy than the initial). The two cases thus correspond to the processes of absorption and emission of radiation respectively.

---

‡ The equations which follow apply equally, with an appropriate interpretation of the matrix elements, to any kind of system: Φ, Ψ, etc., could thus be many-electron state functions. Superscript zeros (unperturbed system) have been discarded.

As in Section 1.8 (p. 29), it is not possible (for fixed ω) to obtain time-proportional transitions, and hence to define a transition probability per unit time, unless one of the two states is a continuum state; this corresponds to the photo-electric effect in which absorption of a photon leads to the ejection of an electron. The first term in (7.67) is then dominant, and we obtain (cf. p. 31) the final result by integrating $|c_f(t)|^2$ over all the states in the vicinity of ω_f to obtain a time-proportional probability of a transition to that group of states. The transition probability per unit time is

$$w = \frac{2\pi}{\hbar^2}\, n(\omega_f)\big|H_{fi}{}'\big|^2 \tag{7.69}$$

where $n(\omega_f)\,d\omega_f$ is the number of free-electron states in the frequency range $d\omega_f$. In terms of the density of states per unit range of *energy* we note $n(\omega_f) = \hbar\rho(E_f)$. This provides an important concrete application of "Fermi's golden rule" (p. 32). More detailed consideration of the free-electron states (see, for example, Bethe, 1965) allows us to evaluate the probability that the ejected electron will emerge in an element of solid angle inclined at angles θ, φ to the incident radiation and hence to evaluate an important characteristic of the system—the "differential cross-section"—in terms of the matrix element $H_{fi}{}'$ defined in (7.67). Further discussions of scattering theory appear elsewhere (Vol. 4).

Polychromatic radiation

When the initial and final levels are both discrete (7.68) leads to time-proportional transitions only when there is a continuous distribution of ω values centred around the peak $\omega = \omega_{fi}$ (considering again the case of absorption). In this case, dropping the second term in (7.68) and making a simple reduction, we obtain

$$|c_f(\omega, t)|^2 = 4\,\frac{|H_{fi}{}'(\omega)|^2}{\hbar^2}\,\frac{\sin^2 \tfrac{1}{2}\theta t}{\theta^2}\quad (\theta = |\omega_{fi} - \omega|)$$

for the growth of c_f due to a single wave train with frequency factor ω. When there is a continuous distribution of incoherent wave trains, all with $\omega \simeq \omega_{fi}$, each makes an additive contribution[‡] and the resultant effect is described by

$$|c_f(t)|^2 = \int |c_f(\omega, t)|^2 n(\omega)\,d\omega$$

[‡] This is a consequence of the assumed incoherence. Strictly, the perturbation $H^{(1)}$ is a sum of contributions; additivity of the contributions to $|c_f|^2$ (which requires proof) depends on the random phases in the vector potential expressions for different wave trains.

where $n(\omega)\,d\omega$ is the number of incoherent wave trains in range $d\omega$. Integration then yields a time-proportional result, corresponding to transition probability per unit time

$$w = \frac{2\pi}{\hbar^2}\, n(\omega)\big|H_{fi}{'}\big|^2 \tag{7.70}$$

which is formally identical with (7.69). The density function now refers, however, not to the number of states but rather to the number of wave trains per unit range in the vicinity of $\omega = \omega_{fi}$.

To complete the discussion, we reduce the matrix element $H_{fi}{'}$ defined in (7.67) by writing the constant factor $\mathbf{A}_0 = \mathbf{A}_0(\omega)$ which determines both amplitude and polarization of the vector potential in the form $\mathbf{A}_0 = A_0 \mathbf{a}_0$. Here $\mathbf{a}_0$ is a unit vector in the sense $\mathbf{a}_0{}^* \cdot \mathbf{a}_0 = 1$, and determines the polarization alone, while $A_0 = \sqrt{(\mathbf{A}_0{}^* \cdot \mathbf{A}_0)}$ is the real scalar amplitude which determines the intensity of radiation of frequency ω according to (7.63). We withdraw the constant $A_0(\omega)$ from the integral in (7.67), together with a factor $\hbar$ from the momentum operator, and substitute in (7.70). On introducing the intensity (7.63), we obtain

$$w = 2\pi\, \frac{I(\omega)}{\varepsilon_0 c \omega^2}\left(\frac{e}{m}\right)^2 |M_{fi}|^2 \tag{7.71}$$

where the matrix element M_{fi} is defined by

$$M_{fi} = \int \Phi_f{}^*\, e^{i\mathbf{k}\cdot\mathbf{r}}\mathbf{a}_0 \cdot \nabla\Phi_i\, d\mathbf{x}. \tag{7.72}$$

This quantity completely determines the transition probability for absorption of radiation, resulting in excitation of the system from initial state Φ_i to final state Φ_f, and therefore implicitly contains the various selection and intensity rules encountered in the various branches of spectroscopy. A similar discussion of the process of emission, associated with the second term in (7.68), shows that in a given radiation field the probability of the *reverse* transition, from state Φ_f to state Φ_i, is identical with that just evaluated. On the other hand, there is a finite probability of transition from a higher state to a lower state *even in the absence* of a radiation field; this is referred to as a probability of *spontaneous emission*, to distinguish it from the probability of *stimulated* emission in the presence of radiation. The probability of spontaneous emission per unit time cannot be calculated by the semi-classical method and requires quantization of the field. It may, however, be inferred by a statistical argument referring to a large number of identical systems in thermal equilibrium with a radiation field. For these and other developments, which provide the theoretical basis of spectroscopy, the reader is referred elsewhere (e.g. Bethe and Jackin, 1968, for many applications of the semi-classical approach; Griffith, 1961, Chap. 3, for a simplified and very readable account of quantization of the field; Heitler, 1954, for a more comprehensive discussion). Some of the generalizations relating to many-electron systems will be taken up in Vol. 3.

REFERENCES

BETHE, H. A. (1964) *Intermediate Quantum Mechanics*, Benjamin, New York; see also BETHE, H. A. and JACKIN (1968) for a revised and enlarged edition.

DAVIES, D. W. (1967) *The Theory of the Electric and Magnetic Properties of Molecules*, Wiley, London and New York.

GRIFFITH, J. S. (1961) *The Theory of Transition-Metal Ions*, Cambridge University Press, London.

HEITLER, W. (1954) *The Quantum Theory of Radiation*, Oxford University Press.

LAMB, W. E. (1941) *Phys. Rev.* **60**, 817.

SLATER, J. C. (1960) *Quantum Theory of Atomic Structure*, Vols. 1 and 2, McGraw-Hill, New York.

APPENDIX 1

UNITS AND DIMENSIONS

It is now customary to express all physical quantities in the internationally agreed SI[‡] system of units. Unfortunately a second system of units, the older cgs system, is also in use and is likely to remain so for many years since it is the one adopted by many of the classic textbooks. Moreover, at the elementary-particle level, it is frequently more convenient to employ yet another system of "atomic units": since these units are provided by naturally occurring quantities (such as the magnitude of the electronic charge, and Planck's quantum of action) they might well be called "natural units".

It is therefore important to be able to convert all kinds of quantity freely from one system to another, and this requires an appreciation of the idea of the *dimensions* of a quantity. In the older cgs system, certain dimensional ambiguities and inconsistencies are present in the representation of electric and magnetic quantities (manifest in the use of a "mixed" system of esu and emu): these are removed by the introduction of SI units. We shall therefore discuss mainly SI units, atomic units, and the relations between them.

Dimensions

In order to be absolutely clear about the significance of units we recall first the most elementary ideas.

The statement $l = 10$ m implies that a unit of length, the metre, has been defined and that the *ratio* of the length (l) to that of the unit is a pure number 10. The number 10 is a *numerical measure* of the length l, but the same length would have a different measure if the unit were defined differently: if the unit were decreased by a factor k the measure would increase by a factor k. For example, changing to the cm as unit would correspond to $k = 100$ and the measure of l would increase by this factor: $l = 1000$ cm. Other quantities would suffer corresponding changes: the measure of an area would be increased by k^2, that of a

---

‡ "Système International d'Unités." The system is essentially the mks system, with a "rationalized" choice of electric and magnetic units.

226

volume by k^3. The index of k indicates the *dimension in length* of the quantity and we express such observations symbolically by writing[‡]

$$[\text{area}] = L^2, \qquad [\text{volume}] = L^3, \qquad \text{etc.}$$

When only "mechanical" quantities are considered, each is characterized by its dimensions in mass, length and time (M, L, T). Thus

$$[\text{force}] = MLT^{-2}$$

means that if the unit of mass is made k times smaller, the measure of the force is made k times larger; but if the unit of time is made k times smaller the measure of the force is multiplied by k^{-2}. From the definition it follows that when quantities are multiplied together their dimensions are similarly multiplied, e.g.

$$[\text{work}] = [\text{force} \times \text{distance}] = MLT^{-2}L = ML^2T^{-2}.$$

Some of the most commonly occurring "mechanical" quantities are listed in column 1 of Table A1.2, along with their dimensions.

In science, as distinct from mathematics, the symbols occurring in an equation normally denote *physical quantities* rather than the *numbers* in which they are measured.[§] Thus, $l_1 + l_2 = l_3$ could represent $5\text{ m} + 10\text{ m} = 15\text{ m}$, and this would imply the equality of the *numbers* of units on each side of the equation, $5 + 10 = 15$: equality of two quantities means equality of their numerical measures in terms of a common unit. Since equality of two physical quantities must be independent of choice of units, it is evident that an equation must be *dimensionally homogeneous*, i.e. the dimensions of each side must be identical—for otherwise, in a change of units, the numerical measures of the quantities on the two sides would be multiplied by different factors, contradicting their equality. Similarly, quantities with different dimensions cannot be added: this means, in particular, that when a function such as $e^x = 1 + x + x^2/2! + \ldots$ occurs in an equation referring to physical quantities, x *must be a pure number* (e.g. a *ratio* of two lengths l_1/l_2 which is obviously invariant against change of units). Dimensional considerations are frequently of great value in checking the internal consistency of equations, as well as in dealing with changes of units.

[‡] In words, "The dimensions of area are el-two", etc.

[§] The term "quantity calculus" has been used to indicate this usage of symbols in science.

TABLE A1.1

Primary atomic units

Quantity and dimensions‡	Atomic unit and name§	SI equivalent‖
Mass M	$m = 1$ em (electron-mass)	$9{\cdot}1091\ 10^{-31}$ kg
Charge Q	$e = 1$ e (electron)	$1{\cdot}60210\ 10^{-19}$ C
Action MLT^{-1} (A)	$\hbar = 1$ P (Planck)	$1{\cdot}05450\ 10^{-34}$ J s
Permittivity $Q^2\,W^{-1}\,L^{-1}$	$\kappa_0 = 4\pi\varepsilon_0$ (no name)	$4\pi \times 8{\cdot}8542\ 10^{-12}$ F m^{-1}

† The dimensions of all quantities are expressible in terms of the more conventional symbols M, L, T, Q. For convenience, A (action) and W (energy) are also frequently employed.

§ The choice of m, e, $\hbar$ and κ_0 as primary units fixes all others. There is no generally accepted set of names for the atomic units, though some (e.g. Planck, Hartree and Bohr) have been adopted fairly widely in the literature (see Shull and Hall, 1959).

‖ The four base units required in these tables are the metre, kilogram, second and ampere (m, kg, s, A), while the other important named units are the joule (J), newton (N), coulomb (C), watt (W), volt (V), farad (F), weber (Wb), henry (Hn) and tesla (T). (The henry is here denoted by Hn to avoid confusion with the Hartree atomic unit (H). For further information the reader is referred to *Quantities, Units & Symbols* (Report by the Symbols Committee of the Royal Society, London, 1971).

SI units. Electric and magnetic quantities

The differences between SI and cgs units are trivial except in the case of electromagnetic units, amounting only to replacement of centimetre, gram, second by metre, kilogram, second, i.e. a change to mks units. Correspondingly the dyne and the erg are replaced by the newton (N) and the joule (J), but the unit of power, the watt (W), remains. Units are indicated using indices corresponding to the dimensions of a quantity: thus, for force (MLT^{-2}) $1\ N = 1$ kg m s^{-2}. It is also frequently convenient to use energy (W) as a secondary dimension; for example, $[f] = MLT^{-2} = ML^2T^{-2}L^{-1} = WL^{-1}$ and the newton may therefore be written in $1\ N = 1$ J m^{-1}. Any dimensionally equivalent statement of units is acceptable. In the present example, the first expression defines the newton as the force (mass × acceleration) which would give an acceleration of 1 m s^{-2} to a mass of 1 kg: while the second defines it as the force which does 1 J of work in moving its point of application 1 m.

When electric and magnetic quantities are admitted, a fourth primary dimension is needed: electric charge is indicated by the dimensional symbol Q. The dimensions of all electric and magnetic quantities then follow directly from their definitions. The fundamental equation[‡]

$$\mathbf{f} = q\mathbf{F} + q\mathbf{v} \times \mathbf{B} \qquad (A1.1)$$

defines the electric field strength $\mathbf{F}$ and the magnetic flux density $\mathbf{B}$ in terms of the force $\mathbf{f}$ experienced by a particle of charge q moving with velocity $\mathbf{v}$. The dimensions of $\mathbf{F}$ and $\mathbf{B}$ are evidently

$$[\mathbf{F}] = \mathrm{MLT^{-2}Q^{-1}}, \qquad [\mathbf{B}] = \mathrm{MT^{-1}Q^{-1}},$$

while those of current, potential difference, etc., follow in a straight-forward way. In fact, it is the unit of *current* (not charge) which is fixed by international agreement as the fourth base unit. The unit of current is the ampere (A) (with dimensions $\mathrm{QT^{-1}}$) and the unit of charge, the coulomb, is thus the amount of charge carried by a current of 1 ampere flowing for 1 second; $1\ \mathrm{C} = 1\ \mathrm{A\ s}$.

The main differences between SI and cgs units arise because, when the charge is defined independently of the force acting between charged particles, Coulomb's law must be written $f = q_1 q_2 / \kappa_0 r^2$ where κ_0 (subscript zero for free space) is a determinable proportionality constant. When κ_0 is written in the "rationalized" form, $\kappa_0 = 4\pi\varepsilon_0$, ε_0 is the "permittivity of free space" and for dimensional consistency it is clear that

$$[\varepsilon_0] = \mathrm{M^{-1}L^{-3}T^2Q^2}.$$

This proportionality constant thus occurs whenever a field (i.e. the force on a "test charge") is related to the charge producing the field. A similar proportionality constant occurs in relating the magnetic flux density to the charge and velocity of a moving particle producing the magnetic field. The two basic equations giving $\mathbf{F}$ and $\mathbf{B}$ at a point $\mathbf{r}$, due to a charge particle at the origin, moving with velocity $\mathbf{v}$, may be written in vector form as

$$\mathbf{F} = \frac{q\,\mathbf{r}}{4\pi\varepsilon_0 r^3}, \qquad \mathbf{B} = \frac{q\,\mathbf{v}\times\mathbf{r}}{4\pi\bar{\mu}_0 r^3}, \qquad (A1.2)$$

where $\bar{\mu}_0$ is the analogue of ε_0. Unfortunately, for historical reasons, the proportionality constant in the second equation is normally written $(\mu_0/4\pi)$ where μ_0 is the "permeability of free space"; hence $\bar{\mu}_0 = \mu_0^{-1}$. The constant $(\mu_0/4\pi)$ has the dimensions $\mathrm{MLQ^{-2}}$ or $\mathrm{MLT^{-2}}\,(\mathrm{QT^{-1}})^{-2}$

---

‡ As we are using F (instead of E) for the electric field, the force is denoted by a lower-case letter. Note also that a vector equation is equivalent to three equations in the components (ordinary scalar quantities) so that exactly similar dimensional arguments may be applied.

(i.e. force $\times$ current^{-2}) and may thus be measured in $N\,A^{-2}$; it is conventionally fixed as

$$(\mu_0/4\pi) = 10^{-7}N\,A^{-2}$$

in defining the unit of current. All the electric and magnetic units are then fixed: the names of the principal units are listed beneath Table A1.1 and are employed in defining various secondary quantities in Tables A1.3 and A1.4. It is also noteworthy that μ_0 and ε_0 are related (p. 220) by

$$\varepsilon_0\mu_0 = 1/c^2,$$

where c is the velocity of light, and that in many textbooks (e.g. Slater, 1960) μ_0 is eliminated from all equations in favour of ε_0.

Atomic or natural units

In the treatment of the hydrogen atom (fixed nucleus approximation; see, for example, Vol. 1, Section 2.3) there occur the quantities[‡]

$$a_0 = \frac{\hbar^2\kappa_0}{me^2}, \qquad E_0 = \frac{me^4}{\kappa_0^2\hbar^2} = \frac{e^2}{\kappa_0 a_0}.$$

We have already adopted these quantities as the atomic units of length and energy, noting that they arise naturally when we choose m, e, and $\hbar$ as the units of mass, charge and action, respectively, provided we fix a fourth unit by taking κ_0 to have the numerical measure unity. By adopting such units it is possible to eliminate the various fundamental constants from all equations. The quantities that appear in the equations may then be re-interpreted as *dimensionless variables* (i.e. as the numerical measures of the physical quantities they represent) and any calculated quantity x must likewise be regarded as x *atomic units*. In other words, we calculate with numbers and supply the correct units at the end by referring to the dimensions of the calculated quantity. An example with which we are already familiar is angular momentum: the natural unit is $\hbar$ (the fundamental constant with dimensions ML^2T^{-1}) and if the dimensionless operator is L then the actual angular momentum operator is $\hbar L$. Similarly, if r has been used to denote the measure of a distance, then the actual distance is understood to be $\rho = ra_0$. In practice, it would be tiresome to use different symbols (cf. ρ and r) for quantities and their numerical measures; the distinction is not therefore usually made and when using atomic units, "distance r" is understood to mean "r atomic units" or ra_0.

[‡] From now on we use the abbreviation $\kappa_0 = 4\pi\varepsilon_0$.

Table A1.2

Derived units (mechanical)

Quantity and dimensions	Atomic unit and name[‡]	SI equivalent[§]
Length L	$\dfrac{\kappa_0 \hbar^2}{me^2} = 1$ B (Bohr, a_0)	$5 \cdot 29157 \ 10^{-11}$ m
Time T	$\dfrac{\kappa_0^2 \hbar^3}{me^4} = \dfrac{\hbar}{E_0} = 1$ as ("atomic second")	$2 \cdot 41889 \ 10^{-17}$ s
Velocity[‖] LT^{-1}	$\dfrac{e^2}{\kappa_0 \hbar} = \dfrac{eE_0}{\hbar} = 1$ B as^{-1}	$2 \cdot 18764 \ 10^6$ m s^{-1}
Force MLT^{-2}	$\dfrac{m^2 e^6}{\kappa_0^3 \hbar^4} = \dfrac{E_0}{a_0} = 1$ H B^{-1}	$8 \cdot 23831 \ 10^{-8}$ N (J m^{-1})
Energy $ML^2 T^{-2}$ (W)	$\dfrac{me^4}{\kappa_0^2 \hbar^2} = \dfrac{e^2}{\kappa_0 a_0} = 1$ H (Hartree, E_0)	$4 \cdot 35944 \ 10^{-18}$ J
Power $ML^2 T^{-3}$	$\dfrac{m^2 e^8}{\kappa_0^4 \hbar^5} = \dfrac{E_0^2}{\hbar} = 1$ H as^{-1}	$1 \cdot 80225 \ 10^{-1}$ W (J s^{-1})

[‡] The names Bohr and Hartree are fairly widely used for the units of length and energy. For the atomic unit of time the name "atomic second" is proposed, with the symbol "as" formed by adding the prefix "a" to the conventional symbol "s". Each unit is expressed first in terms of the primary units ($m, e, \hbar, \kappa_0$) of Table A1.1 and, wherever it is more convenient, in terms of a dimensionally equivalent combination of derived units.

[§] The expressions in parentheses indicate dimensionally equivalent units (e.g. 1 N $= 1$ Jm^{-1}), corresponding to those employed in column 2.

[‖] In these units the velocity of light is $137 \cdot 030$ B as^{-1}, the fine structure constant being $\alpha = e^2/\kappa_0 \hbar c = 1/137 \cdot 030$.

TABLE A1.3

Derived units (electrical)

Quantity and dimensions	Atomic unit and name[‡]	SI equivalent[§]
Electric permittivity $Q^2 W^{-1} L^{-1}$	$\kappa_0\ (=4\pi\varepsilon_0) = 1\ e^2\ H^{-1}\ B^{-1}$	$4\pi \times 8\cdot8542\ 10^{-12}\ F\ m^{-1}\ (C^2\ J^{-1}\ m^{-1})$
Electric current QT^{-1}	$\dfrac{me^5}{\kappa_0^2\hbar^3} = \dfrac{eE_0}{\hbar} = 1\ e\ as^{-1}$	$6\cdot62329\ 10^{-3}\ A$
Electric potential WQ^{-1}	$\dfrac{me^3}{\kappa_0^2\hbar^2} = \dfrac{E_0}{e} = 1\ aV\ (\text{``atomic volt''})$	$2\cdot72108\ 10^1\ V$
Capacitance $Q^2 W^{-1}$	$\dfrac{\kappa_0^2\hbar^2}{me^2} = \dfrac{e^2}{E_0} = 1\ e^2\ H^{-1}$	$5\cdot88774\ 10^{-21}\ F\ (C^2\ J^{-1})$
Electric field strength $(\mathbf{E})\ WQ^{-1}\ L^{-1}$	$\dfrac{m^2e^5}{\kappa_0^3\hbar^4} = \dfrac{E_0}{ea_0} = 1\ aV\ B^{-1}$	$5\cdot14220\ 10^{11}\ V\ m^{-1}$
Electric displacement $(\mathbf{D})\ QL^{-2}$	$\dfrac{m^2e^5}{\kappa_0^2\hbar^4} = \dfrac{e}{a_0} = 1\ e\ B^{-2}$	$5\cdot72142\ 10^1\ C\ m^{-2}$
Electric dipole moment LQ	$\dfrac{\kappa_0\hbar^2}{me} = ea_0 = 1\ e\ B$	$8\cdot47778\ 10^{-30}\ C\ m$
Electric polarizability $L^2Q^2W^{-1}$	$\dfrac{\kappa_0^3\hbar^4}{m^2e^5} = \dfrac{e^2a_0^2}{E_0} = 1\ e\ B\ aV^{-1}$	$1\cdot64857\ 10^{-43}\ F\ m^2$

[‡] Each unit is expressed first in terms of the primary units (m, e, $\hbar$, κ_0) of Table A1.1 and, wherever it is more convenient, in terms of a dimensionally equivalent combination of derived units. The "atomic volt" (aV) is proposed as a convenient unit of potential: 1 Hartree (27·2108 eV) of work is required to move charge e between points differing in potential by 1 aV, and hence 1 aV $= 27\cdot2108$ V.

[§] The expressions in parentheses indicate dimensionally equivalent units (e.g. $1\ N = 1\ Jm^{-1}$), corresponding to those employed in column 2.

It is useful to extend the proposals of Shull and Hall (1959), already adopted by some authors, by introducing the following named units:

$$1 \text{ e } = \text{ unit of charge (magnitude of the electron charge)},$$

$$1 \text{ em } = \text{ unit of mass (electron mass)},$$

$$1 \text{ P(lanck) } = \text{ unit of action, or angular momentum } (\hbar),$$

$$1 \text{ B(ohr) } = \text{ unit of length } (a_0),$$

$$1 \text{ H(artree) } = \text{ unit of energy } (e^2/\kappa_0 a_0).$$

In fact, only e, m, $\hbar$, and κ_0 (all taken as natural units) are required in defining all other quantities. These primary quantities are expressed in SI units in Table A1.1.

The atomic units of many other commonly encountered quantities, together with their SI equivalents, are collected for reference in Tables A1.3–4, columns 2 and 3. In the table a few more named units have been proposed and comments are perhaps necessary.

In the first place the natural unit of time is fixed in terms of the Planck and the Hartree, since [action] = WT : the unit is $1 \text{ PH}^{-1} = \hbar\kappa_0 a_0/e^2$ and is the time taken for an electron in the first Bohr orbit to travel through one Bohr. Atomic analogues of the basic SI units have in a few cases been introduced in the table by means of a prefix "a"; thus $1 \text{ as } = 1 \text{ PH}^{-1}$ denotes the "atomic second". Secondly, in view of the importance of electric and magnetic quantities, it is convenient to have named atomic units of electric potential and magnetic flux density. The "atomic volt" is simply $1 \text{ aV } = 1 \text{ He}^{-1}$ (unit charge e falling through 1 aV loses 1 H of potential energy). The magnetic flux density has a slightly less direct definition: it has dimensions $MQ^{-1}T^{-1}$ and the unit is thus $\text{em e}^{-1} \text{ as}^{-1}$. This unit is the "atomic tesla", $1 \text{ aT } = 1 \text{ em e}^{-1} \text{ as}^{-1}$. The magnitude of the unit may be visualized by reference to the Biot–Savart law (A1.2). For unit charge (e) moving in a circular orbit, the flux density at the centre is

$$\mathbf{B} = \frac{\mathbf{r} \times e\mathbf{v}}{(4\pi\bar{\mu}_0)r^3} = \frac{e(\mathbf{r} \times m\mathbf{v})}{mc^2(4\pi\varepsilon_0)r^3}.$$

For unit angular momentum and unit radius this would yield, since $c = (1/\alpha) \text{ B as}^{-1} = 137 \cdot 030 \text{ B as}^{-1}$, a flux density of magnitude

$$B = (137 \cdot 030)^{-2} \text{ em e}^{-1} \text{ as}^{-1} = (137 \cdot 030)^{-2} \text{ aT}.$$

The atomic tesla is a rather large unit ($1 \text{ aT } = 2 \cdot 35055 \times 10^5$ T); but fields in atoms may evidently be of the order 10^{-4} aT, corresponding to

TABLE A1.4

Derived units (magnetic)

Quantity and dimensions	Atomic unit and name‡	SI equivalents§
Magnetic permeability $ML\,Q^{-2}$	$\dfrac{\kappa_0\hbar^2}{e^4} = \dfrac{ma_0}{e^2} = 1\text{ em B e}^{-2}$	$1\cdot87797\ 10^{-4}\text{ N A}^{-2}$
Magnetic flux WTQ^{-1}	$\dfrac{\hbar}{e} = 1\text{ P e}^{-1} = 1\text{ aV as}$	$6\cdot58199\ 10^{16}\text{ Wb (V s)}$
Magnetic flux density $(\mathbf{B})\ MT^{-1}Q^{-1}$	$\dfrac{m^2e^3}{\kappa_0^2\hbar^3} = \dfrac{\hbar}{ea_0^2} = 1\text{ aT (``atomic Tesla'')}$	$2\cdot35055\ 10^5\text{ T}$
Magnetic inductance WT^2Q^{-2}	$\dfrac{k^2{}_0\hbar^4}{me^6} = \dfrac{\hbar^2}{e^2E_0} = 1\text{aV as}^2\text{ e}^{-1}$	$4\cdot33226\ 10^{-31}\text{ Hn (Wb A}^{-1})$
Magnetizing force $(\mathbf{H})\ QL^{-1}T^{-1}$	$\dfrac{m^2e^7}{\kappa_0^3\hbar^5} = \dfrac{E_0e}{a_0\hbar} = 1\text{ H aT}^{-1}\text{B}^{-3}$	$1\cdot25165\ 10\text{ A}^8\text{ m}^{-1}\ (\text{J T}^{-1}\text{ m}^{-3})$
Magnetic dipole moment ‖ L^2QT^{-1}	$\dfrac{e\hbar}{m} = \dfrac{E_0ea_0}{\hbar} = 1\text{ H aT}^{-1}$	$1\cdot85464\ 10^{-23}\text{ J T}^{-1}$
Magnetic polarizability¶ $L^2Q^2M^{-1}$	$\dfrac{\kappa_0^2\hbar^4}{m^3e^2} = \dfrac{ea_0^2}{m} = 1\text{ H aT}^{-2}$	$7\cdot89023\ 10^{-29}\text{ J T}^{-2}$

‡ Each unit is expressed first in terms of the primary units (m, e, $\hbar$, κ_0) of Table A1.1 and, wherever it is more convenient, in terms of a dimensionally equivalent combination of derived units. The "atomic tesla" (aT) is proposed as a convenient unit of flux density: unit charge (e) moving in a circular orbit of unit radius (1B) would generate a field at the origin of flux density $(137\cdot030)^{-2}$ aT or $12\cdot518$ T.

The expressions in parentheses indicate dimensionally equivalent units (e.g. $1\text{ N} = 1\text{ J m}^{-1}$), corresponding to those employed in column 2.

‖ The Bohr magneton ($\beta = e\hbar/2m$) corresponds to $\tfrac{1}{2}$ atomic unit of magnetic moment: the Zeeman splitting for a free electron ($g = 2$) in a field of flux density 1 aT would be 1 H.

¶ Induced moment/flux density.

20 T or 200,000 gauss. The analogues of the second, the volt and the tesla, namely

$$1 \text{ as } = 1 \text{ PH}^{-1} = \hbar\kappa_0 a_0/e^2 = 2 \cdot 41889 \times 10^{-17} \text{ s,}$$

$$1 \text{ aV } = 1 \text{ He}^{-1} = e/\kappa_0 a_0 = 2 \cdot 72108 \times 10^1 \text{ V,}$$

$$1 \text{ aT } = 1 \text{ em e}^{-1} \text{ as}^{-1} = me/\kappa_0 \hbar a_0 = 2 \cdot 35055 \times 10^5 \text{ T,}$$

have been freely used in the tables.

<h3 style="text-align:center">REFERENCES</h3>

SHULL, H. and HALL, G. G. (1959) *Nature* **184**, 1559.
SLATER, J. C. (1960) *Quantum Theory of Atomic Structure*, Vols. 1 and 2, McGraw-Hill, New York.